INDUSTRIAL APPLICATIONS OF HOMOGENEOUS CATALYSIS

CATALYSIS BY METAL COMPLEXES

INDUSTRIAL APPLICATIONS OF HOMOGENEOUS CATALYSIS

Edited by

A. MORTREUX AND F. PETIT

Laboratory of Heterogeneous and Homogeneous Catalysis,
Laboratory of Applied Organic Chemistry,
University of Lille, France

D. REIDEL PUBLISHING COMPANY

ACADEMIC PUBLISHERS GROUP

DORDRECHT / BOSTON / LANCASTER / TOKYO

Library of Congress Cataloging in Publication Data

Industrial applications of homogeneous catalysis / edited by A. Mortreux and F. Petit.
p. cm. -- (Catalysis by metal complexes)
Updated and expanded versions of lectures given at a meeting held at the University of Lille, September 1985, and sponsored by the Commission of the European Communities.
Bibliography: p.
Includes index.
ISBN 9027725209
1. Catalysis--Congresses. I. Mortreux, A. (Andre), 1943- II. Petit, F. (Francis), 1942- . III. Commission of the European Communities. IV. Series.
TP156.C35I53 1988
660.2′995--dc19 87-32334

Published by D. Reidel Publishing Company,
P.O. Box 17, 3300 AA Dordrecht, Holland.

Sold and Distributed in the U.S.A. and Canada
by Kluwer Academic Publishers,
101 Philip Drive, Norwell, MA 02061, U.S.A.

In all other countries, sold and distributed
by Kluwer Academic Publishers Group,
P.O.Box 322, 3300 AH Dordrecht, Holland.

Printed in the Netherlands

TABLE OF CONTENTS

PREFACE

Catalysts are now widely used in both laboratory and industrial-scale chemistry. Indeed, it is hard to find any complex synthesis or industrial process that does not, at some stage, utilize a catalytic reaction.

The development of homogeneous transition metal catalysts on the laboratory scale has demonstrated that these systems can be far superior to the equivalent heterogeneous systems, at least in terms of selectivity. Thus, there is an increasing interest in this field of research from both an academic and industrial point of view.

In connection with the rapid developments in this area, four universities from the E.E.C (Aachen, FRG; Liège, Belgium; Milan, Italy; and Lille, France) have collaborated to organise a series of seminars for high-level students and researchers. These meetings have been sponsored by the Commission of the E.E.C and state organizations.

The most recent of these meetings was held in Lille in September 1985 and this book contains updated and expanded presentations of most of the lectures given there.

These lectures are concerned with the field of homogeneous transition metal catalysis and its application to the synthesis of organic intermediates and fine chemicals from an academic and industrial viewpoint.

The continuing petroleum crisis which began in the early 1970s has given rise to the need to develop new feedstocks for the chemical industry. Up to the present, interest has focussed on carbon monoxide and methanol, the latter now being available at the competitive price of $100 per ton. Thus, the opening chapter of this book is concerned with new catalytic processes for the synthesis of chemical building blocks from these feedstocks.

The second part of this book describes the use of coordination catalysts for the activation of small molecules like CO, H_2, etc. and fine chemical synthesis.

The next section contains lectures given on the chemistry of processes already used by the chemical industry. For example, alkene oligomerization and polymerization, together with new findings in this field. Included in this section is a treatment of hydrocarbon activation in the alkene metathesis reaction, and is concluded by a report of recent work in the important and challenging field of alkane activation.

The last section contains two unusual examples of catalytic processes, namely photochemically induced catalysis and supported molecular cluster catalysis. The latter is a new and promising area of research, since it combines the advantages of both heterogeneous and homogeneous catalysis.

Professor Keim, who organized the first of these meetings in Aachen in 1982, concludes this book by an overview of homogeneous catalysis and discusses its future: a future in which changing raw material supply, engineering requirements and environmental considerations are all important.

Lille, September 1987.

A. MORTREUX
F. PETIT

ACKNOWLEDGEMENTS

This book develops the lectures given by several speakers who attended a four day seminar on Homogeneous Catalysis held at the University of Lille in September 1985. The meeting was organized by the Laboratoire de Catalyse Hétérogène et Homogène of Lille University, under the auspices and support of the Commission of the European Communities.

The organizors greatly acknowledge the following institutions for their financial and material support:

— Le Centre National de la Recherche Scientifique
— Le Ministère de l'Education Nationale
— Le Ministère de l'Industrie et de la Recherche
— L'Université des Sciences et Techniques de Lille Flandres Artois
— L'Ecole Nationale Supérieure de Chimie de Lille
— L'Institut Français du Pétrole
— La Société Chimique des Charbonnages de France.

M. RÖPER*

CHEMICALS FROM METHANOL AND CARBON MONOXIDE

1. Introduction

Methanol is a versatile, readily available C_1-compound made from synthesis gas. Large scale industrial methanol production from CO/H_2 was started up in 1925 by BASF using ZnO/Cr_2O_3 catalysts. The present methanol production capacity has been reported to be 21 mio t/a, while the actual demand is only in the range of 12 mio t/a. This overcapacity is mainly due to the build-up of new plants in the Midde East, Eastern Europe, New Zealand and Latin America, where surplus natural gas is available at a very low price [1, 2]. The ready supply as well as the low raw material costs will keep the price of methanol low in the near future. This will stimulate methanol demand and will help to introduce new methanol-based processes for motor fuels as well as for organic base chemicals [3].

The present industrial uses of methanol include the syntheses of formaldehyde, methyl esters, methyl animes and methyl halides. In addition, methanol or its derivatives find increasing interest as a substrate for carbonylation and dehydration reactions, which are summarized in Scheme 1. Some of these processes have already been commercialized, such as the synthesis of acetic acid, of acetic anhydride, or of methyl formate. The feasibility of the MTG (methanol-to-gasoline) process has been proved by pilot plant operation and a commerical unit of a capacity of 560 000 t/a of hydrocarbons had started up operation in New Zealand by 1985. In addition there is a variety of carbon monoxide-based reactions, which convert methanol mainly into C_2-oxygenated compounds.

These can potentially replace ethylene-based routes, e.g. in the case of ethanol, acetaldehyde, vinyl acetate, and ethylene glycol. With methanol from cheap natural gas becoming available in the near future, these processes, although uneconomic today, might become industrially attractive.

* Present address: BASF AG, 6700 Ludwigshafen, F.R. Germany

A. Mortreux and F. Petit (Eds.), Industrial Applications of Homogeneous Catalysis, 1—17.

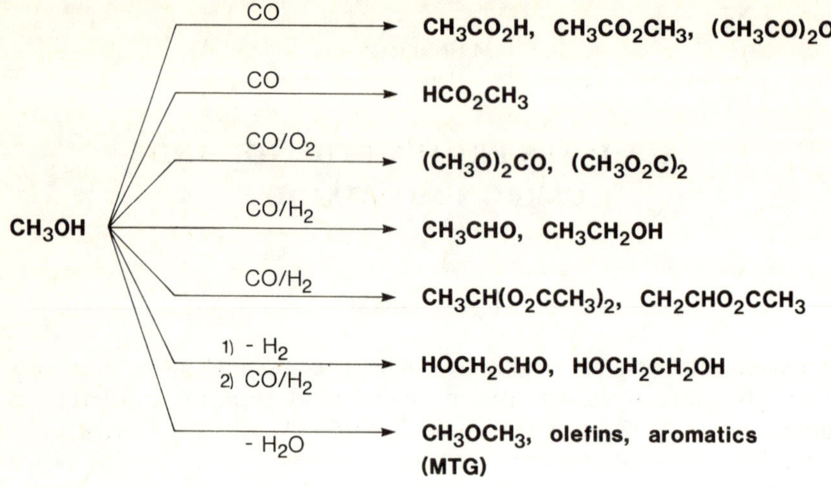

Scheme 1. Chemicals from methanol.

2. Carbonylation of Methanol and of Methanol Derivatives

While the base-catalyzed carbonylation of methanol yields methyl formate, a versatile intermediate for formic acid and formamide synthesis, the transition metal-catalyzed carbonylation involves C—C coupling, giving acetic acid derivatives as C_2 oxygenates.

2.1. TRANSITION METAL CATALYZED CARBONYLATION [4—11]

From the industrial point of view one of the major achievements of homogeneous catalysis has been the introduction of acetic acid processes via the carbonylation of methanol. These processes allow not only the use of methanol as a cheaper feedstock as compared to ethylene, but are also characterized by an extremely high selectivity.

The carbonylation of methanol to give acetic acid is catalyzed by Group VIII transition metal complexes, especially rhodium, cobalt, and nickel:

$$CH_3OH + CO \longrightarrow CH_3CO_2H \qquad (1)$$

Iodine compounds are essential cocatalysts, and the reaction is believed to proceed via methyl iodide, which alkylates the transition metal. At elevated pressures and temperatures both acetic acid and the iodine compounds are highly corrosive. Thus, the development of corrosion-resistant alloys such as Hastelloy C was a prerequisite to commercialization of this reaction. Two industrial processes have been developed: the cobalt-catalyzed BASF process was introduced in the late 1950s, and the rhodium-based Monsanto process followed in the early 1970s. As is evident from Table I, rhodium catalysts operate at very mild conditions and with extremely high selectivities by comparison with cobalt or nickel catalysts. It is therefore no surprise that most commercial plants now use the rhodium-based Monsanto process. Meanwhile, the worldwide capacity for acetic acid from methanol is well over 1 000 000 t/a and is expected to increase further.

TABLE I
Acetic acid by carbonylation of methanol [12—14].

Catalyst	Temperature (°C)	Pressure (bar)	Yield (%)
Rh_2O_3/HI	175	1—15	99
$Co(OAc)_2/CoI_2$	250	680	90
$Ni(CO)_4$/MeI/LiOH	180	70	84

Besides methanol, dimethyl ether and methyl acetate can also be effectively carbonylated with rhodium (or iridium) catalysts to give acetic anhydride:

$$CH_3OCH_3 + CO \longrightarrow CH_3CO_2CH_3 \quad (2)$$

$$CH_3CO_2CH_3 + CO \longrightarrow (CH_3CO)_2O \quad (3)$$

In addition to methyl iodide, phosphines and early transition metal compounds such as $Cr(CO)_6$ are also neccessary as cocatalysts. Typical conditions are 10—100 bar and 150—200 °C and, at conversions of 50—80%, selectivities to acetic anhydride of up to 90% can be achieved [15]. This acetic anhydride route avoids the high energy intermediate ketene and has been commercialized by Halcon SD/Eastman Kodak in a plant with a design capacity of 200 000 t/a [16].

The acetic anhydride produced in this plant is used in the synthesis of cellulose acetate and the acetic acid liberated in this process is recycled via esterification with methanol. A schematic diagram of Eastman Kodak's C_1 complex is shown in Figure 1.

The carbonylation of acetaldehyde dimethyl acetal or the hydrocarbonylation of methyl acetate by use of rhodium catalysts yields ethylidene diacetate, which can be cracked thermally to give vinyl acetate:

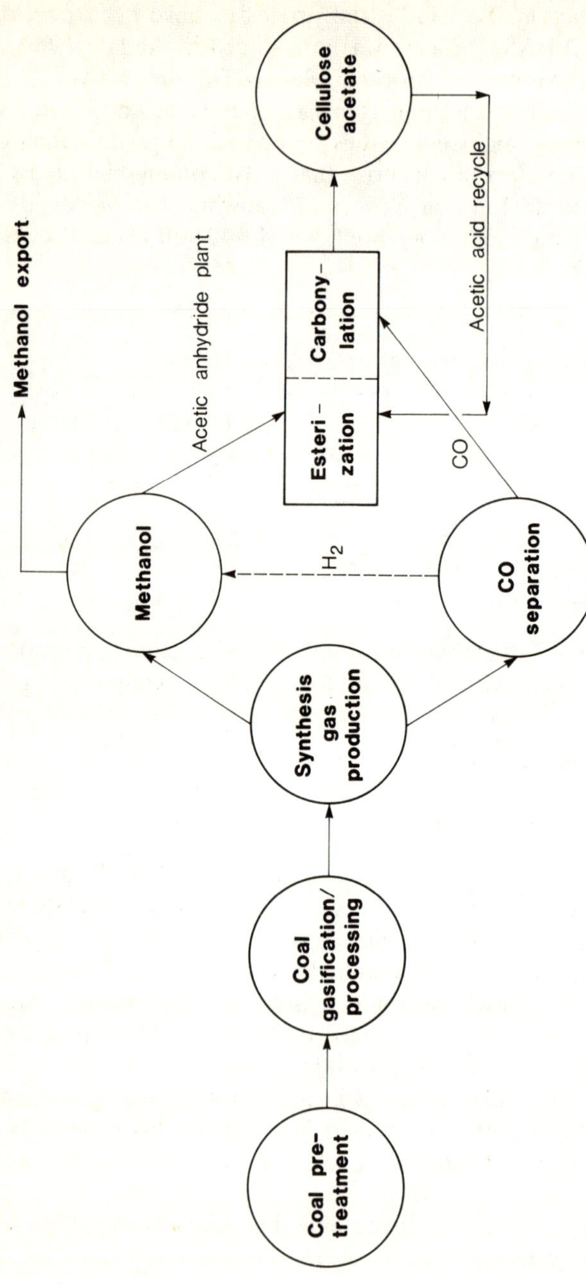

Fig. 1. Tennessee Eastman C_1 complex.

$$CH_3CH(OCH_3)_2 + 2\ CO \longrightarrow CH_3CH(O_2CCH_3)_2 \quad (4)$$

$$CH_3CO_2CH_3 + 2\ CO + H_2 \longrightarrow CH_3CH(O_2CCH_3)_2 \quad (5)$$

$$CH_3CH(OC_2CH_3)_2 \longrightarrow CH_2{=}CH{-}O_2CCH_3 + CH_3CO_2H \quad (6)$$

Modified catalysts of the type $RhCl_3$/MeI have been reported to be active for reactions (4) and (5) [11]. This methanol-based vinyl acetate process has been developed by Halcon SD and is ready for commercialization [16].

2.1.1. *Side reactions*

Typical side reactions of the methanol carbonylation are the formation of methyl acetate (eq. 7), methyl formate (eq. 8), dimethyl ether (eq. 9), and the water gas shift reaction (eq. 10):

$$CH_3OH + CH_3CO_2H \rightleftharpoons CH_3CO_2CH_3 + H_2O \quad (7)$$

$$CH_3OH + CO \rightleftharpoons HCO_2CH_3 \quad (8)$$

$$2\ CH_3OH \rightleftharpoons (CH_3)_2O + H_2O \quad (9)$$

$$H_2O + CO \rightleftharpoons CO_2 + H_2 \quad (10)$$

These reactions are equilibria which can be controlled by reaction conditions, catalyst metal, ligand, promoters, and solvents.

2.1.2. *Rhodium catalysts*

The great advantage of rhodium catalysts is their high selectivity, which is $> 99\%$, based on methanol. Even in the presence of hydrogen, no hydrogenation products such as methane, acetaldehyde or ethanol are observed, in contrast to cobalt-based catalysts. In addition, the high activity allows the use of metal concentrations as low as 10^{-3} M [10]. Besides methanol, a variety of other alcohols can be submitted to the rhodium-catalyzed carbonylation and more detailed data are known for ethanol [17] and benzyl alcohol [18].

The mechanism of this reaction has been elucidated by Forster [10] and Hjortkjaer [19], using kinetic and spectroscopic techniques. Under steady-state conditions, CO pressures above 2 bar, and methanol concentrations above 0.5 mol/l the following expression for the reaction rate was found:

$$r = k\ [Rh]^1\ [I]^1\ [CO]^0\ [CH_3OH]^0 \tag{11}$$

Neither of the reactants affects the rate which is dependent to first order in rhodium and in iodide. This finding has been interpreted by assumption of oxidative addition of methyl iodide to a Rh^I species as the rate determining step, as elucidated in Scheme 2:

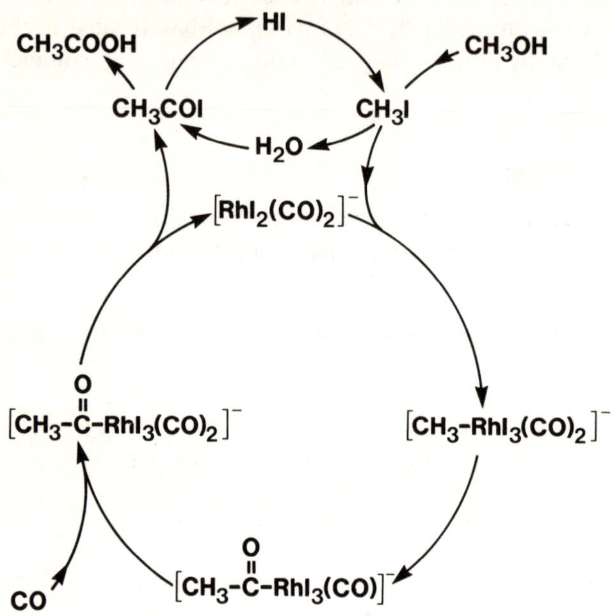

Scheme 2. Mechanism of the Rh-catalyzed methanol carbonylation [10, 19].

The sequence involves the formation of methyl iodide from HI and methanol, oxidative addition to an anionic Rh^I species, CO-insertion, then reductive elimination of acetyl iodide, followed by its hydrolysis to acetic acid and HI.

Iodide is a necessary component in the overall reaction scheme, and any source of iodide will serve as a promoter. Iodide is a good nucleophile, a good leaving group, a good ligand for rhodium, and a very weak proton base. Bromide is much less effective as a promoter, as is pentachlorothiophenoxide, $C_6Cl_5S^-$ [20].

A dimeric intermediate $[Rh_2(COMe)_2(CO)_2I_6]^{2-}$ has been isolated from the reaction, which under CO pressure generates the mononuclear species $[Rh(COMe)(CO)_2I_3]^-$ [21, 22]. More recent ^{13}CO labelling studies indicate that methyl migration to give the acetyl complex is irreversible in this case [23].

Supported Rh catalysts have also been used successfully in methanol carbonylation [24]. The rates of both homogeneous and heterogeneous systems at 250 °C and 1 bar CO differ significantly, as is obvious from Table II.

TABLE II
Methanol carbonylation at 250 °C/1 bar CO by homogeneous and supported Rh catalyst systems [24].

Catalyst	Rh content [wt %]	Rate [g MeOAc/g Rh · h]
$RhCl_3$	homogeneous	3000
$Rh(NO_3)_3/C$	3	25
$RhCl(CO)(Ph_3P)_2$-Al_2O_3	1.39	40
$RhCl_3$-NaX	0.25	50

The low activity of the supported Rh catalysts is explained by assuming that only about 1% of Rh is active in this case. The choice of the carrier seems to have some impact on the catalytic activity, and X-zeolites have attracted special interest [25]. Also, higher alcohols such as ethanol, have been carbonylated over these systems [26].

2.1.3. *Cobalt catalysts*

The lower activity of Co catalysts requires a high catalyst concentration of about 0.1 M. The selectivity reaches 90%, based on methanol. By-products are methane, acetaldehyde, ethanol and ethers. $HCo(CO)_4$ is supposed to be the active species, and is formed according to equations (12) and (13) [13]:

$$2\ CoI_2 + 2\ H_2O + 10\ CO \longrightarrow Co_2(CO)_8 + 4\ HI + 2\ CO_2 \quad (12)$$

$$Co_2(CO)_8 + H_2O + CO \longrightarrow 2\ HCo(CO)_4 + CO_2 \quad (13)$$

The assumption of hydro cobalt carbonyl as the active species is in accordance with the observation that small amounts of H_2 enhance the catalytic activity. The mechanism shown in Scheme 3 has been proposed [13]. The role of iodine is possibly not restricted to the formation of alkyl iodides, which act as alkylating agents (eq. 14),

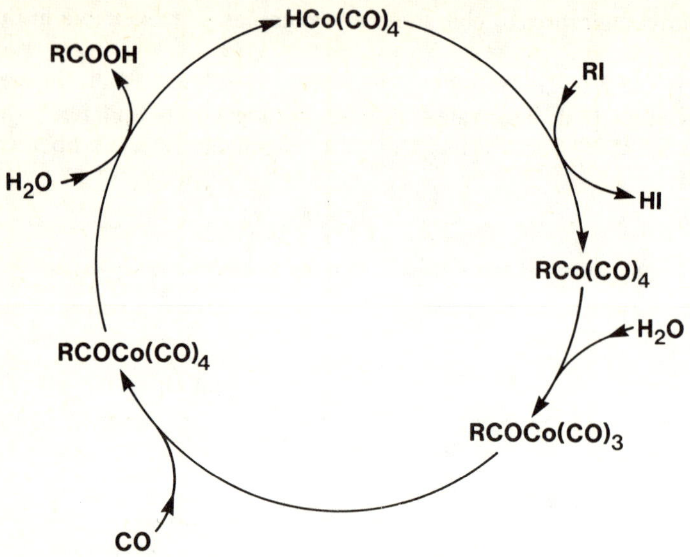

Scheme 3. Mechanism of the Co-catalyzed alcohol carbonylation [13].

$$\mathrm{ROH} + \mathrm{HI} \rightleftharpoons \mathrm{RI} + \mathrm{H_2O} \qquad (14)$$

but also includes catalytic hydrolysis of the acyl intermediate via formation of an acyl iodide [27]. In contrast to the Rh catalyzed alcohol carbonylation the rate depends on the pressure of CO which makes high pressures of 600—700 bar necessary. Lower pressures are possible if metals such as Ru, Ir, Pd, Pt, and Cu are added [28, 29]. Cobalt catalysts can also be used for the carbonylation of higher alcohols as has been exemplified in the case of benzyl alcohols [30].

2.1.4. *Nickel catalysts*

Nickel metal, as well as a variety of nickel compounds, have been used as catalysts in the presence of iodine. $Ni(CO)_4$ is produced from NiI_2 according to equation (15) [18]:

$$\mathrm{NiI_2} + \mathrm{H_2O} + 5\,\mathrm{CO} \longrightarrow \mathrm{Ni(CO)_4} + 2\,\mathrm{HI} + \mathrm{CO_2} \qquad (15)$$

The hydrogen iodide formed in (15) is used to transform the alcohol into an alkyl halide (eq. 14), which, according to Heck, adds oxidatively to Ni^0 [31], Scheme 4.

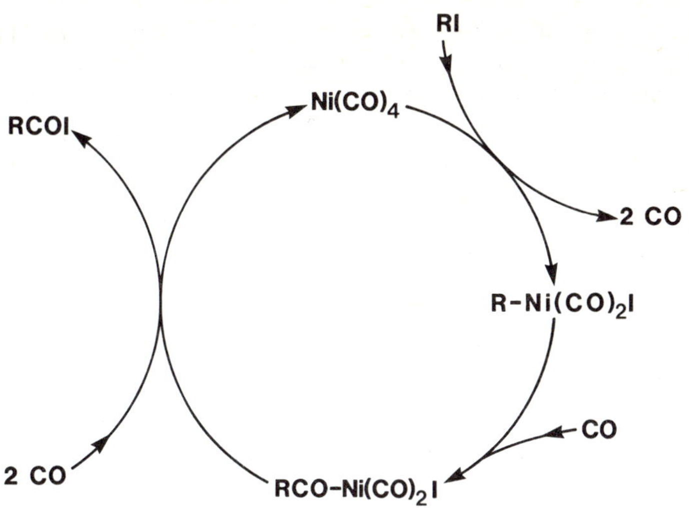

Scheme 4. Mechanism of the Ni-catalyzed alcohol carbonylation [31].

The resulting acyl iodide is hydrolyzed either by water or the alcohol. Usual nickel catalysts require rather high pressures and temperatures [32]. However, if high concentrations of methyl iodide are used, quite mild conditions can be applied [14]. If the molar ratio of CH_3I to CH_3OH is at least 1 : 10, pressures as low as 35 bar can be utilized at 150 °C with a catalyst comprizing $Ni(OAc)_2 \cdot 4\,H_2O$ and Ph_4Sn [33]. Also vapor phase carbonylation of methanol using supported nickel metal catalysts has been reported [34].

2.2. BASE CATALYZED CARBONYLATION

The base catalyzed carbonylation of methanol to give methyl formate is carried out industrially on a large scale to produce formic acid. The reaction proceeds at 70—80 °C and CO pressures above 20 bar with sodium methoxide as the catalyst, and involves nucleophilic attack of methoxide on CO [35].

$$CO + OCH_3^- \longrightarrow CO_2CH_3^- \quad (16)$$

$$CO_2CH_3^- + CH_3OH \longrightarrow HCO_2CH_3 + OCH_3^- \quad (17)$$

Platinum/*N*-ethylpiperidine catalysts have also been used for methyl formate synthesis via methanol carbonylation [36].

Methyl formate is a versatile intermediate, which is currently being used in the synthesis of formic acid and a variety of formamides. Potential future processes may involve the synthesis of pure CO, methanol, dimethyl carbonate, diphosgene, acetic acid, methyl glycolate, and methyl propionate [37]. These reactions either operate at milder conditions as compared to established processes, or avoid the use of high purity CO. An example is the "isomerization" of methyl formate to acetic acid, which proceeds without consumption of additional CO in the presence of Rh or Ir catalysts:

$$HCO_2CH_3 \longrightarrow CH_3CO_2H \quad (18)$$

As in the related methanol carbonylation, methyl iodide is used as the promoter [38].

3. Reductive Carbonylation of Methanol and Methanol-derived Substrates

3.1. METHANOL HOMOLOGATION [11, 39–46]

The reductive carbonylation of methanol yields acetaldehyde or ethanol:

$$CH_3OH + CO + H_2 \longrightarrow CH_3CHO + H_2O \quad (19)$$

$$CH_3CHO + H_2 \longrightarrow CH_3CH_2OH \quad (20)$$

This cobalt catalyzed reaction was first reported by chemists from BASF in the early 1940s and was introduced into the open literature by Wender in 1950. Again, iodine promoters were found to speed up the reaction considerably and, in some cases, ligands such as phosphines are added to improve the selectivity. Whereas acetaldehyde prevails with Co/I-catalysts, high selectivities for ethanol are obtained by addition of a hydrogenation catalyst which, in most cases, is ruthenium. The major problem of methanol homologation is product selectivity, and important side products are methane, methyl ethers, acetates, higher alcohols, and aldehydes. Acetals like 1,1-dimethoxyethane, which are formed especially in the initial phase of the reaction, can be regarded as intermediates which, at higher conversions, are transformed to acetaldehyde/ethanol.

Catalysts which are selective for ethanol formation usually comprize cobalt, phosphines, iodine and ruthenium compounds, and ethanol selectivities in the range of 60—80% can be achieved [42, 44—46].

A high acetaldehyde selectivity and catalytic activity was obtained by Gauthier-Lafaye *et al.* using $Co_2(CO)_8$/MeI/KI (1 : 10 : 205). However, due to the low catalyst concentration, acetaldehyde yields are only in the range of 18% [47].

High yields of acetaldehyde can be achieved, if the reaction is carried out in 1,4-dioxane as the solvent with CoI_2 as the catalyst, as was reported by Walker [48] and Jenner *et al.* [49]. A disadvantage of this system is the low catalytic activity, which is in the range of 25—50 mol acetaldehyde/g atom cobalt. h. As has recently been found in our group, the catalytic activity can be enhanced at least by a factor of 10 to 570—1250 h^{-1}, if ionic promoters are added to the system and the molar ratio I/Co is > 5 [50]. It has been shown by workers at URBK, that this latter catalyst system can be recycled which is a prerequisite for industrial application [51]. Economic studies by Aquilo *et al.* of the Celanese Chemical Co. have shown that acetaldehyde production via reductive methanol carbonylation is superior to the Wacker—Hoechst process, if reinvestment is considered [52]. The same study reveals ethanol production via methanol homologation to be less economic than ethylene hydration.

The mechanism of homologation is still not clear, but the following general scheme can be outlined:

$$CH_3OH + HI \rightleftharpoons CH_3I + H_2O \quad (21)$$

$$CoL_n + CO \rightleftharpoons CoL_n(CO) \quad (22)$$

$$CH_3I + CoL_n(CO) \rightleftharpoons CH_3CoL_n(CO)(I) \quad (23)$$

$$CH_3CoL_n(CO)(I) \rightleftharpoons CH_3COCoL_n(I) \quad (24)$$

$$CH_3COCoL_n(I) + H_2 \rightleftharpoons CH_3CHO + CoL_n + HI \quad (25)$$

Methyl iodide, formed from methanol and hydrogen iodide, adds oxidatively to a cobalt species $CoL_n(CO)$. This is followed by methyl migration, leading to an acetyl species which, in a fast reaction is hydrogenated to give acetaldehyde, hydrogen iodide and the cobalt species CoL_n, which takes up one CO to give $CoL_n(CO)$. In the coordination sphere of this latter complex, CO, I^-, hydrogen, and phosphines/phosphites have to be considered as the ligands, the oxidation state of cobalt being either -1 or $+1$. Depending on the oxidation state and the number and type of ligands this species will be charged. Proposals known from the literature are $Co(CO)_4^-$ [40, 41, 44], $HCo(CO)_3$ [45] or $Co(CO)_xI_y(PR_3)_z$ [53, 54].

For the kinetics of acetaldehyde [55] formation from methanol and syngas with $Co/I/PR_3$-catalysts the following rate law was observed:

$$r = k\,[CH_3OH]\,[\text{Catalyst}]\,P_{CO}\,P_{H_2}^{0.4} \quad (26)$$

This rate law, which in a similar form has also been reported for ethanol formation [56], emphasizes the role of the methylation step of the cobalt species, which is believed to be stabilized by CO.

In the presence of large cations such as PPN^+ ($PPN^+ = (Ph_3P)_2N^+$) the reaction of $Co(CO)_4^-$ with methyl iodide in THF at 0 °C yields almost quantitatively an anionic iodide substituted cobalt acetyl complex (eq. 27):

$$PPN[Co(CO)_4] + CH_3I \longrightarrow PPN[CH_3COCo(CO)_3I] \quad (27)$$

With respect to the strong effect of cations on cobalt/iodine catalysts in methanol homologation, this anionic acetyl complex is of interest as a potential intermediate in the catalytic cycle [57].

Methanol hydrocarbonylation can also be carried out with $Fe(CO)_5/NR_3$ [58] and Ru/I [59] catalysts, although rates are much lower than with cobalt systems. In the case of the iron/amine catalyst CO_2 is formed instead of water as the by product:

$$CH_3OH + 2\,CO + H_2 \longrightarrow CH_3CH_2OH + CO_2 \quad (28)$$

As the rate-determining step the alkylation of an iron species has been postulated:

$$NR_3CH_3^+ + Fe(CO)_4H^- \longrightarrow CH_3Fe(CO)_4H + NR_3 \quad (29)$$

Ruthenium catalysts also have been reported to catalyzed the homologation of methylacetate to ethyl acetate [60].

An indirect method to convert methanol selectively into ethanol has been reported recently by Halcon [61] and Humphreys and Glasgow/BASF [62]. In the first stage methanol is carbonylated according to eq. (1) to give acetic acid, which is subsequently hydrogenated to ethanol as shown in eq. (30):

$$CH_3CO_2H + 2\,H_2 \longrightarrow CH_3CH_2OH \quad (30)$$

While in the Halcon SD process a non-noble metal halide catalyst is used for methanol carbonylation, and acetic acid is converted to methyl acetate prior to hydrogenation, the ENSOL process of Humphreys and Glasgow/BASF utilizes the rhodium/iodide-based Monsanto technology to produce acetic acid which is directly hydrogenated to ethanol. For the ENSOL process an overall thermal efficiency of 50% and a carbon efficiency of 74% is claimed for the conversion of natural gas into ethanol [62].

3.2. HOMOLOGATION OF METHOXY DERIVATIVES

The methoxy moiety of ethers (e.g. dimethyl ether), esters (e.g. methyl formate, methyl acetate), and acetals (e.g. 1,1-dimethoxy ethane) can be hydrocarbonylated to yield acetaldehyde or ethanol [50, 63, 64, 65]. Cobalt/iodide catalysts of the same type as in the hydrocarbonylation of methanol are used and the mechanisms appear to be closely related. While the reactivity of these systems is much lower compared to that of methanol, they offer the advantage of a ready separation of water, e.g. in a preceeding etherization step [50]. The hydrocarbonylation of 1,1-dimethoxymethane can also be directed to yield 1,1,2-trimethoxyethane, a glycol aldehyde derivative [63].

3.3. REDUCTIVE CARBONYLATION OF FORMALDEHYDE

The conversion of derivatives of formaldehyde with synthesis gas in the presence of cobalt catalysts to give ethylene glycol had already been claimed in patents of BASF in 1942. The stoichiometric hydrocarbonylation of monomeric formaldehyde with $HCo(CO)_4$ proceeds at atmospheric pressure and 0 °C almost quantitatively [66]:

$$CH_2O + CO + H_2 \longrightarrow HOCH_2CHO \quad (31)$$

The catalytic hydrocarbonylation to give glycolaldehyde can be achieved with cobalt catalysts at 200—300 bar and 110 °C [67]. With rhodium catalysts such as $RhCl(CO)(Ph_3P)_2$ both formaldehyde conversion and glycolaldehyde selectivity are in the range of 80—90% [68, 69]. The reaction is usually carried out in *N,N*-disubstituted amide solvents or in the presence of bases like phosphines or amines. Glycolaldehyde is easily hydrogenated to give ethylene glycol, which also can be obtained directly from formaldehyde by adjusting the hydrocarbonylation conditions.

4. Oxidative Carbonylation [70]

The oxidative carbonylation of methanol yields dimethyl carbonate or dimethyl oxalate:

$$2\ CH_3OH + CO + 1/2\ O_2 \longrightarrow (CH_3O)_2CO + H_2O \quad (32)$$

$$2\ CH_3OH + 2\ CO + 1/2\ O_2 \longrightarrow (CH_3O_2C)_2 + H_2O \quad (33)$$

As was reported by Romano *et al.*, cuprous chloride is an effective catalyst for selective dimethyl carbonate synthesis. The reaction proceeds at 90—100 °C at > 20 bar of CO and can be either separated into an oxidation and a reduction step, or carried out in a one-pot redox system [71]. Dimethyl carbonate is a versatile reagent which can replace phosgene and dimethyl sulfate as carbonylating and methylating agents, respectively.

A 5000 t/a plant has been operated by ENICHEM since 1983 in Ravenna, Italy [72].

If palladium/copper halide catalysts are used, C—C coupling occurs and dimethyl oxalate is formed. The by product water can be trapped by addition of orthoformates and, instead of oxygen, quinones can also be used as the oxidizing agent. Typical conditions are 70 bar of CO and temperatures of 125 °C. Based on results of Rivetti *et al.*, product formation can be envisioned via alkoxycarbonyl species [73]:

$$PdCl_2(CO)_2 + 2\ OCH_3^- \longrightarrow [PdCl_2(CO_2CH_3)_2]^{2-} \quad (34)$$

$$[PdCl_2(CO_2CH_3)_2]^{2-} \longrightarrow Pd + CO_2CH_3\text{—}CO_2CH_3 + 2\ Cl^- \quad (35)$$

More recently, Ube Industries have published an indirect process to carbonylate methanol oxidatively to dimethyl oxalate utilizing nitrous acid methyl ester as the oxidant. A supported Pd/Fe catalyst is used and methyl nitrite can be generated either *in situ* or in a separate reactor from methanol and NO [74]:

$$2\ CH_3OH + 2\ NO + 1/2\ O_2 \longrightarrow 2\ CH_3ONO + H_2O \quad (36)$$

$$2\ CH_3ONO + 2\ CO \longrightarrow CH_3O\text{-}C(=O)\text{-}C(=O)\text{-}OCH_3 + 2\ NO \quad (37)$$

Dimethyl oxalate is of interest as a solvent, in agriculture (oxalamide), the pharmaceutical industry, and food production. It can also serve as a precursor of ethylene glycol, and the catalytic hydrogenation of dimethyl oxalate has been investigated by Union Carbide/Ube Industries [11].

5. Conclusions

The increasing supply of cheap methanol will stimulate its use as a fuel additive or as a fuel precursor via the MTG (methanol-to-gasoline) process. It will also help to introduce further methanol carbonylation processes for the synthesis of oxygenated C_2 compounds besides the already well established acetic acid and acetic anhydride processes. Present research and development activities concentrate on target molecules like ethylidene diacetate, acetaldehyde, glycolaldehyde, ethylene glycol, dimethyl carbonate, and dimethyl oxalate [75].

Institut für Technische Chemie und Petrolchemie
der Rheinisch-Westfälischen Technischen Hochschule Aachen,
F.R. Germany.

References

1. K. Kobayashi, *Chem. Econ. Eng. Rev.,* **16**(6), 32 (1984).
2. H. Itami, *Chem. Econ. Eng. Rev.,* **16**(4), 21 (1984).
3. K. Kobayashi, *Chem. Econ. Eng. Rev.,* **14**(4), 20 (1982).
4. A. Muller in J. Falbe (ed.), *New Syntheses with Carbon Monoxide*, Springer Verlag, Berlin, 1980, p. 243.
5. J. Falbe, *Synthesen mit Kohlenmonoxid*, Springer Verlag, Berlin, 1977.
6. J. Falbe in F. Korte (ed.), *Methodicum Chimicum*, vol. 5, Georg Thieme Verlag, Stuttgart, 1975.
7. N. V. Kutepow and W. Himmele in *Ullmanns Encyclopädie der technischen Chemie*, vol. 9, 4th edition, Verlag Chemie, Weinheim, 1975, p. 155.
8. F. Piacenti and M. Bianchi in I. Wender and P. Pino (eds.), *Organic Syntheses via Metal Carbonyls*, John Wiley and Sons, New York, 1977, p. 1.
9. T. A. Weil, L. Cassar and M. Foa in I. Wender and P. Pino (eds.), *Organic Syntheses via Metal Carbonyls*, Vol. 2, John Wiley and Sons, New York, 1977, p. 517.
10. D. Forster, *Adv. Organometal. Chem.,* **17**, 255 (1979).
11. K.-H. Keim, J. Korff, W. Keim and M. Röper, *Erdöl, Kohle, Erdgas, Petrochem.,* **35**, 297 (1982).
12. F. Paulik and J. E. Roth, *J. Chem. Soc., Chem. Commun.*, 1578 (1968).
13. N. von Kutepow, W. Himmele and H. Hohenschutz, *Chem. Ing. Techn.,* **37**, 383 (1965).
14. J. Gauthier-Lafaye and R. Perron (Rhone-Poulenc), *Eur. Pat. Appl.* 35458 (1981); *Chem. Abstr.* **96**, 34582 p (1982).
15. R. V. Porcelli and V. S. Bhise (Halcon Research and Development Corp.), *Ger. Pat.* 3.024.353 (29/1/1981); *Chem. Abstr.* **94**, 120902h (1981).
16. J. E. Erler and B. Juran, *Hydrocarbon Processing*, 109 (Febr. 1982).
17. J. Hjortkjaer and J. C. A. Jørgensen, *J. Mol. Catal.,* **4**, 199 (1978).
18. F. E. Paulik, A. Hershman, W. R. Knox and J. F. Roth (Monsanto Co.), *D. E. Offen.* 1.941.449 (1970); *Chem. Abstr.* **72**, 110807 y (1970).
19. J. Hjortkjaer and V. W. Jensen, *Ind. Eng. Chem., Prod. Res. Dev.,* **15**, 46 (1976).
20. K. M. Webber, B. C. Gates and W. Drenth, *J. Catal.,* **47**, 269 (1977).
21. D. Forster, *J. Am. Chem. Soc.,* **98**, 846 (1976).
22. G. W. Adamson, J. J. Daly and D. Forster, *J. Organomet. Chem.,* **71**, C 17 (1974).
23. A. G. Kent, B. E. Mann and C. P. Manuel, *J. Chem. Soc., Chem. Commun.*, 728 (1985).
24. M. S. Scurell, *Platinum Met. Rev.,* **21**, 92 (1977).
25. J. Jamanis, K.-C. Lien, M. Caracotsios and M. W. Powers, *Chem. Eng. Commun.,* **6**, 355 (1981).
26. M. S. Scurell and T. Hauberg, *Appl. Catal.,* **2**, 225 (1982).

27. D. Forster and M. Singleton, *J. Mol. Catal.*, **17**, 299 (1982).
28. K. Nozaki (Shell), *D. E. Offen.* 2.400.534 (1974); *Chem. Abstr.* **81**, 120023 d (1974).
29. N. v. Kutepow and F.-J. Müller (BASF), *D.E. Offen.* 2.303.271 (1973); *Chem. Abstr.* **81**, 135473 z (1974).
30. T. Imamoto, T. Kusomoto and M. Yokoyama, *Bull. Chem. Soc. Jpn.*, **55**, 643 (1982).
31. R. F. Heck, *J. Am. Chem. Soc.*, **85**, 2013 (1963).
32. H. J. Hagemeyer, Jr. (Eastman Kodak Co.), *US Patent* 2.739.169 (1956); *Chem. Abstr.* **50**, 16835 d (1956).
33. A. N. Naglieri and N. Rizkalla (Halcon Int.), *D.E. Offen.* 2.749.954 (1978); *Chem. Abstr.* **89**, 42469 c (1978).
34. K. Fujimoto, T. Shikada, K. Omata and H. Tominaga, *Ind. Eng. Chem., Prod. Res. Dev.*, **21**, 429 (1982).
35. S. P. Tonner, D. L. Trimm, M. S. Wainwright and N. W. Cant, *J. Mol. Catal.*, **18**, 215 (1983).
36. R. A. Head and M. I. Tabb, *J. Mol. Catal.*, **26**, 149 (1984).
37. M. Röper, *Erdöl, Kohle, Erdgas, Petrochem.*, **37**, 506 (1984); T. Ikarashi, *Chem, Econ. Eng. Rev.*, **12**(8), 31 (1980).
38. R. L. Pruett and R. T. Kacmarcik, *Organometallics* **1**, 1693 (1982); M. Röper, E. O. Elvevoll and M. Lütgendorf, *Erdöl, Kohle, Erdgas, Petrochem.*, **38**, 38 (1985).
39. H. Bahrmann and B. Cornils, *Chem. Ztg.*, **104**, 39 (1980).
40. D. W. Slocum, in W. Jones (ed.), *Catalysis in Organic Syntheses*, Academic Press, New York 1980, p. 245.
41. M. E. Fakley and R. A. Head, *Appl. Catal.*, **5**, 3 (1983).
42. H. Bahrmann, W. Lipps and B. Cornils, *Chem. Ztg.*, **106**, 249 (1982).
43. M. Röper and H. Loevenich, in W. Keim (ed.), *Catalysis in C_1-Chemistry*, D. Reidel Publ. Comp., Dordrecht, 1983, p. 105.
44. W. R. Pretzer and T. P. Kobylinski, *Ann. N.Y. Acad. Sci.*, **333**, 58 (1980).
45. I. Wender, *Catal. Rev.-Sci. Eng.*, **14**, 97 (1976).
46. G. Braca and G. Sbrana, in R. Ugo (ed.), *Aspects of Homogeneous Catalysis*, D. Reidel Publ. Comp. Dordrecht, 1984, vol. 5, p. 241.
47. J. Gauthier-Lafaye, R. Perron and Y. Colleuille, *J. Mol. Catal.*, **17**, 339 (1982).
48. W. E. Walker (Union Carbide Corp.), *Eur. Pat. Appl.* 37586 (1981); *Chem. Abstr.* **96**, 68333 y (1982).
49. G. Jenner and P. Andrianary, *J. Mol. Catal.*, **24**, 87 (1984).
50. M. Röper, W. Keim, K.-H. Keim, W. Feichtmeier and J. Korff, *Ger. Offen.* 3.343.519 (1.12.1983); M. Röper, *Habilitationsschrift*, Rheinisch-Westfälische Technische Hochschule Aachen, F.R.G., 1985.
51. J. Korff, K.-H. Keim, W. Keim and M. Röper (MRBK), *Eur. Pat. Appl.* 193.801 (20.2.1986).
52. A. Aquilo, J. S. Alder, D. N. Freeman and R. J. H. Voorhoeve, *Hydrocarbon Proc.*, **62**(3), 57 (1983).
53. J. R. Blackborow, R. J. Daroda and G. Wilkinson, *Coord. Chem. Rev.*, **43**, 17 (1982).
54. M. Röper, H. Loevenich and J. Korff, *J. Mol. Catal.*, **17**, 315 (1982).
55. H. Loevenich and M. Röper, *C_1 Mol. Chem.*, **1**, 155 (1984).
56. P. F. Francoisse and F. C. Thyrion, *Ind. Eng. Chem., Prod. Res. Dev.*, **22**, 542 (1983).

57. M. Röper, M. Schieren and B. T. Heaton, *J. Organomet. Chem.*, **299**, 131 (1986).
58. M. J. Chen, H. M. Feder and J. W. Rathke, *J. Am. Chem. Soc.*, **104**, 7346 (1982).
59. G. Braca, G. Sbrana, G. Valentini, G. Andrich and G. Gregorio, in M. Tsutsui (ed.), *Fundamental Research in Homogeneous Catalysis*, vol. 3, Plenum Press, 1979, 221.
60. G. Braca *et al., Ind. Eng. Chem., Prod. Res. Dev.*, **20**, 115 (1981).
61. B. Juran and R. V. Porcelli, *Hydrocarbon Process.*, 85 (Oct. 1985).
62. C. L. Winter, *Hydrocarbon Process.*, 71 (Apr. 1986).
63. H. Hanrath, *Dissertation*, Rheinisch-Westfälische Technische Hochschule Aachen, F.R. Germany, 1986.
64. R. W. Wegman and D. C. Busby, *J. Chem. Soc., Chem. Commun.*, 332 (1986).
65. R. Bartek, M. M. Habib and W. R. Pretzer, *J. Mol. Catal.*, **33**, 245 (1985).
66. J. A. Roth and M. Orchin, *J. Organomet. Chem.*, **172**, C 27 (1979).
67. T. Yukawa, K. Kawasaki and H. Wakamatsu (Ajinomoto Co.), *Ger. Pat.* 2.427.954 (9/6/1975); *Chem. Abstr.* **82**, 124761 m (1975).
68. A. Spencer, *J. Organomet. Chem.*, **194**, 113 (1980).
69. A. S. C. Chan, W. E. Caroll and D. E. Willis, *J. Mol. Catal.*, **19**, 377 (1983).
70. R. Ugo, in W. Keim (ed.), *Catalysis in C_1-Chemistry*, D. Reidel Publishing Comp., Dordrecht, 1983, p. 135.
71. U. Romano, R. Tesei, M. M. Mauri and P. Rebora, *Ind. Eng. Chem., Prod. Res. Dev.*, **19**, 396 (1980).
72. M. M. Mauri, U. Romano and F. Rivetti, *Quad. Ing. Chim. Ital.*, **21**, 6 (1985).
73. F. Rivetti and U. Romano, *J. Organomet. Chem.*, **154**, 323 (1978).
74. Y. Shiomi, T. Matsuzaki and K. Masunaga (Ube Industries), *Eur. Pat. Appl.* 108359 (1984).
75. J. Haggin, *Chem. Eng. News*, 7 (May 19, 1986).

JEAN GAUTHIER-LAFAYE AND ROBERT PERRON

CARBON MONOXIDE AND FINE CHEMICALS SYNTHESIS

Carbon monoxide has long been considered as a typical petrochemical or basic chemical raw material. Most of its industrial applications and research targets are to be found in areas such as methanol synthesis, Fisher-Tropsch synthetic gasoline production, oxo chemistry, acetic acid synthesis, and so on [1].

It is only very recently that this most versatile molecule has been regarded as a possible starting material for more sophisticated synthesis. In other words, carbon monoxide is nowadays starting to be considered as something else than a mere *raw material* for petrochemicals: it is assuming something like the role of a versatile *reagent* for the fine chemical industry [2].

We shall see hereafter that carbon monoxide can in fact react with and lead to almost any organic functional molecule, thus demonstrating its versatility and potential use in fine organic chemistry.

1. Basis of Carbon Monoxide Chemistry

Carbon monoxide versatility hides an unexpected simplicity of its main chemical mechanisms: as few as four or five principles are sufficient to explain and account for most of the several hundred different reactions published in the literature.

Let us first acknowledge that the molecule is a poorly reactive gas *per se*: metallic or organometallic catalysts are essential in order to activate the molecule and achieve the desired reactions. This activation is generally achieved by coordinating carbon monoxide to the metallic center of the catalyst, i.e. by formation of metal carbonyl species.

$$M + CO \longrightarrow M{\rightarrow}CO$$

All the metals given in Table I are known to yield such species; it is furthermore generally accepted that these metallocarbonyl complexes are more stable when the metal center:

A. Mortreux and F. Petit (Eds.), Industrial Applications of Homogeneous Catalysis, 19—64.

(a) is in the lowest possible oxidation state (e.g.: $Ni^0(CO)_4$, $[Co^{-I}(CO)_4]^-$
(b) contains an even number of electrons (e.g.: $Ni(CO)_4$, $Co(CO)_4^-$, $Ru(CO)_5$, $Cu(CO)_n^+$ . . .).

The maximum stability is obtained with a valence number of either 16 or 18.

TABLE I
Metals known to activate carbon monoxide.

(Ti)	(V)	(Cr)	(Mn)	(Fe)	(Co)	(Ni)	(Cu)
(Zr)	Nb	(Mo)		(Ru)	(Rh)	(Pd)	(Ag)
(Hf)	(Ta)	(W)	(Re)	(Os)	(Ir)	(Pt)	(Au)

The carbon monoxide activation mechanism implies that its reaction with the substrate must take place within the coordination sphere of the catalyst.

In order words the creation of a M—CO bond is necessary for *activation* of carbon monoxide whereas its *reaction* implies the formation of a metal-substrate (M—S) bond. The different mechanisms involved in the substrate activation process depend greatly on the temporal chemical nature of the substrate, for example:

(a) the substrate contains a good leaving group (e.g.: CH_3I, CH_3OTs . . .): it will react following either a classical nucleophilic substitution (*SN*)

$$CH_3—I + Fe(CO)_4^{2-} \longrightarrow CH_3—Fe(CO)_4^- + I^-$$

or an oxidative addition (*OA*) pattern

$$Ar—Br + PdL_3 \longrightarrow Ar—Pd(Br)L_3$$

In both cases the formal oxidation state of the metal is increased (+II). Whereas in the OA mechanism the total valence number is also increased, it is maintained in the *SN* mechanism. For a given reaction the mechanism involved will essentially depend on the catalyst's nature (Mn, Co, Fe usually react following the *SN* mechanism while Rh, Pd, Pt, Ru give oxidative addition reactions) and on the substrate's nature (benzylic or allylic halides obviously favor *SN* type reactions while aromatic or vinylic halides can only react following oxidative additions).

(b) unsaturated substrates (e.g.: $CH_2{=}CH_2$, $R_2C{=}O$) can react either with hydrido complexes:

$$CH_2{=}CH_2 + HCo(CO)_4 \longrightarrow CH_3{-}CH_2{-}Co(CO)_4$$

or indirectly with nucleophiles after an initial activation-coordination step:

$$CH_2{=}CH_2 + Pd^{2+}{-}CO \longrightarrow \left[\begin{matrix} CH_2 \\ \| \\ CH_2 \end{matrix} \rightarrow Pd{-}CO\right] \xrightarrow{Nu^-} \begin{matrix} Nu{-}CH_2 \\ | \\ CH_2 \end{matrix} \diagup Pd^+{-}(CO)$$

The first mechanism is characteristic of CO/H_2 reactions while the second is typical of oxidative carbonylations.

Once the desirable [(Substrate)—M—CO] entity necessary for the reaction to happen is formed, the next step of the carbonylation is an insertion, i.e. the migration of the substrate toward the coordinated carbon monoxide.

$$M \begin{matrix} \swarrow CO \\ \searrow Sub \end{matrix} \rightleftharpoons M{-}\overset{\overset{\displaystyle O}{\|}}{C}{-}Sub$$

This is the key step of the overall synthesis, since it allows the bonding of the carbon atom of the carbon monoxide molecule to the substrate skeleton; hence the product contains one more carbon atom than the substrate.

This insertion step is always an equilibrium reaction and it is very rarely a rate limiting step. It was demonstrated that its actual mechanism is a migration of the substrate into to the M—CO bond and not the reverse, even though the substrate nature, shape and steric hindrance are never an obstacle to the migration: the feasibility and facility of the reaction depend only on the catalyst's nature. The metals known to follow the reaction are those circled in Table I.

The last step of the overall carbonylation reaction consists in the final product synthesis and its decomplexation from the metallic center. This step is generally either:

— an hydrogenolysis $$\mathrm{Sub{-}C({=}O){-}M + H^+ \longrightarrow Sub{-}CHO + M^+}$$

— an hydrogenation $$\mathrm{Sub{-}C({=}O){-}M + H_2 \longrightarrow Sub{-}CHO + HM}$$

— a nucleophilic substitution $$\mathrm{Sub{-}C({=}O){-}M + Nu^- \longrightarrow Sub{-}C({=}O){-}Nu + M^-}$$

— a reductive elimination $$\mathrm{Sub{-}C({=}O){-}M{-}Z \longrightarrow Sub{-}C({=}O){-}Z + M}$$

— a dimerization $$\mathrm{Sub{-}C({=}O){-}M + M{-}Z \longrightarrow Sub{-}C({=}O){-}Z + M{-}M}$$

This oversimplified panorama of carbon monoxide's main chemical mechanisms surely does not account for the whole of carbon monoxide chemistry. It nevertheless allows a ready understanding of most of the reactions, as we shall see hereafter.

It must be considered as an elementary guide for those who are not familiar with catalysis and/or carbon monoxide.

2. Carbonylation of organic halides

The carbonylation of organic halides is useful to prepare [3—8]:

— aldehydes $$\mathrm{RX + CO + H_2 \longrightarrow R{-}CHO + HX}$$
— acids and esters $$\mathrm{RX + CO + ROH \longrightarrow R{-}CO_2R + HX}$$
— amides $$\mathrm{RX + CO + HNR_2 \longrightarrow R{-}CONR_2 + HX}$$
— ketones $$\mathrm{RX + CO + R^- \longrightarrow R{-}COR + X^-}$$
— acid halides $$\mathrm{RX + CO \longrightarrow R{-}COX}$$

The X group can either be a halogen (Cl, Br, I) or a pseudohalide and the catalyst used will generally depend on the precise nature of the substrate (RX):

- if RX is an alkyl, an allyl or a benzylic halide (i.e. X is a good leaving group) then good nucleophiles such as $Co(CO)_4^-$, $NiX(CO)_3^-$, $Fe(CO)_4^{2-}$ are required.
- if RX is an aromatic or a vinylic halide then good oxidative-addition type reagents such as Pd, Pt, Ru, or Rh derivatives are useful.

All the reactions leading to HX co-production should be performed in the presence of bases in order to neutralize the acid formed thus avoiding the catalyst destruction.

2.1. SYNTHESIS OF ALDEHYDES

2.1.1. *Aromatic and vinylic halides*

The results obtained with both aromatic bromides or iodides are usually fairly good when using palladium catalysts under mild conditions (Table II); chlorides do not react unless specially activated.

Aromatic halides can be replaced by acid chlorides, thus enabling the so-called Rosenmund reaction [9]:

$$\text{Ar—COX} + \text{CO/H}_2 \xrightarrow{\text{Pd/Ph}_3} \text{Ar—CHO}$$

2.1.2. *Alkyl halides*

Using alkyl halides the scope of the reaction is limited to benzylic compounds unless one uses stoichiometric amounts of the versatile Collman reagent $Na_2Fe(CO)_4$ (Table II).

TABLE II
Synthesis of aldehydes.

Starting material	Product	Catalyst	Conditions °C	bar	Yield (%)	Ref.
CH_3O–C_6H_4–Br	CH_3O–C_6H_4–CHO	Pd	70	50	84	[9]
1-bromonaphthalene (Br)	1-naphthaldehyde (CHO)	Pd	70	50	82	[9]
N(C_5H_4)–Br	N(C_5H_4)–CHO	Pd	70	50	80	[9]
C_4H_9—CH=CH—Br	C_4H_9 CH=CH—CHO	Pd	70	50	65	[9]
$C_6H_5CH_2$—Cl	$C_6H_5CH_2$—CHO	$Co_2(CO)_8$	100	100	88	[10]
o-Cl $C_6H_4CH_2$—Cl	*o*-Cl$C_6H_4CH_2$—CHO	$Co_2(CO)_8$	100	100	75	[10]
n-C_5H_{11}—Br	*n*-C_5H_{11}—CHO	$Na_2Fe(CO)_4$ Stoichiometric	—	—	99	[11]
n-C_6H_{13}CH(Br)(CH_3)	*n*-C_6H_{13}CH(CHO)(CH_3)	$Na_2Fe(CO)_4$ Stoichiometric	—	—	50	[11]

2.2. SYNTHESIS OF ACIDS AND ESTERS

Carbonylation of RX into acids or esters is one of the best known and most utilized reactions. Its scope is unusually large since it has been applied to vinylic, aromatic, heterocyclic, benzylic, aliphatic and allylic halides or pseudohalides (Table III).

Iodides are always more reactive than bromides; chlorides are, as usual, less reactive, especially in the aromatic series, where high temperatures (hence high pressures) are necessary.

2.2.1. *Aromatic and vinylic halides*

Active under mild conditions, palladium catalysts are the more generally used; they allow the stereospecific synthesis of vinylic esters [12]. Nickel is sometimes used, but only in polar solvents [15]; cobalt is only active under phase transfer conditions combined with activation by irradiation [16].
If needed, aromatic halides can be replaced by diazo compounds:

$$\mathrm{Ar{-}N_2^+X^- + CO + ROH \longrightarrow Ar{-}CO_2R + HX + N_2}$$

2.2.2. *Aliphatic halides*

Almost any aliphatic, benzylic or allylic halide (chlorides as well as bromides and iodides) can yield the corresponding carboxylic ester or acid using cobalt, nickel, iron or palladium catalysts (Table III).

Phase transfer catalysis is often used, especially when water-soluble carboxylates are to be prepared [26]:

$$\mathrm{Z{-}C_6H_4{-}CH_2Cl + CO + 2\ NaOH \xrightarrow[\text{solv./water}]{Co(CO)_4^-} Z{-}C_6H_4{-}CH_2CO_2Na + NaCl}$$

Z = H, *p*-CN, *p*-CH_3 Yield: 90—100%
Z = *o*-CH_3 Yield: 60%

In some cases alkyl halides can be replaced by ammonium salts, for example [27]:

$$\mathrm{C_6H_5CH_2{-}NEt_3^+\ Cl^- + CO + H_2O \xrightarrow{Co_2(CO)_8/CH_3I} C_6H_5{-}CH_2CO_2H + HNEt_3^+\ Cl^-}$$

Yield: 99%

2.3. SYNTHESIS OF AMIDES

Synthesis of amides is only feasible starting from aromatic or vinylic

TABLE III
Synthesis of acids and esters.

Starting Material	Product	Catalyst	Conditions °C	Conditions bar	Yield %	Ref.
$NC-C_6H_4-Br$	$NC-C_6H_4-CO_2n\text{-}Bu$	Pd	100	1	88	[12, 13]
$C_4H_9-C_6H_3(NHCOCH_3)-Br$	$C_4H_9-C_6H_3(NHCOCH_3)-CO_2H$	Pd	110	2	90	[14]
C_6H_5-Br	$C_6H_5-CO_2H$	Ni	100	—	88	[15]
$Cl-C_6H_4-Br$	$Cl-C_6H_4-CO_2H$	$Co_2(CO)_8$ $h\nu$	65	1	95	[16]
$Ar-CH=CH-Br$	$Ar-CH=CH-CO_2Bu$	Pd	70	110	80	[12, 13]
$Ar-CH=CH-Br$	$Ar-CH=CH-CO_2Bu$	Pd	70	110	58	[12, 13]
$CH_2=CH-Cl$	$CH_2=CH-CO_2Me$	$Pd-SnCl_2$	120	150	83	[17]
$Cl-CH_2CO_2Me$	$CH_2(CO_2Me)_2$	$Co_2(CO)_8$	70	20	98	[18, 19]

TABLE III (Continued)

Starting Material	Product	Catalyst	Conditions °C	Conditions bar	Yield %	Ref.
$Cl—CH_2COCH_2CO_2Et$	$CH_2(CO_2Et)(COCH_2CO_2Et)$	$Co_2(CO)_8$	70	20	60	[21]
 2-thienyl-CH_2Cl	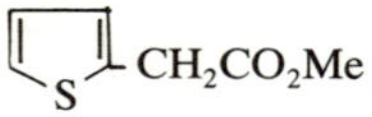 2-thienyl-CH_2CO_2Me	$Co_2(CO)_8$	70	20	95	[22, 23]
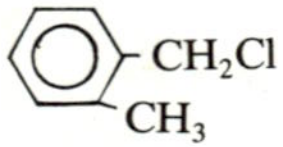 o-$CH_3C_6H_4CH_2Cl$	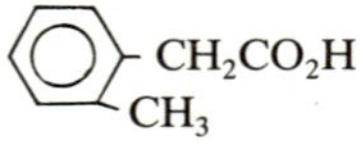 o-$CH_3C_6H_4CH_2CO_2H$	$Fe(CO)_3NO$	60	5	94	[24]
n-$C_8H_{17}—Cl$	n-$C_8H_{17}—CO_2t$-Am	Fe—CRACO	65	1	72	[25]

halides (I > Br ≫ Cl); it is then as general as the preceding ester synthesis (Table IV).

Use of stoichiometric amounts of $Na_2Fe(CO)_4$ and oxidation of the resulting intermediate by iodine in the presence of an amine extends the reaction scope to aliphatic halides (Table IV).

2.4. SYNTHESIS OF KETONES

Ketones are prepared by the carbonylation of aromatic or aliphatic halides in the presence of tetra-alkyl stannous compounds.

$$RX + CO + SnR'_4 \xrightarrow{Pd} R\text{—}CO\text{—}R' + XSn\text{-}r'_3$$

The reaction is general as far as the halide substrate is concerned (Table V) but it seems, on the other hand, to be limited to organostannous derivatives.

An interesting propargylic ketone synthesis has been published by T. Kobayashi [34] which avoids these stannous compounds:

$$Ar\text{—}I + CO + Ar\text{—}C{\equiv}CH \xrightarrow[base/120\,°C/20bar]{Pd} Ar\text{—}CO\text{—}C{\equiv}C\text{—}Ar$$

Yield = 85%

Use of the stoichiometric Collman reagent allows dialkyl ketone synthesis via:

(a) double CO/alkene insertion into an initial alkyl halide [35]

$$n\text{-}C_5H_{11}\text{—}Br + CO + CH_2{=}CH_2 \xrightarrow[2\,—\,addition\ HX]{1\,—\,Na_2Fe(CO)_4} n\text{-}C_5H_{11}\text{—}CO\text{—}CH_2CH_3$$

Yield = 91%

(b) double RX/R′X′ reaction on $Na_2Fe(CO)_4$ [36]:

$$Fe(CO)_4^{=} \xrightarrow[-X^-]{RX} RCO\text{—}Fe(CO)_4^- \xrightarrow[-X'^-]{R'X'/CO} R\text{—}CO\text{—}R' + Fe(CO)_5$$

$RX = n\text{-}C_8H_{17}Br$
$R'X' = C_2H_5I$ Yield = 99%

2.5. SYNTHESIS OF ACID HALIDES

The methyl chloride carbonylation into acetyl chloride has been known

TABLE IV
Synthesis of amides.

Starting Material	Product	Catalyst	Conditions °C	Conditions bar	Yield %	Ref.
$MeO-C_6H_4-Br$	$MeO-C_6H_4-CONHCH_2Ar$	Pd	100	5	76	[28]
$O_2N-C_6H_4-Br$	$O_2N-C_6H_4-CONHCH_2Ar$	Pd	100	5	57	[28]
C_4H_3S-Br	$C_4H_3S-CONHCH_2Ar$	Pd	100	5	63	[28]
C_6H_5-I	$C_6H_5-CONHAr$	Pd on C	100	2	70	[29]
$Ar-CH=CH-Br$	$Ar-CH=CH-CONHCH_2Ar$	Pd	100	5	81	[28]
$CH_2=C(Cl)CH_3$	$CH_2=C(CONHAr)CH_3$	Pd	100	5	74	[13]
$C_6H_5-COCH_2-Cl$	$C_6H_5-COCH_2-CONHAr$	$Na_2Fe(CO)_4$	—	—	44	[30]

TABLE V
Synthesis of ketones.

Starting Material		Product	Catalyst	Pressure bar	Yield %	Ref.
 C_6H_5–I	and $SnBu_4$	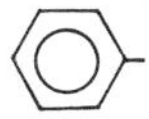 C_6H_5–CO—Bu	Pd	10	73	[31]
 ClC_6H_4–N_2^+, X^-	and $SnMe_4$	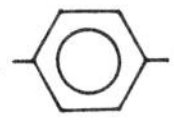 ClC_6H_4–CO—Me	Pd	10	90	[33]
$C_6H_2CH(Br)CH_3$	and $SnMe_4$	$C_6H_5(CO—CH_3)CH_3$	Pd	10	70	[32]
$CH_3CH(Br)CO_2Et$	and $SnMe_4$	$CH_3CH(CO—CH_3)CO_2Et$	Pd	10	62	[32]

since 1920 [37]. The poor results then obtained (Yield = 30%) were recently improved [38] using rhodium catalysts in heptane solvent; at 150—185° under 45 to 75 bar the selectivity reaches 100% with a space productivity of 300 g/h.l.

$$CH_3Cl + CO \xrightarrow{RH/CH_3I} CH_3COCl$$

Allylic and aromatic acid halides have been obtained using palladium catalysts [39, 40]. In the presence of sodium fluoride aromatic chlorides can directly be converted into acid fluorides but yields and selectivities are low despite severe reaction conditions [41].

$$C_6H_5—Cl + CO + NaF \xrightarrow{Pd/AlCl_3} C_6H_5—COF + NaCl$$

Yield = 21%

2.6. SYNTHESIS OF KETO-ACIDS AND KETO-AMIDES

The double carbonylation reaction (i.e. α-insertion of two carbon monoxide molecules on a single substrate molecule) was discovered by R. Perron in 1975 [42]. Phenylpyruvic acid derivatives were then directly prepared from benzyl halides using cobalt catalysts (Table VI).

The reaction has recently been extended, with palladium catalysts, to aromatic and vinylic α-keto-amides syntheses starting from aromatic or vinylic halides (Table VI).

TABLE VI
Synthesis of keto-acids and keto-amides.

Starting Material	Product	Catalyst	Conditions °C	bar	Yield %	Ref.
C_6H_5–CH_2Cl	C_6H_5–CH_2CO—CO_2H	Co	60	5	75	[42, 4]
F–C_6H_4–CH_2Cl	F–C_6H_4–CH_2CO—CO_2H	Co	60	5	81	[42]
Cl–C_6H_4–CH_2Cl	Cl–C_6H_4–CH_2CO—CO_2H	Co	60	5	73	[42]
o-F–C_6H_4–CH_2Cl	o-F–C_6H_4–CH_2CO—CO_2H	Co	60	5	73	[42]
m-CH_3–C_6H_4–CH_2Cl	m-CH_3–C_6H_4–CH_2CO—CO_2H	Co	60	5	79	[42]
C_6H_5–Br	C_6H_5–CO—$CONEt_2$	Pd	100	50	74	[44, 45]
MeO–C_6H_4–Br	MeO–C_6H_4–CO—$CONEt_2$	Pd	100	50	20	[45]
2-thienyl–I	2-thienyl–CO—$CONEt_2$	Pd	100	50	66	[45]
$C_6H_5CH{=}CH$—Br	$C_6H_5CH{=}CH$—CO—$CONEt_2$	Pd	100	50	55	[44, 45]

2.7. SYNTHESIS OF ANHYDRIDES

Symmetric and asymmetric anhydrides are readily affordable [46, 170] starting from aliphatic or benzylic halides (and pseudo-halides) and carboxylates, using cobalt catalysts (Table VII).

$$RX + CO + R'CO_2Na \xrightarrow{Co(CO)_4^-} R\text{—}CO\text{—}OCOR' + NaX$$

3. Carbonylation of Alcohols

The carbonylation of alcohols is useful to prepare:

— alcohols and aldehydes $ROH + CO/H_2 \longrightarrow RCH_2OH$ or $RCHO$
— carboxylic acids $ROH + CO \longrightarrow RCO_2H$

TABLE VII
Synthesis of anhydrides.

Starting material	Product	Catalyst	Conditions °C	bar	Yield	Ref.
CH_3I and CH_3CO_2Na	CH_3—CO—$OCOCH_3$	Co	80	50	75	[46]
CH_3I and $C_6H_5CO_2Na$	CH_3—CO—$OCOC_6H_5$	Co	80	50	73	[46]
n-C_4H_9I and CH_3CO_2Na	n-C_4H_9—CO—$OCOCH_3$	Co	80	50	60	[46]

Phenols are not reactive: the reactions are limited to aliphatic and benzylic alcohols. The reactivities of the different aliphatic alcohols are in the order [47]: Primary < Secondary < Tertiary, with the noticeable exception of methanol, which is comparable to a tertiary alcohol.

The catalysts employed in aldehyde or alcohol synthesis are always based on cobalt while rhodium and nickel are used for the synthesis of acids; strong acidic media are employed for tertiary alcohol carbonylation.

Oxalates are obtained when oxidative carbonylations are performed using palladium catalysts:

$$2\ ROH + 2\ CO + O_2 \xrightarrow{Pd} ROCO\text{—}COOR$$

3.1. SYNTHESIS OF ALCOHOLS AND ALDEHYDES

Most of the literature concerns methanol hydrocarbonylation [167]. The reaction has been extended to benzylic alcohols [48, 49]:

$$C_6H_5\text{-}CH_2OH + CO + 2\ H_2 \xrightarrow[130\ °C\ —\ 50\ bar]{Co/Ru/NaI} C_6H_5\text{-}CH_2CH_2OH + H_2O$$

Yield = 73%

In a comparable reaction, methyl esters can be homologated into the corresponding ethyl esters [50, 168]:

$$R\text{—}CO_2CH_3 + CO + 2\ H_2 \xrightarrow{Co/Ru/I} R\text{—}CO_2CH_2CH_3 + H_2O$$

$R = CH_3$ Selectivity = 90%
$R = C_6H_5$ Selectivity = 90%

3.2. SYNTHESIS OF CARBOXYLIC ACIDS

The rhodium catalyzed carbonylation of methanol into acetic acid has been in commerical operation for more than 15 years (Monsanto process [51]).

$$CH_3OH + CO \xrightarrow{Rh/CH_3I} CH_3CO_2H$$

In a similar way ethanol yields propionic acid, but the rates are slower; higher aliphatic alcohols are not reactive. Benzyl alcohol can be readily converted into phenyl acetic acid using nickel catalysts [52, 53].

Carbonylation of tertiary alcohols is effective using strong acid catalysis; rearranged tertiary carboxylic acids are obtained [54]:

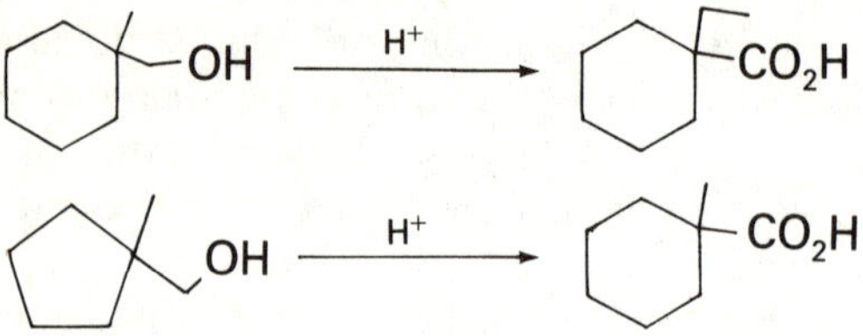

OH $\xrightarrow{H^+}$ CO_2H

OH $\xrightarrow{H^+}$ CO_2H

These reactions have been extended to esters, thus yielding the corresponding anhydrides: for example methyl acetate is carbonylated into acetic anhydride using rhodium, nickel or cobalt catalysts [55—60, 169]:

$$CH_3CO_2{-}CH_3 + CO \xrightarrow{\text{Rh or Ni or Co}} CH_3CO_2{-}COCH_3$$

The reaction is now commercially operated by Eastman-Kodak (240 000 t/a) [61]).

3.3. SYNTHESIS OF OXALATES AND CARBONATES

The palladium catalyzed oxidative carbonylation of alcohols yields the corresponding oxalates [171]:

$$2\ ROH + 2\ CO + O_2 \xrightarrow{Pd} ROCO{-}COOR + H_2O$$

The reaction was first developed by D. M. Fenton [75] in the presence of oxygen (or air) and copper cocatalysts; unfortunately it was slow and scarcely selective.

Addition of different water scavengers and different bases improved the results, but not enough for economic commercialization.

The best results have been obtained by Ube [76, 77] who replaced the classical $CuCl_2/O_2$ redox couple by a $RONO/O_2$ couple, the reaction thus becomes:

$$2\ RONO + 2\ CO \xrightarrow{Pd} ROCO{-}COOR + 2\ NO$$
$$2\ NO + \tfrac{1}{2}\ O_2 \longrightarrow N_2O_3$$
$$N_2O_3 + 2\ ROH \longrightarrow 2\ RONO + H_2O$$

The three steps can be performed in the same reactor and the overall reaction conditions are mild. In the liquid phase *n*-butanol gives dibutyl oxalate in excellent yields using a simple palladium on charcoal catalyst; furthermore, the water/butyl nitrite azeotrope allows facile separation of the water formed.

The preparation of methyl and ethyl oxalates can be achieved in the gas phase under mild conditions (100—120 °C under 1 to 5 bar) but the reaction has to be run in two different steps: the nitrites are first prepared starting from the alcohols and NO/O_2; they are carbonylated afterwards.

Dibutyl oxalate, which is the starting material for oxalic acid and oxamide (6000 and 600 t/year, respectively) synthesis, is prepared nowadays using such a carbonylation route.

The copper catalyzed oxidative carbonylation of methanol yields dimethyl carbonate instead of the oxalate [78]:

$$2\ ROH + 2\ CO + O_2 \xrightarrow{Cu} CO(OR)_2 + H_2O$$

The reaction has been industrialized in Italy. It has been extended to other alcohols such as 1,1,1-trifluoroethanol [79] and phenol [80] using palladium/manganese catalysts.

4. Carbonylation of Nitro Compounds

Carbonylation of nitro compounds yields:

— isocyanates $Ar{-}NO_2 + 3\ CO \rightarrow Ar{-}NCO + 2\ CO_2$
— carbamates $Ar{-}NO_2 + 3\ CO + ROH \rightarrow Ar{-}NHCO_2R + 2\ CO_2$
— ureas $Ar{-}NO_2 + 3\ CO + HNR_2 \rightarrow Ar{-}NHCONR_2 + 2\ CO_2$
— formamides $Ar{-}NO_2 + 3\ CO + H_2 \rightarrow Ar{-}NHCHO + 2\ CO_2$

These syntheses are limited to aromatic nitro compounds.

4.1. ISOCYANATES, CARBAMATES AND UREAS

The direct synthesis of toluene diisocyanate (TDI) starting from dinitrotoluene has been extensively studied since 1960 using sophisticated multimetallic palladium or rhodium based catalysts [62—64]. The best results obtained to date are not sufficient to insure its industrial application.

The reaction has been extended to other phenyl isocyanates. Performed in alcoholic solvents it yields the corresponding carbamates, with better yields using milder conditions, whereas ureas are obtained in presence of amines (Table VIII).

TABLE VIII
Synthesis of isocyanates, carbamates and ureas.

Starting Material	Product	Catalyst	Conditions °C	bar	Yield %	Ref.
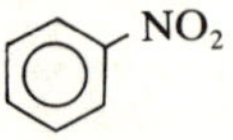 NO_2	 NCO	Rh—Mo	25	50	90	[57, 6]
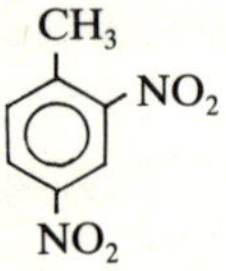 CH_3, NO_2, NO_2	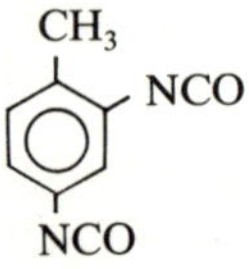 CH_3, NCO, NCO	Pd—Cu—Mo	210	210	73	[62—64]
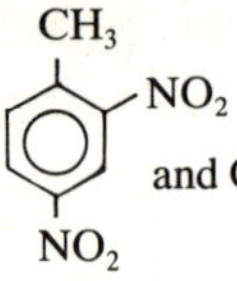 CH_3, NO_2, NO_2 and C_2H_5OH	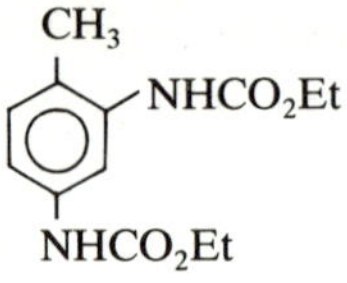 CH_3, $NHCO_2Et$, $NHCO_2Et$	Pd—$FeCl_3$	180	100	95	[69—71]
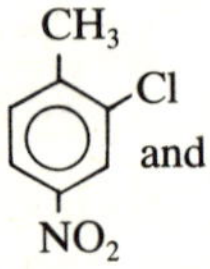 CH_3, Cl, NO_2 and $(CH_3)_2NH$	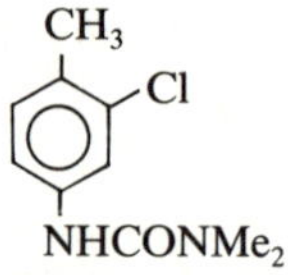 CH_3, Cl, $NHCONMe_2$	S	110	50	96	[66]

In all these reactions the expensive palladium based catalysts can be replaced by elementary selenium [65] (even sulfur [66] in some cases). The yields are good but industrialization of this new type of catalyst might be difficult because of the toxicity and the volatility of selenium.

Some heterocycles have been prepared from orthosubstituted nitro compounds [72, 73]:

$$NO_2,\ OH \quad +3\,CO \longrightarrow \quad H,\ N,\ O \quad C{=}O + 2\,CO_2$$

$$NO_2,\ NH_2 \quad +3\,CO \longrightarrow \quad H,\ N,\ N,\ H \quad C{=}O + 2\,CO_2$$

4.2. SYNTHESIS OF FORMAMIDES

Hydrocarbonylation of aromatic nitro compounds yields the corresponding aromatic formamides [74]:

$$Ar{-}NO_2 + CO/H_2 \xrightarrow{Ru} Ar{-}NHCHO$$

Yield = 52 to 95%

Ar = C_6H_5, *p*-ClC_6H_4, *p*-$MeOC_6H_4$, *p*-$C_6H_5CO{-}C_6H_4$

5. Carbonylation of Amines

The carbonylation of amines is useful to prepare:

— formamides	$R_2NH + CO \longrightarrow$	$R_2N{-}CHO$
— isocyanates	$RNH_2 + CO + M^{2+} \longrightarrow$	$R{-}NCO + 2\,H^+ + M$
— carbamates	$RNH_2 + CO + O_2 + ROH \longrightarrow$	$R{-}NHCO_2R + H_2O$
— symmetrical ureas	$2\,RNH_2 + CO + O_2 \longrightarrow$	$R{-}NHCONH{-}R + H_2O$

5.1. SYNTHESIS OF FORMAMIDES

The synthesis of formamides was discovered by W. Reppe in the '40s using iron, nickel or cobalt catalysts under high pressure. Use of bases (NaOH, NaOMe, etc.) in alcoholic medium has recently allowed much milder conditions; hence dimethylformamide can, for example, be prepared at room temperature under only 5 to 10 bar with excellent yields (Table IX).

TABLE IX
Synthesis of formamides.

Starting Material	Product	Catalyst	Conditions °C	bar	Yield %	Ref.
$(CH_3)_2NH$	$(CH_3)_2N—CHO$	Co	220	200	60	[81]
$(CH_3)_3N$	$(CH_3)_2N—CHO$	Co	220	200	100	[82]
$(CH_3)_2NH$	$(CH_3)_2N—CHO$	OH^-	30	10	98	[83]
(piperidine) NH	(piperidine) N—CHO	Cu Zeolite	300	100	95	[84]

5.2. SYNTHESIS OF ISOCYANATES, CARBAMATES AND UREAS

Isocyanates are readily affordable by oxidative carbonylation of the corresponding amines using palladium/copper (or iron) catalysts; for example using stoichiometric amounts of Pd(II) [85].

$$\text{(Me, } CH_2NH_2\text{, Me, } CHMe_2 \text{ substituted hydrophenanthrene)} + CO + Pd^{2+} \longrightarrow \text{(Me, } CH_2NCO\text{, Me, } CHMe_2 \text{ substituted hydrophenanthrene)} + Pd^0 + 2\,H^+$$

These isocyanates are in fact very rarely isolated, since they react immediately with the starting amine to yield the corresponding symmetrical ureas; carbamates are obtained when alcoholic solvents are used (Table X).

Interesting results have recently been obtained using tellurium catalysts [90]; due to the intrinsic instability of tellurium-dihydride a direct "dehydrogenative carbonylation" is feasible:

$$2\,RNH_2 + CO + Te \longrightarrow RNHCONHR + TeH_2$$
$$TeH_2 \longrightarrow Te + H_2$$

6. Carbonylation of Alkenes

The carbonylation of alkenes is used to prepare:

— aldehydes and alcohols $RCH{=}CH_2 + CO/H_2 \longrightarrow RCH_2CH_2CHO/RCH_2CH_2CH_2C$

— carboxylic acids and esters $RCH{=}CH_2 + CO + ROH \longrightarrow RCH_2CH_2CO_2R$

— amines $RCH{=}CH_2 + CO + H_2O + HNR_2 \longrightarrow RCH_2CH_2CH_2NR_2$

— ketones $RCH{=}CH_2 + CO + RX \longrightarrow RCH{=}CH—CO—R + HX$

TABLE X
Synthesis of carbamates and ureas.

Starting Material	Product	Catalyst	Conditions °C	Conditions bar	Yield %	Ref.
n-BuNH$_2$	BuNH—CO—NHBu	Pd—Cu	40	1	90	[86]
		Se	100	1	100	[87]
$C_6H_{11}NH_2$	C_6H_{11}NH—CO—NHC_6H_{11}	$Mn_2(CO)_{10}$	200	130	58	[88]
		Te	140	30	12	[90]
C_6H_5—NH_2	C_6H_5—$NHCO_2Et$	Pd—Cu	150	150	90	[89]
CH_3, NH_2, NO_2 (benzene ring)	CH_3, $NHCO_2Et$, $NHCO_2Et$ (benzene ring)	Pd—$FeCl_3$	170	100	85	[69]

Almost any alkene can be used in these reactions: terminal or internal alkenes, mono or polyfunctionnal, simple or conjugated.

Whatever the catalyst used (Co, Rh, Pd, Ru . . .) the active catalytic species are usually metallic hydrido species and the first step of the overall process is an insertion of the C═C double bond into the hydrido M—H bond; two different final products are therefore always possible:

$$\text{M—H} + \text{RCH=CH}_2 \rightarrow \text{M–CH}_2\text{CH}_2\text{R} \rightarrow \text{M–C(=O)–CH}_2\text{CH}_2\text{R} \rightarrow \text{R–CH}_2\text{CH}_2\text{–C(=O)–Z}$$

$$\text{M—H} + \text{RCH=CH}_2 \rightarrow \text{M–CH(R)–CH}_3 \rightarrow \text{M–C(=O)–CH(R)CH}_3 \rightarrow \text{R(CH}_3\text{)CH–C(=O)–Z}$$

The reaction's selectivity will depend on the nature of both the metal and the ligand as well as on the experimental conditions (temperature, pressure, solvent).

6.1. SYNTHESIS OF ALDEHYDES AND ALCOHOLS

The so-called "oxo reaction" is of tremendous industrial importance [91]: butyraldehyde and several higher aliphatic aldehydes and alcohols (detergents) are prepared industrially all over the world by such hydrocarbonylation reactions using either cobalt (Ruhrchemie, BASF, Shell, Atochem

. . . processes) or rhodium (Union Carbide, Mitsubishi, Ruhrchemie—Rhone—Poulenc . . . processes).

The reaction has therefore been thoroughly studied (Table XI); both the operating conditions and the final selectivities depend to a high degree on the exact nature of the catalytic system used:

— cobalt is active at high temperature under high pressure; linear aldehydes are produced (selectivities ~ 80%).
— addition of alkyl phosphines (PBu_3 for example) changes the selectivity toward linear alcohols (selectivities ~ 90%); the better thermal stability of the resulting catalyst allows lower pressures (100 to 150 bar) despite the higher temperatures used. Internal olefins can be used to prepare linear terminal alcohols.
— rhodium is more active than cobalt; it is therefore used in much lower concentrations. Good selectivities are obtained at high pressure (over 200 bar)
— addition of triphenylphosphine does not alter the rhodium activity and enhances the catalyst's selectivity at low pressure; 90% linear butyraldehyde is produced under only 10 to 30 bar when operating at high to very high (20 to 100) phosphorous/rhodium ratios.

Other catalysts are known but they are generally less active.

Asymmetric hydroformylation is possible. Starting from vinyl acetate optical yields of 50% enantiomeric excess (ee) have been obtained with chiral rhodium catalysts [105]; a 95% optical yield has even been described with styrene using a chiral platinum—tin catalyst [109].

If hydrosilanes $HSiR_3$ are used in these reactions instead of hydrogen, silyl enol ethers will be formed [110]:

$$\text{cyclohexene} + CO + HSiMeEt_2 \xrightarrow{Co} \text{cyclohexylidene}{=}CH{-}OSiMeEt_2$$

Yield = 89%

Mixtures of carbon monoxide and water can be used instead of CO/H_2. It has been demonstrated that the reaction mechanism is significantly different from that of the classical CO/H_2 reaction but, from a formal point of view, everything goes on "as if" an internal water-gas-shift reaction occurred thus generating the hydrogen necessary for the hydroformylation step:

$$RCH{=}CH_2 + 2\ CO + H_2O \longrightarrow RCH_2CH_2{-}CHO + CO_2$$

TABLE XI
Synthesis of aldehydes and alcohols by hydroformylation

Starting Material	Product	Catalyst	Conditions °C	Conditions bar	Yield %	Ref.
$CH_3CH{=}CH_2$	$CH_3CH_2CH_2{-}CHO$	Co	150	200	80	[92]
		Rh	100	30	90	[91]
$C_3H_7CH{=}CH_2$	$C_5H_{11}{-}CHO$	Co/PBu_3	150	40	90	[93]
$C_6H_{13}CH{=}CH_2$	$C_9H_{19}{-}CHO$	Rh/PPh_3	90	10	80	[94]
$C_3H_7CH{=}CH{-}CH_3$	n-$C_6H_{13}{-}CH_2OH$	Co/PBu_3	150	40	68	[96]
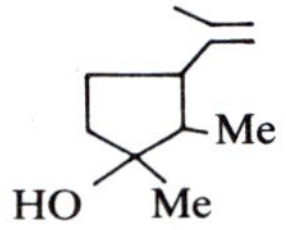 Me, HO, Me	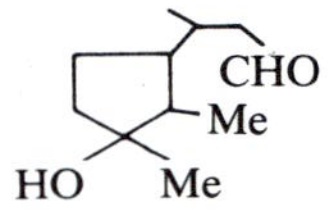 CHO, Me, HO, Me	Rh	100	80	65	[95]
 O	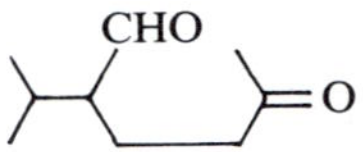 CHO, O	Rh	70	650	85	[91]
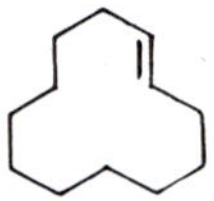	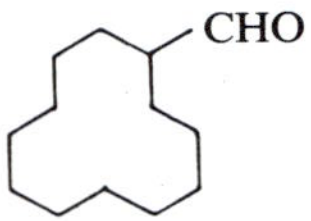 CHO	Rh	60	60	100	[97]
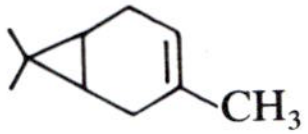 CH_3	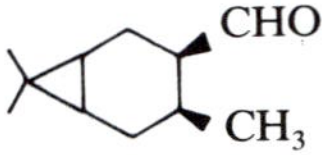 CHO, CH_3	Rh	100	60	ND	[98]
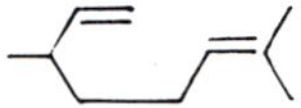	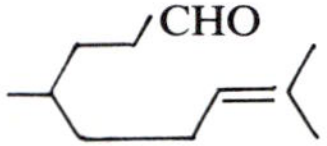 CHO	Rh	70	10	90	[99]

TABLE XI (Continued)

Starting Material	Product	Catalyst	Conditions °C	Conditions bar	Yield %	Ref.
C_6H_5—CH=CH_2	C_6H_5—CH(CHO)CH_3	Rh	25	1	90	[100]
C_6F_5—CH=CH_2	C_6F_5—CH(CHO)CH_3	Rh	90	100	98	[101]
2-thienyl—CO—C_6H_4—CH=CH_2	2-thienyl—CO—C_6H_4—CH(CHO)CH_3	Pd—HCl			90	[102]
CH_2=CH—CH_2OH	HO—$(CH_2)_3$—CHO	Rh	80	50	75	[103]
CH_2=C(CH_3)CH_2OH	HO—CH_2CH(CH_3)CH_2—CHO	Rh	80	80	85	[104]
CH_2=CH—CO_2Me	MeOCO—CH(CHO)CH_3	Rh	110	80	85	[91]
N-succinimidyl—CH=CH_2	N-succinimidyl—CH(CHO)CH_3	Rh	60	35	100	[106]

TABLE XI (Continued)

Starting Material	Product	Catalyst	Conditions °C	Conditions bar	Yield %	Ref.
N–$COCH_3$ (cyclic, ring with C=C)	N–$COCH_3$ (cyclic), –CHO	Rh	60	35	100	[106]
CF_3—CH=CH_2	CF_3—CH(CHO)(CH_3)	Rh	100	150	72	[107]
$(MeO)_2P(=O)$—CH=CH_2	$(MeO)_2P(=O)$—CH(CHO)(CH_3)	Rh	80	600	80	[108]

The most active catalysts are ruthenium [111, 112] and iron [113, 114] (note that these metals are not good catalysts for the classical oxo reaction). Hydrogen is usually not detected during the initial course of the reaction, thus providing good evidence for a specific mechanism. This new "oxo" reaction is known as the Reppe reaction; it is often performed in the presence of bases (sodium hydroxide, tertiary amine, ...) but acidic catalysis is also known.

When performed in the presence of primary or secondary amines the Reppe reaction yields the homologated amine resulting from the hydrogenation of the imino intermediate formed by condensation of the starting amine with the oxo product:

$$R{-}CH{=}CH_2 + 3\ CO + H_2O + HNR'_2 \longrightarrow RCH_2CH_2{-}CH_2NR'_2 + 2\ CO_2$$

This is the so-called aminomethylation reaction (Table XII) [115, 130].

6.2. SYNTHESIS OF CARBOXYLIC ACIDS

Hydroxycarbonylation is very similar in scope to the preceding hydroformylation: almost any alkene can be used as starting material. The best results are obtained with palladium catalysts in acidic medium (and not with rhodium as in the oxo reaction). Cobalt is only active in basic medium at high pressure; it is then capable to hydroxycarbonylate an internal olefin into a terminal ester (Table XIII).

The reaction has been extended to amide and thioester synthesis [119]:

$$C_5H_{11}{-}CH{=}CH_2 + CO + HSEt \xrightarrow{Pd/SnCl_2/PR_3} n\text{-}C_7H_{15}{-}COSEt$$

Yield = 78%

In the presence of carbon tetrachloride, β-trichlomethyl esters are formed [120]:

$$\text{(3,4-methylenedioxyphenyl)}{-}CH_2CH{=}CH_2 + CO + CCl_4 + EtOH \xrightarrow[50\,^\circ C\,-\,50\ bar]{Pd/PR_3} \text{(3,4-methylenedioxyphenyl)}{-}CH_2{-}CH(CO_2Et){-}CH_2{-}CCP_3$$

In strong acidic medium (H_2SO_4, HF) metallic catalysts are not necessary; rearranged carboxylic acids are often obtained [121, 122]:

$$CH_2{=}C(CH_3)(CH_2Br) + CO + H_2O \xrightarrow{H_2SO_4/SO_3} BrCH_2{-}C(CH_3)_2{-}CO_2H$$

Yield: 90%

TABLE XII
Synthesis of amines by aminomethylation

Starting Material	Product	Catalyst	Conditions °C	Conditions bar	Yield %
C_3H_7—CH=CH_2 + HN O (morpholine ring)	*n*-C_5H_{11}—CH_2—N O (morpholine ring)	Ru—Fe	150	100	94
(cyclohexene ring) + $HNMe_2$	(cyclohexane ring)—$CH_2N\ Me_2$	Rh	140	100	90
$(CH_3)_2$C=CH_2 + HN (piperidine ring)	$(CH_3)_2$CH—CH_2—CH_2—N (piperidine ring)	Rh	—	—	70

TABLE XIII
Synthesis of carboxylic acids and esters.

Strategic Material	Product	Catalyst	Conditions °C	Conditions bar	Yield %	Ref.
$CH_2{=}CH_2$	$CH_3CH_2{-}CO_2H$	Rh/Br	150	30	99	[116]
$n\text{-}C_5H_{11}{-}CH{=}CH{-}C_5H_{11}$	$n\text{-}C_{12}H_{25}{-}CO_2Me$	Co/NR_3	170	200	75—80	[117]
N O —$CH{=}CH_2$	N O —CH(CH_3)CO_2Me	Pd/HCl	100	150	90	[118]

$$\text{(octalone)} + CO + H_2O \xrightarrow{HF/SbF_5} \text{(decalone carboxylic acid)}$$

Yield: 90%

6.3. SYNTHESIS OF KETONES

Some specific ketone syntheses have recently been developed starting from dienes and aliphatic halides (Table XIV).

$$\text{diene} + CO + RX \xrightarrow[-Hx]{Co} \text{diene–CO—R}$$

Addition of supplementary nucleophile allows an easy double functionnalisation of dienes (Table XIV):

$$\text{diene} + CO + RX + CH^{-}\langle^{Z}_{Z'} \xrightarrow{Co} R-\overset{O}{\overset{\|}{C}}\text{–CH}_2\text{CH=CHCH}_2\text{–}CH\langle^{Z}_{Z'} + X^{-}$$

6.4. OXIDATIVE CARBONYLATION OF ALKENES

Depending on the carbon monoxide pressure and on the nature of the reaction mixture (acidic or basic) the oxidative carbonylation of alkenes can be directed toward unsaturated esters, diesters or ether-esters (Table XV).

$$RCH{=}CH_2 + R'OH + CO/O_2 \longrightarrow \begin{cases} R{-}CH{=}CH{-}CO_2R' \\ RCH(CO_2R'){-}CH_2CO_2R' \\ RCH(OR'){-}CH_2CO_2R' \end{cases}$$

The formation of these different products can be explained mechanistically.

In acidic or neutral medium the alcohol reacts with the coordinated olefin thus yielding an alkoxy palladium intermediate; subsequent carbonylation leads to the corresponding ether-ester.

$$X_2Pd(CO)\leftarrow\| + HOR \xrightarrow[-HX]{} X\text{–}Pd(CO)\text{–}CH_2CH_2OR \xrightarrow{CO} X_2Pd\text{–}CO\text{–}CH_2CH_2OR \xrightarrow[RX]{ROH} Pd + ROCH_2CH_2COOR$$

TABLE XIV
Synthesis of ketones from dienes and aliphatic halides.

Starting Material	Product	Catalyst	Conditions °C	bar	Yield %	Ref.
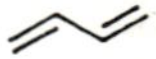 and CH_3I	$CH_3—CO—CH{=}CH—CH{=}CH_2$	Co	70	5	56	[125]
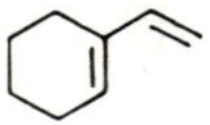 and CH_3I	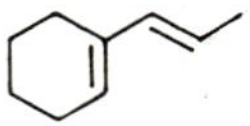 $CO—CH_3$	Co	20	1	ND	[126]
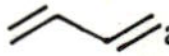 and $BrCH_2CO_2Et$ and $CH^-(CO_2Me)_2$	$EtO_2CCH_2—CO—CH_2—CH{=}CH—CH_2—CH(CO_2Me)_2$	Co	20	1	47	[127]
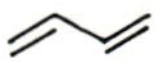 and CH_3I and $CH^-(CO_2Et)_2$	$CH_3CO—CH_2—CH{=}CH—CH_2—CH(CO_2Me)$	Co	20	1	49	[128]

TABLE XV
Oxidative carbonylation of alkenes.

Starting Material	Product	Catalyst	Conditions °C	Conditions bar	Yield %	Ref.
Ar—CH=CH$_2$	ArCH=CH—CO$_2$CH$_3$	Pd—Cu	30	1	80	[126]
	CH$_3$OCO CO$_2$CH$_3$	Pd—Cu	100	120	80	[127]
Ar—CH=CH$_2$	Ar—CH(CO$_2$CH$_3$)—CH$_2$CO$_2$CH$_3$	Pd—Cu	50	15	70	[126]
	CO$_2$Me, CO$_2$Me	Pd—Cu	50	10	85	[128]

In basic medium, the alcoholate anion reacts with the coordinated carbon monoxide thus yielding a carboxy palladium intermediate which then inserts the olefin to yield the carboxyalkyl palladium species.

X—Pd(X)(C≡O) ← ‖ + $\bar{O}$R $\xrightarrow{-X^-}$ Pd(X)(C(=O)—OR) ← ‖ ⟶ Pd(X)—CH$_2$CH$_2$—CO$_2$R $\xrightarrow{-HX}$ =/COOR + Pd

If the carbon monoxide pressure is sufficient an acyl palladium intermediate will be formed which will lead to the expected β-diester.

X—Pd—CH$_2$CH$_2$—CO$_2$R + CO ⟶ X—Pd—CO—CH$_2$CH$_2$—CO$_2$R $\xrightarrow[-HX]{ROH}$ Pd + RO$_2$C—CH$_2$CH$_2$—CO$_2$R

If, on the contrary, the carbon monoxide pressure is low, a β-elimination reaction will take place, yielding the unsaturated ester.

7. Carbonylation of Alkynes

The alkyne carbonylation reactions are used to prepare:

— unsaturated acids $RC{\equiv}CH + CO + H_2O \longrightarrow R{-}CH{=}CH{-}CO_2H$

— α, α' unsaturated diacids $RC{\equiv}CH + 2\,CO + 2\,H_2O + \frac{1}{2}\,O_2 \longrightarrow R{-}C(CO_2H){=}CH{-}($

— dicarboxylic acids $RC{\equiv}CH + 2\,CO + 2\,H_2O \longrightarrow RCH(CO_2H){-}CH_2CO$

— hydroquinones $2\,RC{\equiv}CH + 2\,CO + H_2 \longrightarrow$ HO-C$_6$H$_2$R$_2$-OH (hydroquinone ring with two R substituents)

All these reactions were extensively studied during and just after the last war; acrylic acid today is prepared industrially by acetylene carbonylation (BASF in Germany, Dow—BASF and Rohm and Haas in the U.S., Toagosei in Japan; 350 000 t/a worldwide).

Unexpectedly, however, they have not been thoroughly investigated recently; considerable progress is therefore to be expected in this area in the near future.

7.1. SYNTHESIS OF UNSATURATED ACIDS

This is the best known alkyne carbonylation reaction. It has been industrialized in the special case of acrylic acid synthesis [129].

$$HC{\equiv}CH + CO + H_2O \xrightarrow{Ni/Cu} CH_2{=}CH{-}CO_2H$$

The reaction has been extended to unsymmetrical alkynes (Table XVI).

Usually performed at high temperature using nickel (or palladium) catalysts, the reaction is effective at room temperature using stoichiometric quantities of nickel tetracarbonyl:

$$4\,RC{\equiv}CH + 4\,H_2O + Ni\,(CO)_4 + 2\,HCl \longrightarrow 4\,RCH{=}CH{-}CO_2H + NiCl_2 + H_2$$

Use of cobalt instead of nickel catalysts favors succinic derivatives (Table XVI). Use of an alcoholic solvent allows the synthesis of the corresponding esters but yields are generally lower and palladium catalysts seem then preferable.

Unsaturated diesters can be directly prepared under mild conditions using palladium catalysts in the presence of thiourea (Table XVI).

7.2. SYNTHESIS OF HYDROQUINONES

Hydroquinone derivatives are prepared by hydrocarbonylation of acetylenic compounds:

TABLE XVI
Synthesis of unsaturated acids from alkynes.

Starting Material	Product	Catalyst	Conditions °C	Conditions bar	Yield %	Ref.
$HC{\equiv}CH$	$CH_2{=}CH{-}CO_2H$	Ni–Cu	200	50	90	[129]
$n\text{-}C_6H_{13}{-}C{\equiv}CH$	$n\text{-}C_6H_{13}{-}CH{=}CHCO_2H$	Ni–Cu	200	50	50	[129]
$ClCH_2{-}C{\equiv}CH$	$CH_2{=}C(CO_2CH_3)(CH_2CO_2CH_3)$	Pd	50	100	70	[131]
$HC{\equiv}CH$	$MeOCO{-}CH{=}CH{-}CO_2CH_3$	Pd/thiourea	30	50	90	[132]
$HC{\equiv}CH$	$MeOCO{-}CH_2{-}CH_2{-}CO_2CH_3$	Co	90	150	85	[133]

$$2\,RC{\equiv}CH + 2\,CO + H_2 \xrightarrow{\text{Fe or Ru}} \text{2,5-R}_2\text{-hydroquinone (OH, OH, R, R)}$$

The reaction can either be performed:

— under mild conditions (60 °C — 20 bar) using stoichiometric quantities of iron pentacarbonyl [134]
— under more severe conditions using catalytic quantities of iron [135] or ruthenium [136, 137] catalysts.

Hydroquinone has thus been obtained with a 73% yield at 250 °C under 200 bar of carbon monoxide using a ruthenium catalyst [138].

The reaction stoichiometry imposes the presence of hydrogen but it is very often conducted with CO/water mixtures instead of CO/H_2:

$$2\,RC{\equiv}CH + 3\,CO + H_2O \longrightarrow \text{2,5-R}_2\text{-hydroquinone (OH, OH, R, R)} + CO_2$$

Once again, everything behaves "as if" an initial water-gas-shift reaction were taking place.

8. Carbonylation of C—H Bonds

The C—H carbonylation reactions can be used to prepare:

— aldehydes $RH + CO \longrightarrow RCHO$
— acids $RH + CO + \frac{1}{2}\,O_2 \longrightarrow RCO_2H$
— ketones $RH + CO + R'H + \frac{1}{2}\,O_2 \longrightarrow RCOR' + H_2O$

They are restricted to aromatic or activated aliphatic C—H bonds. We shall see that all types of catalysis, organometallic but also meer acidic and basic catalysis, have been used.

8.1. SYNTHESIS OF ALDEHYDES

The direct carbonylation of toluene into *p*-tolualdehyde has received much attention. Good results have been obtained using HF/BF_3 catalysts under low pressure [139].

$$C_6H_5CH_3 + CO \xrightarrow{HF/BF_3} p\text{-}CH_3C_6H_4CHO$$

The first step is the stoichiometric complexation of toluene by HF/BF_3:

$$C_6H_5CH_3 + HF + BF_3 \longrightarrow C_6H_5CH_3\text{—}HF, BF_3$$

The intermediate toluene/HF/BF_3 complex is then carbonylated (20 to 50 °C under 1 to 5 bar) into the corresponding *p*-tolualdehyde complex:

$$C_6H_5CH_3\text{—}HF, BF_3 + CO \longrightarrow p\text{-}CH_3C_6H_4CHO\text{—}HF, BF_3$$

which is subsequentely decomposed into the final free *p*-tolualdehyde by simple heating:

$$p\text{-}CH_3C_6H_4CHO\text{—}HF, BF_3 \longrightarrow p\text{-}CH_3C_6H_4CHO + HF + BF_3$$

Excellent yields (92 to 97% and regioselectivities (*para/ortho* = 94%) have hitherto been obtained by Mitsubishi Gas Chem. The three chemical steps are actually performed separately. The reaction has been extended to other substrates (Table XVII).

Formylation of heterocyclic compounds (Table XVII) is usually performed in alcoholic medium using basic conditions; intermediate formation of formate species is probable:

$$CO + ROH \underset{}{\overset{RO^-}{\rightleftharpoons}} HCOOR$$

$$\text{(2-H-heterocycle, X)} + HCOOR \xrightarrow{RO^-} \text{(2-CHO-heterocycle, X)} + ROH$$

TABLE XVII
Synthesis from aldehydes by carbonylation of C—H bonds

Starting Material	Product	Catalyst	Conditions °C	bar	Yield %	Ref.
	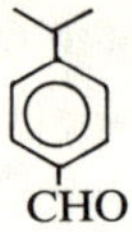 CHO	HF/BF_3	50	10	82	[140]
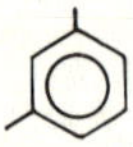	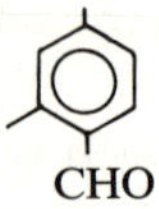 CHO	HF/BF_3	50	10	96	[140]
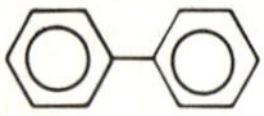	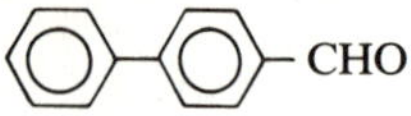 CHO	$AlCl_3$	45	2	73	[141]
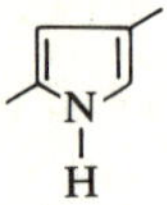 N H	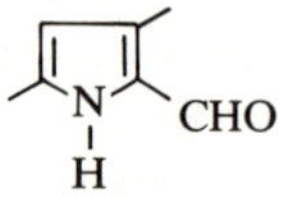 N H CHO	NaOEt				[142]
C_6H_5—CH_2—CN	C_6H_5—CH(CN)(CHO)	NaOEt	40	50	92	[143]

The same reaction conditions are used to formylate activated aliphatic compounds (Table XVII).

8.2. SYNTHESIS OF CARBOXYLIC ACIDS

The reaction is an oxidative carbonylation since oxygen is necessary to account for the stoichiometry:

$$RH + CO + \tfrac{1}{2}\,O_2 \longrightarrow RCO_2H$$

Different oxydants can be used, depending on the reaction conditions, catalysts and starting materials.

8.2.1. *Synthesis of aromatic carboxylic acids*

Most of the aromatic C—H carbonylations have been performed under mild conditions using stoichiometric amounts of palladium(II) salts (Table

XVIII); these salts act as both the carbonylation catalyst and the oxidant; inactive palladium(0) is recovered at the end of the reaction:

$$Ar{-}H + CO + PdX_2 \xrightarrow{ROH} ArCO_2R + Pd(0) + 2\,HX$$

Few attempts have been made to render the reaction truly catalytic in palladium by adding some external oxidants such as $CuCl_2$, $FeCl_3$, O_2.

This difficulty can be obviated starting from organometallic derivatives of the initial aromatic compounds. For example [148, 149] aromatic organothallium compounds, which are readily obtained from Ar—H and thallium salts, are smoothly carbonylated to give aromatic esters using catalytic quantities of palladium (Table XVIII).

$$ArH + Tl(OCOCF_3)_3 \longrightarrow Ar{-}Tl(OCOCF_3)_2 + CF_3CO_2H$$
$$Ar{-}Tl(OCOCF_3)_2 + CO + ROH \longrightarrow Ar{-}CO_2R + Tl(OCOCF_3) + CF_3CO_2H$$

Phenol substrates allow completely different reaction conditions (Table XVIII); organometallic catalysts are not necessary and the reactions can be directly performed using sodium or potassium carbonates:

$$Ar{-}H + Na_2CO_3 + CO \longrightarrow Ar{-}CO_2Na + HCO_2Na$$

One interesting arylacetic synthesis has been proposed recently by UBE, starting from benzene and formaldehyde [152]:

$$C_6H_6 + CO + HCHO \xrightarrow{Rh/HBr} C_6H_5{-}CH_2CO_2H$$

The reaction is catalyzed by rhodium; bromide co-catalysts are used.

8.2.2. *Synthesis of aliphatic carboxylic acids*

Aliphatic acids are prepared in strong acidic media (H_2SO_4, HF, etc.) using either an olefin or an alcohol as “oxidant” [147].

$$RH + CO + {>}C{=}C{<} + H_2O \xrightarrow{H^+} R{-}CO_2H + {>}CH{-}CH{<}$$

$$RH + CO + R'OH \xrightarrow{H^+} R{-}CO_2H + R'H$$

These special “oxidants” are of course partially carbonylated into acidic derivatives during the course of the main reaction; complex mixtures will accordingly always result, and the type of olefin (or the alcohol) must therefore be chosen with care. Because of the strong acidic medium used, rearranged products are, furthermore, usually obtained; for example [147]:

TABLE XVIII
Synthesis of carboxylic acids by carbonylation of C—H bonds.

Starting Material	Product	Catalyst	Conditions °C	bar	Yield %	Ref.
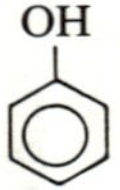 OH (phenol)	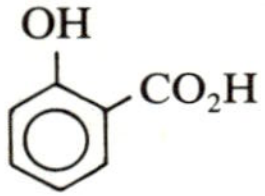 OH, CO_2H	$PdCl_2$ Stoichiometric	20	1	53	[146]
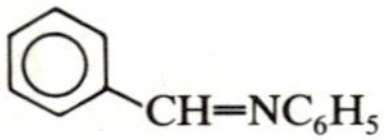 $CH{=}NC_6H_5$	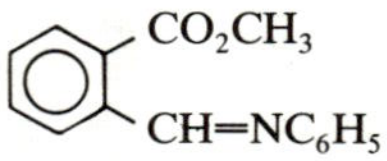 CO_2CH_3, $CH{=}NC_6H_5$	$PdCl_2$ Stoichiometric	50	1	55	[145]
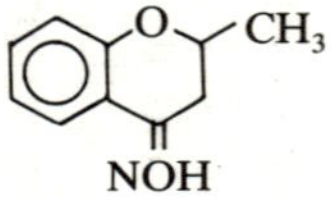 O, CH_3, NOH	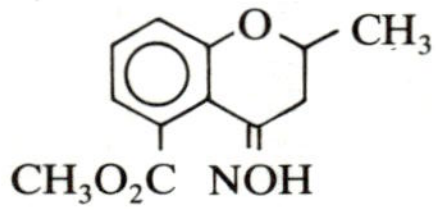 O, CH_3, CH_3O_2C NOH	$PdCl_2$ Stoichiometric	50	1	40	[144]
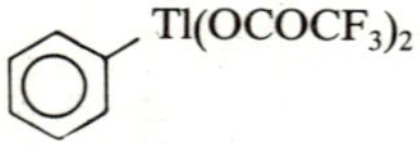 $Tl(OCOCF_3)_2$	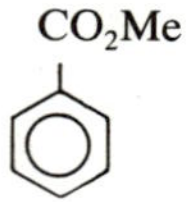 CO_2Me	Pd	20	1	75	[148, 149]
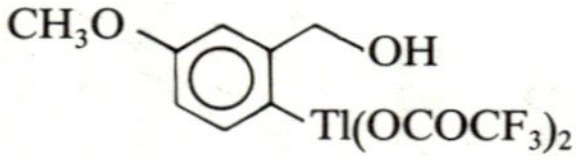 CH_3O, OH, $Tl(OCOCF_3)_2$	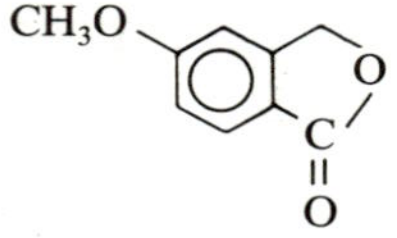 CH_3O, O, C=O	Pd		1	89	[148, 149]

TABLE XVIII (Continued)

Starting Material	Product	Catalyst	Conditions °C	bar	Yield %	Ref.
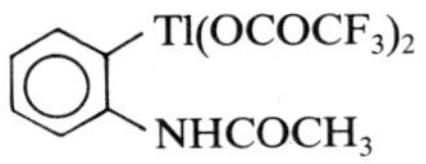 $Tl(OCOCF_3)_2$, $NHCOCH_3$ (ortho-substituted benzene)	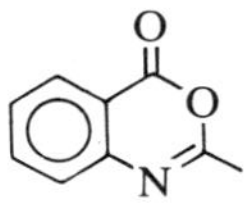 O, O, N (benzoxazinone ring, 2-methyl)	Pd	20	1	89	[148, 149]
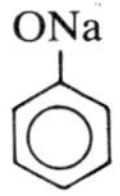 ONa (sodium phenoxide)	 ONa, CO_2Na (para-substituted benzene)	Na_2CO_3	260	60	90	[150]
		K_2CO_3	350	100	80	[151]

$$(CH_3)_2CH{-}CH_2CH_3 + \text{cyclohexene} \longrightarrow \begin{cases} CH_3CH_2{-}C(CH_3)_2{-}CO_2H & \text{Yield} = 37\% \\ + \text{cyclohexane} \\ + \text{1-methylcyclopentanecarboxylic acid } (C_5H_8(CH_3)CO_2H) & \text{Yield} = 50\% \end{cases}$$

Strained alicyclic compounds yield acyclic carboxylic acids; in this special reaction no extra oxidant is needed since the starting molecule acts both as the substrate and the oxidant [153]:

$$\text{cyclo-}(CH_2{-}CH_2{-}CH_2) + CO + H_2O \xrightarrow{H^+} (CH_3)_2{-}CH{-}CO_2H$$

Yield = 80%

8.3. SYNTHESIS OF KETONES

The direct transformation of a C—H bond into a ketone is not a general reaction; it is usually restricted to benzene, which is transformed into benzophenone or anthraquinone [154]; and cyclopropane, which is transformed into cyclobutanone [155].

The former reaction is an oxidative carbonylation:

$$2\, C_6H_5{-}H \xrightarrow[-2\,[H]]{+CO} C_6H_5{-}CO{-}C_6H_5 \xrightarrow[-2\,[H]]{+CO} \text{anthraquinone}$$

It is performed with palladium catalysts in the presence of stoichiometric quantities of $CuCl_2$ or $FeCl_3$, at 250 °C under 60 bar.

The reaction conditions can be considerably improved when starting from mercury derivatives [156, 157]:

$$2\,Z\text{-}C_6H_4\text{-}HgCl + CO \xrightarrow[25\,^\circ C - 1\,bar]{Rh} Z\text{-}C_6H_4\text{-}C(=O)\text{-}C_6H_4\text{-}Z$$

Z	Yield
H	99%
p-CH_3	87%
p-CH_3O	80%

A quite unexpected aryl-alkyl ketone synthesis has recently been described, starting from *p*-di-*tert*-butylbenzene [158]:

$$p\text{-}(t\text{-}Bu)_2C_6H_4 + CO \xrightarrow[25^\circ C - 1\,bar]{AlCl_3} p\text{-}t\text{-}Bu\text{-}C_6H_4\text{-}CO\text{-}tBu$$

Yield = 86%

9. Synthesis of Amino Acids

In the presence of primary amides, aldehydes can be directly carbonylated into amino acids:

$$R\text{—}CHO + CO + R'CONH_2 \xrightarrow{Co} R\text{—}CH(NHCOR')\text{—}CO_2H$$

Discovered by H. Wakamatsu in 1970 [159], the reaction has been extended:

— to olefins [159—161]:

$$R\text{—}CH{=}CH_2 + 2\,CO + H_2 + R'CONH_2 \longrightarrow RCH_2CH_2\text{—}CH(NHCOR')\text{—}CO_2H$$

— to benzyl halides [162, 163].

$$Ar\text{—}CH_2Cl + CO + H_2 + R'CONH_2 \longrightarrow Ar\text{—}CH_2CH(NHCOR')\text{—}CO_2H$$

The reaction is catalyzed by cobalt; hydrogen is necessary whatever the starting material, which can be aldehydes, olefins or benzyl halides. The yields are generally good (50% to 98%) and the selectivities are excellent.

P. Pino [164] has shown that the rather drastic reaction conditions (100 to 150 °C under 200 to 300 bar) initially used by H. Wakanatsu were not necessary; good (or even better) results are obtainable under much milder conditions: 70 to 100 °C under 100 to 150 bar.

The reaction mechanism has been studied by P. Pino [164] and is summarized here:

$$R{-}CHO + H_2N{-}COR' \rightleftharpoons RCH(OH)(NHCOR') \underset{}{\overset{-H_2O}{\rightleftharpoons}} RCH(NHCOR')_2$$

$$RCH(OH)(NHCOR') \xrightarrow{HCo(CO)_4} RCH(Co(CO)_4)(NHCOR') \xrightarrow{CO} RCH(COCo(CO)_4)(NHCOR')$$

(A): $RCH(COCo(CO)_4)(NHCOR') \longrightarrow RCH(COOH)(NHCOR')$

(B): $RCH(COCo(CO)_4)(NHCOR') \longrightarrow RCH(C(=O){-}Co(CO)_4)(N{=}C(OH){-}R') \longrightarrow RCH\langle C(=O){-}O{-}C(R){=}N \rangle \xrightarrow{+H_2O} RCH(COOH)(NHCOR')$

It is not at present possible to know which of the elementary steps A and/or B are actually followed; both are possible.

The intermediacy of the *gem* amido-alcohol has recently been demonstrated by Ajinomoto [165], who succeeded in preparing oxazolones starting from such an amido-alcohol in anhydrous medium:

$$\text{R—CH}\begin{matrix}\diagup \text{OH} \\ \diagdown \text{NHCOR}'\end{matrix} + \text{CO} \xrightarrow[\text{sieve 4 Å}]{\text{Co}_2(\text{CO})_8} \text{RCH}\begin{matrix}\diagup \text{CO}-\text{O} \\ \quad\quad\; | \\ \diagdown \text{N}{=}\text{CH}-\text{R}'\end{matrix}$$

Starting from olefins (or benzyl halides) the first step consists of the *in situ* synthesis of the necessary aldehyde:

$$\text{RCH}{=}\text{CH}_2 + \text{CO} + \text{H}_2 \xrightarrow{\text{Co}} \text{RCH}_2\text{CH}_2\text{—CHO}$$

$$\text{RCH}_2\text{CH}_2\text{CHO} + \text{CO} + \text{H}_2\text{NCOR}' \xrightarrow{\text{Co}} \text{RCH}_2\text{CH}_2\text{—CH}\begin{matrix}\diagup \text{CO}_2\text{H} \\ \diagdown \text{NHCOR}'\end{matrix}$$

Amino acids can also be prepared by a stoichiometric condensation of the Collman reagent $Na_2Fe(CO)_4$ with a starting mixture of acyl chloride and imidoyl chloride [166]:

$$\text{R}_1\text{COCl} + \text{R}_2\underset{\text{Cl}}{\underset{|}{\text{C}}}{=}\text{NR}_3 + \text{Na}_2\text{Fe(CO)}_4 \longrightarrow \text{(oxazolium-5-olate: ring with } \text{R}_2\text{, } \overset{+}{\text{N}}\text{–R}_3\text{, C–R}_1\text{, O, C–O}^-\text{)}$$

$$\text{(oxazolium-5-olate with } \text{R}_2\text{, N–R}_3\text{, R}_1\text{, }{}^-\text{O)} + \text{EtOH} \longrightarrow \text{R}_2\text{—}\underset{\text{CO}_2\text{Et}}{\underset{|}{\text{CH}}}\text{—NR}_3\text{COR}_1$$

Conclusion

We have given some examples of the extraordinary versatility of carbon monoxide in fine organic chemicals synthesis. It is therefore nowadays possible to consider carbon monoxide as a genuine versatile *Reagent* for the fine chemicals industry, thus renouncing to its old definition as a mere raw material solely of the petrochemical industry.

This redefinition of the significance of carbon monoxide is of considerable importance. It leads to fundamental modifications in its research methodology, to a complete re-thinking of its role and interest in organic chemistry and to the necessity to develop new ways to prepare and handle it.

Such important notions as the reagent price, the recycling of the catalyst, the reaction selectivity and the overall process definition must be reconsidered:

— the price of carbon monoxide is a fundamental parameter governing its basic chemical applications: it can be considered of secondary importance only in fine chemicals synthesis.
— good, but not quantitative, yields and selectivities can be sufficient to insure the commercial interest of sophisticated fine chemicals.
— a perfect catalyst recycling is obviously a *sine qua non* condition for petrochemical processes; even if desirable, this condition is not always absolutely necessary for expensive fine chemicals.
— fine chemical processes are often batch or semi-continuous processes, whereas continuous processes are the rule in petrochemical industries; furthermore the fine chemical processes are usually performed in multipurpose plants, and not in reaction-specific ones.

Much work, both in chemical research and in chemical engineering, has surely to be done in order to develop this new chemical tool to a significant industrial level. No doubt these necessary investigations are already on the way in most of the important chemical companies.

Rhône-Poulenc Recherches
Centre de Recherches de Saint-Fons
B.P. 62, 69192 SAINT-FONS Cédex.

References

1. J. Gauthier-Lafaye *et al., Actual. Chim.*, (9), 11, (1983).
2. Y. Colleuille *et al., Techniques de l'Ingénieur* (3), J5780—1, (1985).
3. F. Calderazzo, *Angew. Chem. Int. Ed.*, **16**, 299, (1977).
4. L. Cassar *et al., Synthesis* (9), 509, (1973).
5. K. Kudo *et al., J. Organomet. Chem.*, **52**, 431, (1973).
6. R. F. Heck, *Advances in Organometallic Chemistry*, Vol. 4, 243, Academic Press Inc., (1966).
7. R. F. Heck *et al., J. Am. Chem. Soc.*, **85**, 1220 and 2779 (1963).
8. R. F. Heck *et al., J. Am. Chem. Soc.*, **84**, 2499, (1962) and **83**, 1097 (1961).
9. A. Schoenberg *et al., J. Am. Chem. Soc.* **96**, 7761 (1974).
10. T. Takano *et al.* (Sumitomo) *EP* 34 430 (1980).
11. M. P. Cooke Jr., *J. Am. Chem. Soc.*, **92**, 6080 (1970).
12. A. Schoenberg *et al., J. Org. Chem.*, **39**, 3318 (1974).
13. Y. Ban *et al., J. Synth. Org. Chem.*, **37**, 430 (1979).

14. Valentine Jr. *et al., J. Org. Chem.,* **46**, 4614 (1981).
15. L. Cassar *et al., J. Organomet. Chem.,* **51**, 381 (1973).
16. J. J. Brunet *et al., Tet. Lett.* **22**, 1013 (1981).
17. J. F. Knifton (Texaco), *US Patent* 3.991.101 (1973).
18. R. F. Heck, *FR* 1.313.360 (1960).
19. Montedison, *FR* 2.055.331 (1969).
20. I. Kibayashi *et al.* (Denki Kagaku Kogyo), *J.* 54.128.512, *J.* 55.027.155, and *J.* 55.53241 (1978).
21. K. J. Boosen *et al.* (Lonza), *EP* 47.514 (1980).
22. A. Moro *et al.* (Montedison), *BE* 865.107 (1976).
23. B. Falk *et al.* (Dynamit Nobel), *DT* 2.704.191 (1977).
24. BASF, *FR* 2.310.992 (1975).
25. J. J. Brunet *et al., J. Org. Chem.* **46**, 3147 (1981).
26. L. Cassar, *Ann. N.Y. Acad. Sci.* **333**, 208 (1980).
27. S. Gambarotta *et al., J. Organomet. Chem.* **194**, C 19 (1980).
28. A. Schoenberg *et al. J. Org. Chem.* **39**, 3327 (1974).
29. T. Ito *et al., Bull. Chem. Soc. Jpn.* **48**, 2091 (1975).
30. Y. Watanabe *et al., Bull. Chem. Soc. Jpn* **52**, 1869 (1979).
31. M. Tanaka, *Tet. Lett.* 2601 (1979).
32. T. Kobayashi *et al., J. Organomet. Chem.* **205**, C 27 (1981).
33. K. Kikukawa *et al., Chem. Lett.* 35 (1982).
34. T. Kobayashi *et al., J.C.S., Chem. Comm.* 333 (1981).
35. M. P. Cooke *et al., J. Am. Chem. Soc.* **97**, 6863 (1975).
36. J. P. Collman *et al., J. Am. Chem. Soc.* **94**, 1788 (1972).
37. I. G. Farben, *FR* 671 938 (1928).
38. Hoechst A. G., *EP.* 39413 (1980).
39. D. Medena *et al., Inorg. Chim. Acta* **3**, 255 (1969).
40 National Distillers, *FR* 1.497.185 (1965).
41. W. W. Prichard *et al.* (Du Pont), *US* 3.632.643 (1967).
42. R. Perron (Rhone-Poulenc Ind.), *BE* 837.401 (1975).
43. H. Des Abbayes *et al., J.C.S., Chem. Comm.* 1090, (1978).
44. T. Kobayashi *et al., J. Organomet. Chem.* **233**, C 64 (1982).
45. F. Ozawa *et al., Tet. Lett.* **23**, 3383, (1982).
46. J. Gallucci *et al.* (Rhone-Poulenc), *EP* 89.908 (1982).
47. F. Piacenti and M. Bianchi *Org. Synth. Met. Carbonyls*, Vol. 2, 1 (1977). (Edited by I. Wender and P. Pino) Wiley; New-York. N.Y.
48. I. Wender *et al., J. Am. Chem. Soc.* **73**, 2656, (1951).
49. M. B. Scherwin *et al.* (Chem Systems), *EP* 4.732 (1978).
50. (a) J. Gauthier-Lafaye *et al.* (Rhone-Poulenc Ind), *EP* 31.784 (1979).
    (b) R. Perron *et al.* (Rhone-Poulenc Ind.), *EP* 46.128 (1980).
    (c) J. Gauthier-Lafaye *et al.* (Rhone-Poulenc Ind.), *EP* 46.129 (1980).
52. D. Forster, *Adv. Organometal. Chem.* **17**, 255, (1979).
52. R. Perron *et al.* (Rhone-Poulenc Ind.), *EP* 18.927 (1979).
53. J. Gauthier-Lafaye *et al.* (Rhone-Poulenc Ind.), *EP* 35.458 (1980).
54. Y. Souma *et al., Bull. Chem. Soc. Jpn.* **46**, 3237, (1973).
55. Halcon Int., *FR* 2.242.362 (1973).
56. Hoechst AG, *FR* 2.289.480 (1974).
57. J. Gauthier-Lafaye *et al.* (Rhone-Poulenc Ind.), *EP* 55.970 (1980).
58. R. Perron *et al.* (Rhone-Poulenc Ind.), *EP* 55.192, (1980).

59. J. Gauthier-Lafaye *et al.* (Rhone-Poulenc Ind.), *EP* 67.777 (1981).
60. Cl. Doussain *et al.* (Rhone-Poulenc), *EP* 70.787 (1981).
61. Chem. Week, **126**, 40 (16 Jan. 1980).
62. Mitsui Toatsu, *J.* 56.68.653 (1979).
63. R. Tsumura *et al.* (Mitsui Toatsu), *J.* 54.079.251 (1977).
64. Mitsui Toatsu, *J.* 56.61.342 (1979).
65. Mitsubishi Chem., *J.* 57.175.157 (1981).
66. M. S. Majhail (Quimico GmbH) — *GB* 2.083.812 (1980).
67. K. Unverferth *et al., React. Kinet. Catal. Lett.* **6**, 231 (1977).
68. K. Unverferth, *J. Prakt. Chem.* **321**, 928 (1979).
69. K. Miyata *et al.* (Mitsui Toatsu), *EP* 815 (1977).
70. J. Zajacek *et al.* (Atlantic Richfield), *US* 3.993.685 (1974).
71. A. Lapidus, *Izv. Akad. Nauk SSSR Ser. Khim.*, 1654 (1981).
72. T. Hirashima *et al.* (Showa Chem.), *FR* 2.485.012 (1980).
73. Osaka Prefecture, *J.* 57.102.876 (1980).
74. H. Alper *et al., J. Am. Chem. Soc.,* **103**, 6514 (1981).
75. D. M. Fenton *et al., J. Org. Chem.* **39**, 701 (1974).
76. T. Yasui, *Shokubai,* **23**, 23 (1981).
77. K. Nishimura *et al., Prepr., Div. Pet. Chem., Am. Chem. Soc.,* **24** (1) 355, (1979).
78. U. Romano *et al., Ind. Eng. Chem. Prod. Res. Dev.* **19**, 396 (1980).
79. General Electric, *NL* 7.908.995 (1978).
80. General Electric, *Ger. Offen.* 2.738.519 and 2.738.520 (1976).
81. M. Orchin, *J. Am. Chem. Soc.* **75**, 3148 (1953).
82. Shell, *US* 3.047.231 (1966).
83. (a) D. Smathers (Du Pont), *US* 4.101.577 (1977).
    (b) H. Bellis (Du Pont), *US* 4.302.598 (1980).
    (c) H. Bellis (Du Pont), *EP* 49.581 (1980).
84. B. K. Nefedov *et al., Izv. Akad. Nauk. SSSR Ser. Khim.* (9), 2119 (1974).
85. J. J. Lindberg *et al., Finn. Chem. Lett.* 153 (1980).
86. YU. Sheludyakov, *SU* 732.252 (1977).
87. N. Sonoda *et al., J. Am. Chem. Soc.* **93**, 6344 (1971).
88. B. D. Dombek *et al., J. Organometal. Chem.* **134**, 203 (1977).
89. N. Kambe *et al., Angew. Chem.* **91**, 582 (1979).
90. N. Kambe *et al., Bull. Chem. Soc. Jpn.* **54**, 1460 (1981).
91. B. Cornils, *React. Struct.: Concepts Org. Chem.* **11** (New Synth. Carbon Monoxide), 1 (1980).
92. R. L. Pruett, *Adv. Organometal. Chem.* **17**, 1 (1979).
93. L. H. Slaugh *et al., J. Organometal Chem.* **13**, 469 (1968).
94. R. L. Pruett *et al. J. Org. Chem.* **34**, 327 (1969).
95. R. T. Gray *et al.* (Shell), *Ger. Offen.* 2.753.644 (1976).
96. A. Hershaman *et al., Ind. Eng. Chem., Prod. Res. Dev.* **7**, 226 (1968).
97. D. B. Kazarnovskaya, *SU* 696.001 (1976).
98. H. Siegel *et al., Angew. Chem.* **92** 182 (1980).
99. F. J. McQuillin (Bush, Boake, Allen), *GB* 1.555.331 (1975).
100. G. Wilkinson *et al., J. Chem. Soc.* (A), 2753 (1970).
101. Sagami, *GB* 2.072.168 (1980).
102. Mitsubishi Petro, *J.* 57.062.275 and *J.* 57.032.279 (1980).
103. C. Botteghi *et al., J. Org. Chem.* **37**, 1835 (1972).
104. M. Tamura *et al., Chem. Econ. Eng. Rev.* **12**, 32 (1980).

105. C. F. Hobbs *et al., J. Org. Chem.* **46**, 4422 (1981).
106. Y. Becker *et al., J. Org. Chem.,* **45**, 2145 (1980).
107. Sagami, *J.* 56.161.340 (1980).
108. BASF, *BE* 865.788 (1977).
109. C. Pittman Jr., *et al., J.C.S., Chem. Comm.*, 473 (1982).
110. S. Murai, *Angew. Chem. Int. Ed.,* **18**, 837 (1979).
111. P. Ford *et al. Adv. Chem. Ser.*, 173 (*Inorg. Compds Unusual Prop.* — II), 81 (1979).
112. J. C. Bricker *et al., J. Am. Chem. Soc.* **104**, 1444 (1982).
113. K. Cann *et al., J. Am. Chem. Soc.* **100**, 3969 (1978).
114. R. G. Pearson *et al., J. Am. Chem. Soc.* **104**, 500 (1982).
115. (a) R. Laine (SRI), *US* 4.292.242 (1978).
(b) R. Laine, *J. Org. Chem.* **45**, 3370 (1980).
116. F. Paulik *et al.* (Monsanto), *US* 3.944.604 (1973), *US* 3.989.748 (1968).
117. P. Hofmann *et al., Ind. Eng. Chem. Prod. Res. Dev.* **19**, 330 (1980).
118. Mitsubishi Petro, *J.* 56.010.190 (1979).
119. A. Mullen, *React. Struct.: Concepts Org. Chem.* **11** (New Synth. Carbon Monoxide), 243 (1980).
120. J. Tsuji *et al., Tet. Lett.*, 893 (1982).
121. K. E. Moller, *Brenn. Chem.* **47**, 10 (1966).
122. J. M. Coustard *et al., J. Chem. Res.* (S), 280 (1977).
123. R. F. Heck *et al., J. Am. Chem. Soc.* **85**, 3383 (1963).
124. H. Alper *et al., Tet. Lett.* 2665 (1979).
125. L. S. Hegedus *et al., J. Am. Chem. Soc.* **104**, 4917 (1982).
126. G. Cometti *et al., J. Organomet. Chem.* **181**, C 14 (1979).
127. Atlantic Richfield, *J.* 7.909.221 (1977).
128. J. K. Stille *et al., J. Org. Chem.* **44**, 3474 (1979).
129. K. Weissermel, *Industrial Organic Chemistry* Verlag Chemie, Weinheim, (1978).
130. F. Jachimowicz, *J. Org. Chem.* **47**, 445 (1982).
131. T. Nogi, *Tetrahedron* **25**, 4099 (1969).
132. L. Cassar, *et al., Synthesis* 509 (1973).
133. A. Reppe, *US* 2.604.490 (1951).
134. BASF, *Ger. Offen.* 870.698 (1953).
135. BASF, *Neth. Appl.* 6.602.880 (1965).
136. P. Pino *et al.* (Lonza), *US* 3.459.812 and *GB* 1.167.687 (1967).
137. P. Pino, *Chem. Ind.* (London) 1732 (1968).
138. P. Pino and G. Braca, *Org. Synth. Met. Carbonyls*, Vol. 2, 419 (Edited by I. Wender and P. Pino) Wiley, New-York, (1977).
139. (a) Mitsubishi Gas Chem., *BE* 815.533 (1973).
(b) Mitsubishi Gas Chem., *BE* 823.739 (1973).
140. (a) Mitsubishi Gas, *BE* 887.021 (1980).
(b) S. Fijiyama, *Hydrocarbon Process.* **57** (11), 147 (1978).
141. Bayer, *DT* 2.413.592 (1974).
142. A. Treibs *et al., Liebigs Ann. Chem.* 11 (1979).
143. Dynamit Nobel, *BE* 872.357 (1977).
144. T. Izumi *et al., Bull. Chem. Soc. Jpn.* **51**, 3407 (1978).
145. J. M. Thompson *et al., J. Org. Chem.* **40**, 2667 (1975).
146. J. E. Hallgren (General Electric), *Ger. Offen.* 2.738.519 (1976).
147. H. Koch, *Liebigs Ann. Chem.* 122 (1960).
148. R. C. Larock *et al., J. Org. Chem.* **45**, 363 (1980).

149. R. C. Larock *et al., J. Am. Chem. Soc.* **104**, 1900 (1982).
150. Y. Yasuhara *et al.*, (Toray Ind.), *J.* 71.024.693 (1968).
151. Y. Yasuhara, *J. Org. Chem.* **33**, 4512 (1968).
152. Ube Ind., *J.* 55.111.441 (1979).
153. J. Falbe, *Chem. Ber.* **97**, 3088 (1964).
154. G. Arzoumanidis *et al., J. Mol. Catal.* **9**, 335 (1980).
155. R. Aumann *et al., Chem. Ber.* **111**, 3429 (1978).
156. N. A. Bumagin *et al., Izv. Akad. Nauk. SSSR Ser. Khim.* 221 (1982).
157. Hercules, *US* 3.745.173 (1965).
158. D. Farcasiu *et al., J. Org. Chem.* **47**, 151 (1982).
159. H. Wakamatsu *et al., J. Chem. Soc.* D., 1540 (1971).
160. Ajinomoto, *US* 3.766.266 (1970).
161. Inst. Français Du Petrole, *FR* 2.395.252 (1977).
162. Ajinomoto, *J.* 54.117.403 (1974).
163. T. Yukawa *et al.*, (Ajinomoto), *J.* 50.149.610 (1974).
164. J. J. Parnaud *et al., J. Mol. Catal.* **6**, 341 (1979).
165. K. Izawa *et al.* (Ajinomoto), *J.* 53.084.967 (1976).
166. H. Alpert *et al., J. Am. Chem. Soc.* **101**, 4245 (1979).
167. J. Gauthier-Lafaye and R. Perron in *Methanol et Carbonylation*, (Edited by Technip) Paris, 39 (1986).
168. J. Gauthier-Lafaye and R. Perron in *Methanol et Carbonylation*, (Edited by Technip) Paris, 97, (1986).
169. J. Gauthier-Lafaye and R. Perron in *Methanol et Carbonylation*, (Edited by Technip) Paris, 150 (1986).
170. J. Gauthier-Lafaye and R. Perron in *Methanol et Carbonylation*, (Edited by Technip) Paris, 191 (1986).
171. J. Gauthier-Lafaye and R. Perron in *Methanol et Carbonylation*, (Edited by Technip), Paris, 233, (1986).

A. F. NOELS AND A. J. HUBERT

TRANSITION METAL CATALYZED REDUCTIONS OF ORGANIC MOLECULES BY MOLECULAR HYDROGEN AND HYDRIDES, AN OVERVIEW

Interest in catalyzed reduction reactions remains high. The general aim is to develop catalysts for selective reductions under mild conditions. Greater product selectivity has an important impact on energy and resource utilization in terms of reduced process energy requirements for product separation and purification, and in terms of low-value by-products.

The energy crisis has contributed to an upsurge in the study of hydrogenation catalysis: indeed, a greater utilization of synthesis gas requires successful methods for selective "oligomeric reductions" of carbon monoxide. In fact, fine examples of the versatility of transition metal catalysis are found in the literature of organic reduction chemistry, still a challenging and demanding area, particularly when working with compounds having several reducible moieties. An astute choice of reducing agent and catalyst can often lead to sufficient chemoselectivity for practical syntheses, without resorting to complex masking procedures.

There is an extensive literature devoted to the subject of selective reductions. The object of the present review is to give a general overview of the current methodologies and mechanisms (when known) associated with transition-metal catalysed reductions (to the exclusion of asymmetric reactions) where hydrogen is being provided by either molecular hydrogen or non-transition-metal hydrides, typically $NaBH_4$ or $LiAlH_4$. The addition of transition metal catalysts to hydrides has added new dimensions to their versatility. This point will be emphasized to some extent.

Two relatively recent and powerful methodologies, namely hydrozirconation and hydrosilylation will also be considered. Indeed, advantage can be taken of the ready formation of Si— or Zr—carbon bonds to selectively reduce (e.g. by protonolysis) or functionalize unsaturated molecules.

1. Activation of Molecular Hydrogen

The metal-catalyzed homogeneous hydrogenation of unsaturated organic

A. Mortreux and F. Petit (Eds.), Industrial Applications of Homogeneous Catalysis, 65—92.

substrates has been the most widely studied of all homogeneous catalytic systems. The binding and activation mechanisms of the overall hydrogen transfer (as hydride or H atom) to the substrate, and the release of the saturated product have been elucidated.

The reductions are typically multistep processes involving discrete hydrido- and organometallic intermediates. Basically, three main mechanistic pathways have been evidenced for the addition of H_2 to unsaturated molecules [1]:

— oxidative additions
— homolytic additions
— heterolytic additions

The list is by no means comprehensive, nor are the above mechanisms mutually exclusive. For example, a further variation has been proposed recently for reductions catalyzed by early transition metals and actinides, namely hydrogenolysis of metal—alkyl bonds via a four-center transition state.

1.1. H_2 ACTIVATION BY OXIDATIVE ADDITION (OA)

This type of catalysis is exemplified by Wilkinson's catalyst, $RhCl(PPh_3)_3$ (1965). Much effort has gone into characterizing the catalytic cycle, the currently accepted one being sketched in Figure 1.

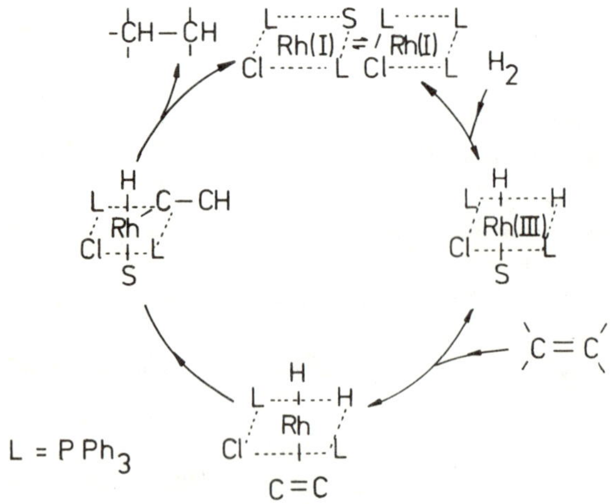

Fig. 1.

The mechanism encompasses a sequence of oxidative addition, migratory insertion, and reductive elimination steps. Actually, what appears at a schematic level as a fairly simple mechanistic scheme corresponds to an extremely complex reaction system involving a large number of species in simultaneous and consecutive reaction steps [2].

For example, in the case of $RhCl(PPh_3)_3$, the active catalytic species appears to contain two tertiary phosphine ligands but how such a species is actually formed from the tritertiary phosphine complex is still open to question. The rate-determining step seems to be the hydride migration to the unsaturated coordinated olefin, a process that can be visualized as a nucleophilic attack of a hydride ligand on an activated double bond.

This particular system is efficient and very specific under mild conditions (room temperature, 1 atm.) for the reduction of olefins. Moreover, it is very sensitive to steric hindrance, as might be expected with a fairly crowded catalyst. Increasing the bulk of the ligands (triarylphosphines) reduces the activity of the catalyst. Internal olefins are reduced much more slowly than terminal ones. Moreover, the system does *not* catalyze the reduction of most common organic groups (keto, cyano, nitro, aryl) and is therefore very chemoselective.

By this way it is possible to selectively hydrogenate biologically active substrates like steroids. An elegant industrial application of Rhodium catalysis is the asymmetric hydrogenation of cinnamic acid derivatives to L-Dopa [3], a drug effective against Parkinson's disease (Monsanto). The induction of asymmetry is achieved through the use of a cationic Rhodium diene complex that bears optically active ligands. Cationic diene complexes of Rhodium such as $[Rh(diene)(PR_3)_2]^+$ operate probably via the same cycle as Wilkinson's catalyst excepted that the olefin coordinates before the oxidative addition of hydrogen, which is the rate determining step.

1.2. H_2 ACTIVATION BY HOMOLYSIS

This type of activation is typically promoted by the pentacyanocobaltate anion, which is very selective for catalyzing the hydrogenation of activated alkenes:

$$H_2 + Co(CN)_5^{\ominus} \longrightarrow 2\ HCo(CN)_5^{\ominus} \text{ or } (H^{\cdot})\,Co(II)(CN)_5^{\ominus}\ ?$$

Non-activated double bonds remain unaffected, thus conjugated dienes are selectively reduced to monoenes. The initial interaction with the activated olefin is viewed as occurring via a four-centre transition state, in which the H atom is transferred to the carbon β to the activating group, seemingly without olefin precoordination.

A
\+ $HCo(CN)_5^{\ominus}$
A
$[(CN)_5 Co \cdots H]^{\ominus}$
A "HCo" A "HCo" A $\ominus$
H H H $(CN)_5 Co$ H

Free radical intermediates were detected in reductions of activated olefins while covalently bonded organocobalt derivatives are formed in the reduction of conjugated olefins (π-allylic systems). Butadiene is selectively reduced to a monoolefin. The stereochemistry of the monoene depends on the concentration of the cyanide anions present in the reaction medium.

1.3. H_2 ACTIVATION BY HETEROLYTIC ADDITION

The real nature of the catalytic cycle is less documented but the H ligand appears to react very much as H^-

$RuCl_2L_3$ $L = PPH_3$
H_2 H_3C-CH_2R
$Ru(II)H(Cl)L_3 + H^+ Cl^-$
$H_2C=CHR$ H_2
CHR
$Ru(II)H(Cl)L_2 + L$ $L_3ClRu-CH_2$
$HC(R)=CH_2$

Formally, no change in the oxidation state of Ru occurs. Although hydrogenolysis of Ru(II)-alkyl to give the alkane (rate determining step) could also occur via heterolytic cleavage of H_2, it more probably occurs via an oxidative addition—reductive elimination sequence.

Actually, $RuCl_2(PPh_3)_3$ is one of the most active homogeneous catalyst for the hydrogenation of terminal olefins (they are hydrogenated at a rate about 10^4 times that of the corresponding internal ones). It is also very efficient for the reduction of $—NO_2$ to $—NH_2$ and aldehydes to alcohols. Nitriles, esters or halides remain unaffected.

Another system where heterolytic cleavage has been evidenced is based on anionic Pt complexes. The commercially available H_2PtCl_2 and $SnCl_2 \cdot 2\,H_2O$ react in methanol to form deep red solutions that contain species such as (A).

$$H_2 + \underset{A}{[Pt(SnCl_3)_5]^{\ominus}} \rightleftharpoons H^{\oplus} + \underset{B}{[HPt(SnCl_3)_4]^{\equiv}} + SnCl_3^{\ominus}$$

The anionic hydrido species (B) is then believed to coordinate an olefin which is then inserted into the Pt—H bond. Presumably, protonolysis promotes the cleavage of the metal—alkyl bond.

These solutions catalyze hydrogenation of simple linear olefins and have been extensively studied for hydrogenation of vegetable oils to remove excessive unsaturation.

1.4. THE CASE OF PENTAMETHYLCYCLOPENTADIENYL (Cp*) Rh AND Ir COMPLEXES

Dinuclear Cp*Rh(III) and Ir(III) complexes show high activity at room temperature and atmospheric pressure for the reduction of olefins. Arenes are also reduced at higher pressures. The process is cocatalyzed by bases and the mechanism seems to involve both heterolytic and oxidative addition sequences [4]:

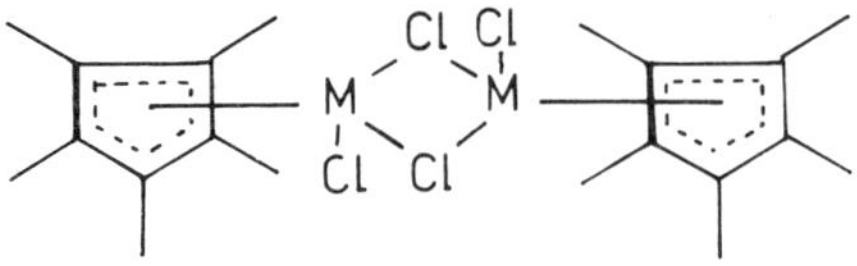

Cl Cl M M Cl Cl = $[M_2Cp^*_2Cl_4]$

M = Rh, Ir

$$[M_2Cp_2Cl_4] \overset{L}{\rightleftharpoons} 2\,MCpCl_2.L \qquad (1)$$

$$MCpCl_2.L + H_2 \rightleftharpoons MCpH(Cl).L + HCl \qquad (2)$$

$$\mathrm{M\,Cp\,H(Cl).L + H_2 \rightleftharpoons M\,Cp\,H_2.L + HCl} \tag{3}$$

$$\mathrm{M\,Cp\,H_2.L + olefin \xrightleftharpoons{slow} M\,Cp\,H_2(olefin)} \tag{4}$$

$$\mathrm{M\,CpH_2(olefin) \xrightarrow{fast} M\,Cp.L_3 + alkane} \tag{5}$$

$$\mathrm{M\,Cp\,L + HCl \rightleftharpoons MCpH(Cl).L} \tag{6}$$

L = solvent

In steps 2 and 3, heterolytical activation takes place to give the dihydride species, probably the real active species. The latter reacts with the olefin and hydrogen in a cycle similar or close to that which has been proposed for olefin hydrogenations catalyzed by $RhCl(PPh_3)_3$, included reductive elimination (step 5) and oxidative addition (step 6) reactions.

1.5. ORGANOLANTHANIDES AND ACTINIDES AS CATALYSTS FOR OLEFIN HYDROGENATION

Hydrides of actinides and lanthanides showed surprisingly high activity for olefin hydrogenation [5]. In fact, organolanthanides of the type$(Cp_2^*MH)_2$ are among the most active homogeneous catalysts for olefins hydrogenation yet reported. Mechanistically these reactions are not well understood and seem to proceed through mechanistic pathways different from the ones admitted with the conventional Group VIII or late transition-metal catalysts. In particular, the metal is formally in a relatively high oxidation state and may not possess an accessible oxidation state for oxidative addition-reductive elimination processes (especially in the lanthanide case). The complexes also possess significant metal-ligand bond polarity [6].

A plausible hypothesis for olefin hydrogenation appears to be a non oxidative hydrogenolysis of the metal alkyl bond (formed by alkene insertion into a preformed M—H bond) via a four-center transition state. Such a pathway is well documented in main group chemistry, the alkyls of Li and Al being particularly relevant examples.

The originality of the system comes from the coupling of an olefin/hydride insertion step with a “four-center” hydrogenolysis of the resulting metal—carbon bond. It is not clear whether olefin insertion or hydrogenolysis is rate controlling. However, the olefin insertion exhibits a strong steric sensitivity and the rate falls significantly for substituted olefins.

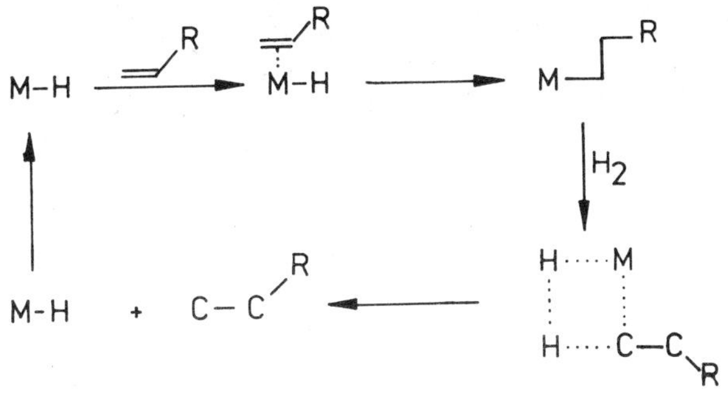

The significant metal—ligand bond polarity of organoactinides and lanthanides is also evidenced by the rapid reactions of their hydrides with ketones, aldehydes and alcohols, which confirms their "hydridic" character.

2. Some Recent Developments in Hydrogenation: Activation of Hydrides by Transition Metal Derivatives

Activation of hydride reagents such as NaH, $LiAlH_4$, $NaBH_4$, etc. by TM complexes has attracted much attention in recent years and considerably extended the scope of their use. Not only are such reagents attractive because of their ease of handling and low costs, but also because of their versatility and the many variations possible. By careful choice of catalyst system, a remarkable degree of selectivity can be attained, often under mild conditions.

Systems are known which will reduce most of the organic functions (amides, carboxylic acids, nitriles, nitro, halides, double and triple bonds, aromatic rings, for instance). In most cases, the true catalyst is formed *in situ* by reduction of the metal with the hydride. Actually, very little is known of the mechanisms of those transition metal-assisted hydride reductions. Catalyzed reductions with $NaBH_4$ have been reviewed [7].

The following examples illustrate the versatility and potentialities of such processes.

2.1. EXAMPLES

2.1.1. *$LiAlH_4$ with First Row Transition Metal Halides*

The reactions of alkenes and alkynes with $LiAlH_4$ in admixture with catalytic amounts of first row TM halides ($CoCl_2$ or $TiCl_3$) are effective in reducing the unsaturated molecule to alkane in high to quantitative yield [8]. Even trisubstituted olefins are reduced quantitatively.

CH_3 — Li Al H_4 / Co or Ni → CH_3

→ C_2H_5

Ph-C≡C-Ph —Li Al H_4 / Ni→ Ph Ph

Co(II) and $NaBH_4$ display an extremely high steric selectivity in the reduction of alkenes, trisubstituted olefins are virtually inert to these reducting conditions. Alkynes are readily reduced to alkenes [9]. However, the reaction is not catalytic in TM and the system is not selective towards ketones or aldehydes. Esters are unaffected under such conditions. $(Ph_3P)_3CoCl$ is a selective reagent for reactions in which an alkyne is selectively reduced to an alkene (mixture of isomers) [10].

R—C₆H₄—I + Na B H_4 —Cp_2TiCl_2→ R—C₆H₄—H

2.1.2. *$LiAlH_4$ with Hard Lewis Acids*

The coupling of $LiAlH_4$ with a hard Lewis acid such as Ce(III) promotes the formation of a powerful reducing system. The actual reducing species may be a low-valent cerium derivative. Organic halides, including fluorine compounds, are smoothly dehalogenated with this reagent [11]. Phosphine oxides (even when crowded) are also reduced to phosphines in excellent yields.

Very interesting are the results of recent investigations on the mechanisms of Co(II) mediated reductions of nitriles, alkenes and alkyl halides by $LiAlH_4$ and $NaBH_4$. Those studies have unambiguously identified borides and aluminides of cobalt as catalysts in all three reductions, a finding clearly at odds with commonly held notions about the mechanisms of such processes and which could also be relevant to other transition-metal—hydride systems [12].

Cp_2TiCl_2 catalyzes the reduction of aromatic halides with $NaBH_4$ in DMF. Yields are high in the presence of air only, which suggest a radical mechanism [13].

2.1.3. *$NaBH_4$ with Ni or Co Salts in MeOH*

The chemoselective borohydride reduction of a double bond conjugated with a keto group can be carried out with Ni or Co salts in methanol [14]:

$$R_1-\underset{\substack{\|\\O}}{C}-CH=C\begin{matrix}R_2\\SR_3\end{matrix}\xrightarrow[NiCl_2]{NaBH_4}R_1-\underset{\substack{\|\\O}}{C}-CH_2-CH_2-R_2$$

2.1.4. *Hydroboration with $NaBH_4$*

Although $NaBH_4$ does not react on its own with unsaturated carbon—carbon bonds, alkenes and alkynes can now be hydroborated with $NaBH_4$ (or $LiBH_4$). A crown ether is needed with $NaBH_4$, a cheaper but less soluble reagent than $LiBH_4$.

The process avoids the handling of diborane. The reaction is efficiently catalyzed by Ti derivatives such as Cp_2TiCl_2 [15] or $TiCl_3$ [16].

The alkylborohydrides thus produced are converted to alcohols on oxidation, and alkenylboranes to alkanes by protonolysis. The hydroboration occurs preferentially in an anti-Markownikoff fashion. In contrast to what occurs with diborane, the Ti catalyzed hydroboration affords almost exclusively monohydroborated products (mixture of isomers) with alkynes or alkadienes, even in the presence of an excess of $NaBH_4$.

$$Ph-CH=CH_2 + NaBH_4 \xrightarrow[18\text{-}C6]{TiCl_4} \xrightarrow[CH_3ONa]{H_2O_2} \begin{matrix}Ph-CH_2-CH_2-OH\\Ph-CHOH-CH_3\end{matrix} \quad 90\%\ (80:20)$$

92% (95 : 5)

HO

42 %

OH 77%

Another system, based on $CoCl_2$—$NaBH_4$ is also efficient for the hydroboration or hydrogenation of alkenes [17]. It was possible to direct the reaction to give the hydroboration product by stirring the $NaBH_4$—cobalt mixture in THF before addition of the olefin. Two moles of alkenes are hydroborated per mole of cobalt. Triply substituted double bonds do not react.

+ $NaBH_4$ $\xrightarrow{CoCl_2}$ OH 85%

OH 90%

OH 70%

2.1.5. *Reduction of Acid Chlorides and Nitro Groups*

Acid chlorides are reduced to the corresponding aldehydes by using $NaBH_4$—$CdCl_2$ in DMF, tributyltin hydrides in the presence of soluble Pd catalyst [18], or preferably, with the stable and easily prepared copper borohydride $Cu(BH_4)(PPh_3)_2$ [19]. The latter complex is very selective and was found to be unreactive towards most organic functions. It did however effectively reduce a large set of substituted aromatic nitro compounds and aliphatic nitro groups attached to a tertiary carbon (75—80% yield) [20].

O_2N NO_2 $\xrightarrow{Cu(BH_4)L_2}$ H_2N NH_2

OAc NO_2 ⟶ OAc NH_2

2.1.6. *Vanadium Chloride and Lithium Hydride*

VCl_3 + LiH(1 : 1) reduces aldehydes, ketones, and esters to the corresponding alcohols in high yields. Terminal olefins are also reduced but *not* alkynes or internal olefins [21].

2.1.7. *Complex Reducing Agents*

The so-called Complex-Reducing-Agents (CRA), formed by reacting NaH + alkoxide + metal salt, show interesting chemoselectivities and are sources of new atmospheric-pressure, heterogeneous hydrogenation catalysts which are cheap and easily prepared. Alk—X (X = I, Br, Cl) are efficiently converted to the corresponding hydrocarbons: benzyl—, vinyl— and allyl—X are also reduced with NiCRA. The reagents are inert toward a number of other functional groups, so that selective reductions can be achieved [22]:

O
X
X = Br, Cl
Na H, Na OAlk
Ni(II)/W(VI)
O
75%

id.
Ni(II)
80 %

O
O
H H
HOH
Ni-CRA 98% -
Zn-CR A - 98 %

2.2. UNUSUAL CHEMOSELECTIVITY

The following examples show how *unusual chemoselectivities* can be attained:

2.2.1. *Reversal of Normal Reduction Sequences are Observed with Lanthanide—$NaBH_4$ Systems*

Typically, selective reductions of ketones in the presence of aldehydes have been achieved without the need for any stable, isolable protecting group [23]. The mild conditions have little effect on other functionalities. The selectivity can be inter- or intramolecular.

$$\text{R-CH=CH-CHO} \xrightarrow[\text{BH}_4^-]{\text{Ce(III)}} \text{R-CH=CH-CH}_2\text{OH} + \text{R'CHO}$$

$$\text{2-oxo-5-formylcyclopentyl-(CH}_2)_n\text{CO}_2\text{CH}_3 \xrightarrow{\text{id.}} \text{2-hydroxy-5-formylcyclopentyl-(CH}_2)_n\text{CO}_2\text{CH}_3$$

The inertness of the aldehyde group has been attributed to selective ketalization (in alcohol—water mixtures) catalyzed by the cerium cation, thus protecting the group from reduction.

2.2.2. *Selective Hydrogenation of Unsaturated Aldehydes and Ketones*

Application of a similar procedure to α, β-unsaturated ketones or aldehydes yields the corresponding allylic alcohol [23, 24].

The unusual selective hydrogenation of α, β-unsaturated aldehydes to unsaturated alcohols has also been accomplished using $[RhCl(CO)_2]_2$ in the presence of tertiary amines under oxo-conditions. Wilkinson's catalyst reduces the alkene double bond under the same conditions [25].

2.3. REDUCTION OF α, β-UNSATURATED NITRILES

Because of the oft-encountered overreduction to the amine, reduction to saturated nitriles is a difficult problem. A mixture prepared from CuBr, $NaAl(OCH_2—CH_2—OCH_3)_2H_2$ (vitride) and sec-butyl alcohol do that work nicely. In contrast to the Mg—CH_3OH method, no coupling was observed [26]. The mechanism of the reactions is largely speculative.

Moreover, in analogy with the alkyl—Cu chemistry, hydrido—Cu reagents might provide selective conjugate addition of hydride and compatibility with most functional groups. Two Cu complexes seem to be efficient for that purpose [27]:

(Mg)

Dimers

"Cu"

- Cu H

99 %

51 %

$$2\ LiAlH(OMe)_3 + CuBr \xrightarrow[0\,°C]{THF} \text{'Li complex'}$$

$$NaAlH_2(OCH_2{-}CH_2OCH_3)_2 + CuBr \longrightarrow \text{'Na complex'}$$

Selective reduction of the olefin unit in conjugated ketones and esters was observed. However, the reactions are often accompanied by side reactions and the reactivity of the complexes toward common functional groups limits their applicability.

"Cu H"

"Cu D"

2.4. HYDROGENATION OF AROMATIC NUCLEI

This area is likely to become increasingly important if coal, which contains polyaromatic compounds, is utilized more for production of petrochemicals (*cf.* coal liquefaction).

Because of unfavorable thermodynamics, partial reduction of aromatics is exceedingly difficult to achieve. The selective reduction of benzene to cyclohexene by tetrahydroborate can be carried out with the help of dicationic η^6-benzene complex $[C_nMe_nM(C_6H_6)]^{++}$ (M = Ir, Rh, $n = 5$; M = Ru, $n = 6$).

The overall reaction consists of addition of two hydrides followed by two protons to coordinated benzene. In this last step, the two protons reoxidize the metal and liberate free cyclohexene [28]. The reaction is thus catalytic in the platinum metal even though it is stoicheiometric in hydride and acid.

The very simple system $NaBH_4$—$RhCl_3$ in hydroxylic solvents has proved to be useful for the reduction of aromates to the saturated cycles under mild conditions [29].

tBu–C₆H₄–R —($RhCl_3$ / $NaBH_4$)→ tBu–C₆H₁₀–R (axial R) + tBu–C₆H₁₀–R (equatorial R)

The reaction is compatible with carboxylic acids, esters and amides which remain unaffected by the reduction system. Even ketones are only partially reduced, though olefinic bonds are completely reduced simultaneously.

This reaction affords more active systems than classical Raney Ni and is convenient at the experimental point of view. However it is *not* catalytic in Rh, which severely limits its practical use.

C_6H_5–nC_6H_{13} ⟶ C_6H_{11}–nC_6H_{13} 100 %

C_6H_5–CH_2–CH(NH–CO–CH_3)–COOH ⟶ C_6H_{11}–CH_2–CH(NH–CO–CH_3)–COOH 94%

pyridine–COOEt ⟶ piperidine–COOEt 60 %

One has to keep in mind that stable and extremely active catalysts have been developed for the homogeneous reduction of aromatic molecules by molecular hydrogen. For instance, under 50 bars of H_2, a cationic Ru(II) complex will only reduce the benzene ring and double C—C bonds to the exclusion of many organic functionalities, as examplified here [30]:

[Ru(II)]⊕ ; H_2, 50 bar

Another methodology that shows great promise for selective reductions is Phase Transfer Catalysis (PTC) under one atmosphere of H_2. Under very gentle conditions, with a quaternary ammonium salt as the PT agent, 1 mol % of $[RhCl(1,5\text{-hexadiene}]_2$ catalyzes the reduction of aromatic rings; yields vary to some extent with the phase transfer agent [31].

H_2, 1 Bar ; 20 °C

OCH_3 → OCH_3

CH_2-C(=O)-CH_3 → CH_2-C(=O)-CH_3

	C_6H_6	C_6H_{12}
CH_2-C(=O)-CH_3	54	100 %
CH_2-CH(OH)-CH_3	14	-

73 %

Arenes are also catalytically reduced under PTC in a very similar system also based on Rhodium, with aliquat-336 as the phase transfer agent.

H_2 ; 1 bar

100%

CO_2CH_3 → CH_2OH + CO_2CH_3 + CH_2OH

1 89 10 %

3. Hydrosilylation (H—Si)

Reduction of unsaturated bonds can be readily achieved by reaction with a Si—H functionality [32].

$$\mathrm{>C{=}C<} \; + \; R_3SiH \longrightarrow -\overset{|}{\underset{H}{\underset{|}{C}}}-\overset{|}{\underset{SiR_3}{\underset{|}{C}}}-$$

R = alkyl, aryl, OAlk, halogen

Since its discovery in the late 1940, hydrosilylation has seen its importance growing steadily. The reaction is catalyzed by a variety of transition metal derivatives. However the catalyst of choice is usually chloroplatinic acid $H_2PtCl_6 \cdot 6\,H_2O$, the so-called Speiers' catalyst (ratio of catalyst to substrate is 10^{-5} to 10^{-7}!!). Hydrosilylation is compatible with a variety of functional groups. The SiR_3 group adds to the less crowded end of the olefin and isomerisation of internal olefins takes place. Optically active silyl groups add with retention of configuration. In addition to its use in organic syntheses, the hydrosilylation reaction is used in the manufacture of silicone polymers. The broadest application is the "curing" of silicone rubbers, a step that converts a syrupy polymer to a gum rubber or plastic polymer to give a hard material (dental cement). The toughening process is accomplished by crosslinking polymer chains. Typically a Si—H group of one chain is added to a vinyl group of another. The vinylsilanes are often conveniently prepared by adding acetylene to Si—H groups:

$$(-O)_2Si(R)H \; + \; CH_2{=}CH{-}Si(R)(O-)_2 \longrightarrow (-O)_2Si(R){-}CH_2CH_2{-}Si(R)(O-)_2$$

$$R_3Si{-}H \; + \; -C{\equiv}C- \longrightarrow R_3Si{-}CH{=}CH_2$$

Another industrial applications of hydrosilylation is the synthesis of solvent-resistant silicone rubbers which are used as liners in self-sealing fuel tanks in aircrafts. Resistance to jet fuel or gasoline is attained by addition of polar cyanoethyl groups:

$$CH_2{=}CH-CN + HSi(CH_3)Cl_2 \longrightarrow Cl_2(CH_3)Si-CH_2-CH_2-CN$$

$$\xrightarrow{R_2Si(OH)_2} \left[-\underset{R}{\overset{R}{Si}}-O-\underset{CH_2-CH_2-CN}{\overset{CH_3}{Si}}-O- \right]_n$$

Accumulated evidence suggests that H-silylation occurs by oxidative addition of Si—H to a low-valent Pt complex, in a mechanism very similar to that of hydrogenation (*cf.* $RhCl(PPh_3)_3$). Alkoxysilanes react more slowly than chlorosilanes, no olefin isomerization being observed.

Many catalysts, other than those based on Pt, are presently used in H-silylation reactions: $Co_2(CO)_8$, Ziegler catalysts (Fe, Co, Ni) and most notably Rh complexes.

Recently, the readily available $[C_5Me_5RhCl_2]_2$ was shown to catalyze the formation of vinyl and allyl silanes from terminal olefins. The ratio of isomers were determined by reaction conditions. Under optimum conditions, selectivity as high as 69% towards the transvinylsilane could be achieved [33]:

$$n\text{-}C_4H_9-CH{=}CH_2 \xrightarrow[Rh]{HSi(C_2H_5)_3} \underset{69\%}{n\text{-}C_4H_9-CH{=}CH-Si(C_2H_5)_3}$$

$$+\ n\text{-}C_4H_9-CH_2-CH_2-Si(C_2H_5)_3 + n\text{-}C_6H_{14}$$

3.1. EXTENSIONS OF HYDROSILYLATION REACTIONS

3.1.1. *Ring closure*

$$RHSi-(CH_2)_n-CH{=}CH_2 \xrightarrow{Pt} \text{HPt}(R_2Si)\cdots(CH_2)_n\text{-olefin} \longrightarrow \text{A (3-methylsilacyclopentane, } SiR_2\text{)} + \text{B (silacyclohexane, } SiR_2\text{)}$$

In fact, some hydrosilylation reactions appear to be sensitive to the relative concentration in metal(catalyst). For example, styrene gives different products of hydrosilylation according to the ratio of catalyst (Rh) to Si [34].

3.1.2. *Hydrosilylation of conjugated dienes*

Not intensively studied, but usually occurs by 1,4-addition, and only at high temperature. Pd catalysts are active with conjugated olefins but *not* with non conjugated ones.

1.8 $$\text{CH}_2{=}\text{CH}{-}\text{CH}{=}\text{CH}_2 + (\text{EtO})_3\text{SiH} \xrightarrow{\text{Pd}} (\text{EtO})_3\text{Si}{-}\text{CH}_2{-}\text{CH}{=}\text{CH}{-}\text{CH}_3 \quad (10\,\%, E+Z)$$
$$+ (\text{EtO})_3\text{Si}{-}\text{CH}_2{-}\text{CH}{=}\text{CH}{-}\text{CH}_2{-}\text{CH}_2{-}\text{CH}{=}\text{CH}{-} \quad (90\,\%)$$

The yield of the 1 : 1 adduct rises up to 85% at 100 °C with π-allyl PdCl as catalyst. Moreover, the reaction is highly stereoselective as the pure Z isomer is obtained by treatment of butadiene with R_3SiH in the presence of $Cr(CO)_6$ under UV irradiation [35].

3.1.3. *Hydrosilylation of acetylenes*

The ratio of (a) to (b) varies with R_3Si, yields are usually good.

$$R_3\text{SiH} + C_4H_9{-}C{\equiv}C{-}H \longrightarrow \underset{\text{(a)}}{R_3\text{Si}{-}\text{CH}{=}\text{CH}{-}C_4H_9} + \underset{\text{(b)}}{\text{CH}_2{=}C(\text{SiR}_3){-}C_4H_9}$$

3.1.4. *Reduction of C=O*

— *to alcohols*: not widely used (because of the many other useful methods available) except for promoting asymetric inductions [36].

$$PhC(=O)CH_3 + H_2SiPh_2 \xrightarrow{M^*} Ph(CH_3)C(H)OSiHPh_2 \xrightarrow{H_3O^{\oplus}} Ph(CH_3)C(H)OH$$

$M = Rh^*, Pt^*$

This reaction is, however, of interest and preferable to catalytic hydrogenation when the resulting alcohol is prone to hydrogenolysis. Moreover, application of H—Si permits highly stereospecific reductions of certain ketones to any stereoisomeric alcohol.

— *to alkanes* (*and alcohols to alkenes*): nonconjugated aldehydes and ketones are reduced by HSi upon addition of Bronsted or Lewis Acids. In general, only aryl ketones and aldehydes with electron donating substituents give synthetically useful yields of completely deoxygenated products [37].

$$R_2C{=}O + BF_3\,(gas) \xrightarrow{R'_3SiH} R'_3SiF + R_2CH{-}O{-}BF_2$$

$$R_2CH{-}O{-}BF_2 \xrightarrow{-BOF_2} R_2CH^+ \xrightarrow{R'_3SiH} R_2CH_2$$

An important application of silicon hydrides is the selective synthesis of (*E*)-crotylsilanes [38].

$$H_3C{-}CH{=}CH{-}CH_2{-}MgBr \xrightarrow[{[M]}]{HSiR_3} H_3C{-}CH{=}CH{-}CH_2{-}SiR_3$$

In this particular case, the most efficient catalyst is dichloro-[1,1′-bis(diphenylphosphino)Ferrocene]-Nickel(II). Yields are quite variable according to R but the ratio of *E* to *Z* isomers is very high. It has been observed that the *Z* isomers react with aldehydes to yield erythro adducts [39].

R—CH=CH—$Si(CH_3)_3$ + R′CHO (E) —$TiCl_4$→ erythro: 95 % + [threo]

3.1.5. *Reduction of α, β-unsaturated carbonyl compounds*

Homogeneous HSi catalyzed by TM complexes is a unique and effective method for the selective reduction of α, β-unsaturated carbonyl molecules *either* to the corresponding saturated ketones *or* to allylic alcohols:

— *monosilanes* bring about the selective hydrogenation of carbon—carbon double bonds conjugated to a carbonyl group via silylenolether whereas *di-* and *trihydrosilanes* effect the selective reduction of carbonyl groups via allylic silyl ether [40].

Rh(I); R_3SiH → CH—C=C—$OSiR_3$ → CH—CH—C=O (c)

Rh(I); R_2SiH_2 → C=C—CH—$OSiHR_2$ → C=C—CH—OH (d)

The carbonyl reduction displays an exceedingly higher selectivity than usual metal hydrides:

	c/d	yield %
Et_3SiH	99 : 1	95
Ph_2SiH_2	1 : 99	95

	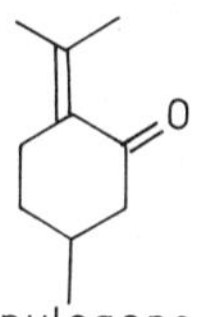 pulegone	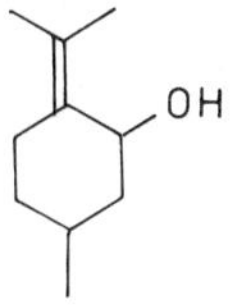 OH	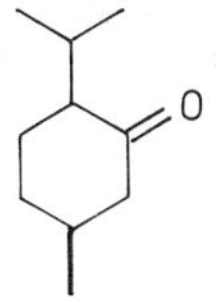 O	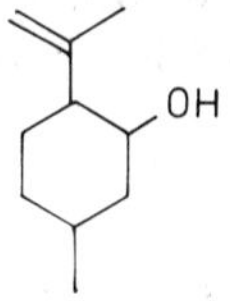 OH
Et_2SiH_2 Rh(I)	-	100	-	-
Ph_2SiH_2 Rh(I)	-	100	-	-
$LiAlH_4$	49	51	-	-
$LiAl(O\,t\text{-}Bu)_3H$	39	43	18	-
$LiBH_4$	7	93	-	-
$NaBH_4$	18	36	-	46

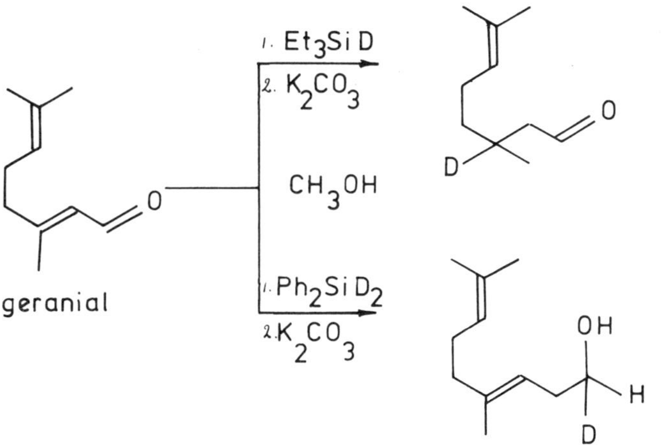

geranial

1. Et_3SiD
2. K_2CO_3

CH_3OH

1. Ph_2SiD_2
2. K_2CO_3

O

D

OH

H

D

No scrambling of D was observed. Introduction of chirality is also observed in the H—Si products of conjugated ketones [41].

From a mechanistic point of view, hydrosilylation of simple carbonyl compounds catalyzed by Rh(I) complexes seems to involve an initial Si migration from the adduct resulting from oxidative addition to the C=O of the coordinated substrate:

$$ClRhL_3 + XR_2SiH \xrightarrow[OA]{-L} L_2Rh(H)(Cl)SiR_2X \longrightarrow XR_2SiRhHL_2Cl$$

π-allyl

?

RE

1,2-add.

1,4-add.

HO

O

3.1.6. *Hydrosilylation of C=N bonds*

Whereas H—Si of alkenes, alkynes and carbonyl compounds has been much investigated only a few publications deal with the corresponding reduction of CN bonds [42] (Schiff bases, oximes, etc.).

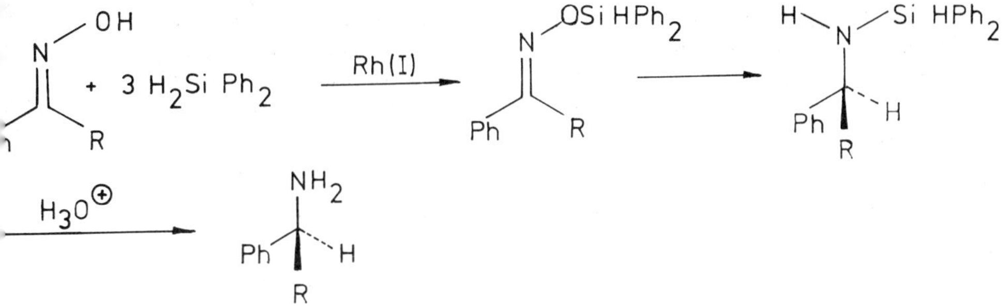

The reaction can be controlled enantioselectively using optically active catalysts and by means of this novel reaction, oximes can be converted into optically active primary amines and schiff bases into optically active secondary amines. Up to now, optical yields are rather poor.

Another point of interest is the reduction of *acetals* to ethers, the reaction is catalyzed by trimethylsilyltriflate [43].

$$R_2C(OCH_3)_2 + (CH_3)_3Si\,H \xrightarrow{(CH_3)_3Si\,OTf} R_2CH{-}O{-}CH_3$$

On the other hand, introduction of functional groups is always possible via hydrosilylation [44].

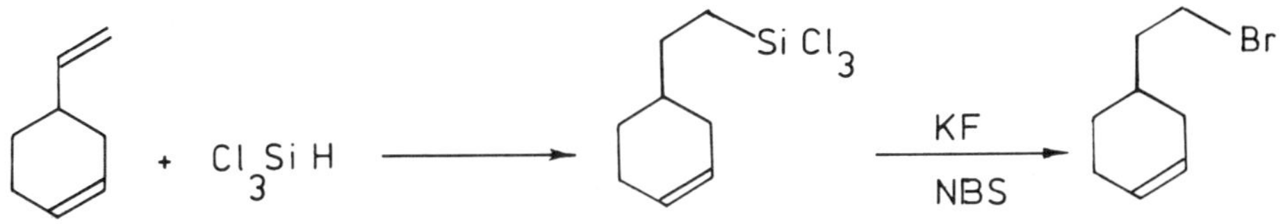

4. Hydrozirconation

Hydrozirconation (H—Zr) has recently been developed as a procedure that appears to have broad potential for reducing and functionalizing alkenes, alkynes and 1,3-dienes [45].

The reagents involved are inexpensive, readily prepared and the yields are high. Hydrozirconation converts unsaturated C—C bonds (even non activated ones) via π-bonded intermediates to σ-bonded Zr derivatives.

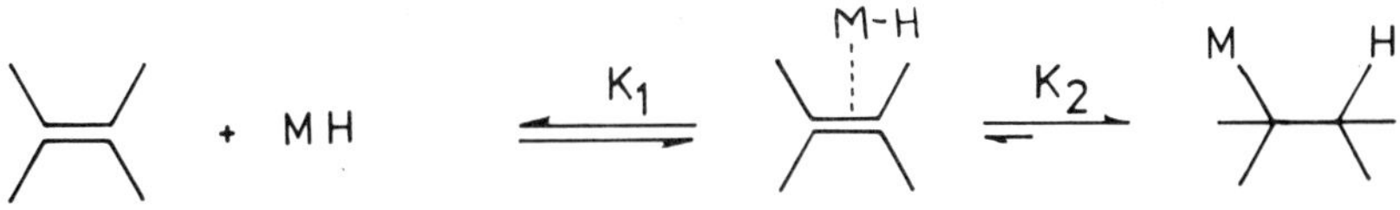

The reaction is cleanly achieved by heating the unsaturated molecule with $H(Cl)ZrCp_2$. Formally this complex is a d^0, high valent complex, two factors favorable for the stability of M—alkyl bonds (K_2 large). Moreover, it is readily prepared from Cp_2ZrCl_2 and $LiAlH_4$ or $NaAlH_2(OR)_2$ in THF.

$Cp_2ZrH(Cl)$ = ([Zr]-H) reacts with unsaturated C—C bonds to give isolable addition products (in opposition to boranes and alanes, organo-Zr are rather stable in dry air) that can be hydrolyzed with diluted acidic solutions. The scope of hydrozirconation with respect to chemoselectivity and substrate structure lies between those of hydroalumination and hydroboration. The facile isomerization of secondary alkylzirconium derivatives into primary alkyl derivatives occurs very readily (even below room temperature, note the difference with Al and B!).

+ Cp_2ZrHCl ≡ [Zr]H → $ZrCp_2Cl$
or

[Zr]
$H_3O^{\oplus}$ → alkane

—[Zr]

Tetrasubstituted olefins fail to react, as do some trisubstituted cyclic olefins. Whereas acetylenes react 70—100 times faster than the corresponding olefins.

Another major difference with hydrosilylation and organo-boranes, Alkyl-Zr do *not* react with functional groups such as aldehydes, ketones and alkylating agents [46]. Carbon—carbon bond formation is nevertheless possible through transfer or carbonylation reactions [45].

The reactions are very regio- and stereospecific (100% *cis*-addition, anti-Markownikov), the less-hindered positional isomer is isolated, even at room temperature.

t-Bu–C≡C–H → (1. [Zr]H; 2. D_2O) → (E)-t-BuCH=CHD; → (1. [Zr]D; 2. D_2O) → t-BuCH=CD₂

Although purified organo-Zr vinylic compounds do not rearrange, isomerization is catalyzed by an excess of [Zr]—H and probably proceeds through an elusive dimetallated molecule.

1,3-dienes react by 1,2-addition to the sterically less-hindered double bond. However 2,4-dienes do not behave analogously; they react with several equivalents of organo-Zr.

4.1. FUNCTIONAL GROUP COMPATIBILITY

Ethers and halogens do not interfer in hydrozirconation. However zirconium hydrides can reduce several types of carbonyl groups to alcohols. Nitriles are reduced to aldehydes. Consequently, these functions have to be protected.

A number of examples of the application of organozirconium hydride transfer reaction [47] has been reported. Homogeneous hydrogenations using H_2ZrCp_2 are known. The proposed catalytic cycle is close to that suggested for hydrogenation catalyzed with organolanthanides and may involve hydrozirconation followed by hydrogenolysis of the resulting zirconium—carbon bond [48].

The low valent Zr species prepared by reduction of $ZrCl_4$ with Na amalgam in the presence of a 1,3-diene and of a phosphine are highly efficient catalysts for the reduction of mono and disubstituted olefins and acetylenes to the corresponding alkanes. The same species also catalyze the metathesis of 1,3-cyclohexadiene into cyclohexane and benzene [49], a reaction that is also very efficiently achieved with cobalt-based Ziegler-type catalysts [50].

More and more, organozirconium chemistry is emerging as a powerful tool in organic synthesis [45, 46, 47]. In addition to the Zr-catalyzed Ziegler—Natta polymerization, the carbon—zirconium bond can be functionalized in a number of ways. Organozirconium derivatives are useful intermediates for the preparation of organic halides, alcohols, heterosubstituted derivatives as well as for carbon—carbon bond formation through carbonylation and acetylation. Of major importance are the recent developments of procedures for Pd- or Ni-catalyzed cross coupling by Neghishi [45] and for Ni-catalyzed conjugate additions by Schwartz [51]. The synthetic interest of organozirconium chemistry is further enormously increased by the facility of promoting transmetallation reactions, i.e. the replacement of Zr by other metals such as Al, Cu, Zn, Sn and Hg to yield organometallic compounds that display a rich chemistry of their own.

Laboratoire de Chimie Macromolécutaire et de Catalyse Organique, and
Laboratoire de Synthèse Organique et de Catalyse,
Institut de Chimie, B. 6,
Université de Liège,
B-4000 Sart-Tilman, Belgium.

References

1. F. J. Mc Quillin, *Homogeneous Hydrogenation in Organic Chemistry*, D. Reidel (1976); C. Master, *Homogeneous Transition-Metal Catalysis*, Chapman and Hall (1981).
2. J. Halpern, *Inorg. Chim. Acta,* **50**, 11 (1981).
3. G. W. Parshall, *J. Mol. Catal.,* **4**, 243 (1978).

4. S. L. Grundy, A. J. Smith, H. Adams and P. M. Maitlis, *J. Chem. Soc. Dalton Trans.*, 1747 (1984).
5. P. J. Fagan, J. M. Manriquez, E. A. Maatta, A. F. Seyam and T. J. Marks, *J. Am. Chem. Soc.,* **103**, 6650 (1981): G. Jeske, H. Lanke, H. Mauermann, H. Schumann and T. J. Marks, *J. Am. Chem. Soc.,* **107**, 8111 (1985).
6. K. W. Bagnall, *Essays in Chemistry,* **3**, 39 (1972).
7. R. C. Wade, *J. Mol. Catal.,* **18**, 273 (1983).
8. E. C. Ashby and J. J. Lin, *J. Org. Chem.,* **43**, 2567 (1978).
9. S. K. Chung, *J. Org. Chem.,* **44**, 1014 (1979).
10. B. Steinberger, M. Michman, H. Schwarz and G. Höhne, *J. Organometal. Chem.,* **244**, 283 (1983).
11. T. Inamoto, T. Takeyama and T. Kusumoto, *Chem. Lett.*, 1491 (1985).
12. J. O. Osby, S. W. Heinzman and B. Ganem, *J. Am. Chem. Soc.,* **108**, 67 (1986).
13. B. Meunier, *J. Organometal. Chem.,* **204**, 345 (1981).
14. T. Nishio and Y. Omote, *Chem. Lett.*, 1223 (1979).
15. H. S. Lee, K. Isagawa and Y. Otsuji, *Chem. Lett.,* 363 (1984).
16. H. S. Lee, K. Isagawe, H. Toyoda, *Chem. Lett.*, 673 (1979).
17. N. Santyanarayana and M. Periasamy, *Tetrah. Lett.,* **25**, 2501 (1984).
18. R. Johnstone and R. Telford, *J. Chem. Soc., Chem. Comm.*, 354 (1978). P. Four and F. Guibe, *J. Org. Chem.,* **46**, 4439 (1981).
19. T. N. Sorrel and P. S. Pearlman, *J. Org. Chem.,* **45**, 3449 (1980).
20. J. A. Cowan, *Tetrah. Lett.*, 1205 (1986).
21. E. C. Ashby and S. A. Noding, *J. Org. Chem.,* **45**, 1041 (1980).
22. Vanderesse R., Brunet J. J. and Caubere P., *J. Org. Chem.,* **46**, 1270 (1981).
23. J. L. Luche and A. L. Gamal, *J. Am. Chem. Soc.,* **101**, 5848 (1979). K. Kashima and Y. Yamamoto, *Chem. Lett.*, 1285 (1978).
24. A. Gamal and J. L. Luche, *J. Am. Chem. Soc.,* **103**, 5454 (1981).
25. T. Mizoroki, K. Seki, S. Meguro and A. Osaki, *Bull. Chem. Soc. Jap.*, 2148 (1977).
26. M. E. Osborn, J. F. Pegues and L. A. Paquette, *J. Org. Chem.,* **45**, 167 (1980).
27. M. F. Semmelhack, R. D. Stauffer and A. Yamashita, *J. Org. Chem.,* **42**, 3180 (1977).
28. S. L. Grundy, A. J. Smith, H. Adams and P. M. Maitlis, *J. Chem. Soc., Dalton Trans.*, 1747 (1984).
29. T. Satoh *et al., Tetrah. Lett.*, 193 (1982).
30. M. A. Bennet, T. W. Tuney, *J.C.S. Chem. Commun.*, 312 (1979).
31. K. R. Januszkiewicz and H. Alper, *Organometallics* **2**, 1055 (1983). *Tetrah. Lett.*, 4139 (1983).
32. J. L. Speier, *Adv. in Organometal. Chem.,* **17**, 407 (1979). G. W. Parshall, *Homogeneous Catalysis*, Wiley-Interscience p. 71 (1980).
33. A. Millan, P. Bentz and P. M. Maitlis, *J. Mol. Catal.,* **26**, 89 (1984).
34. A. Onopchenko, E. T. Sabovrin and D. L. Beach, *J. Org. Chem.,* **48**, 5101 (1983).
35. M. S. Wrighton and M. A. Schroeder, *J. Am. Chem. Soc.,* **96**, 6235 (1984).
36. H. Brunner and G. Riepl, *Angew. Chem. Int., Ed. Engl.*, 377 (1982).
37. J. L. Fry, *J. Org. Chem.,* **43**, 374 (1978).
38. T. Hayashi, K. Kabeta and M. Kumada, *Tetrah. Lett.*, 1499 (1984).
39. T. Hayashi, K. Kabeta, I. Hamachi and M. Kumada, *Tetrah. Lett.*, 2865 (1983).
40 I. Ojima and T. Kogure, *Organometallics,* **1**, 1390 (1982).
41. T. Kogure and I. Ojima, *J. Organometal. Chem.,* **234**, 249 (1982).
42. H. Brunner and R. Becker, *Angew. Chem., Int. Ed. Engl.,* **23**, 222 (1984).

43. T. Tsunoda, M. Suzuki and R. Noyori, Tetrah, Lett., 4679 (1979).
44. M. Kumada *et al., J. Am. Chem. Soc.,* **100**, 290 (1978).
45. (a) J. Schwartz and J. A. Labinger, *Angew, Chem. Int. Ed. Engl.,* **15**, 333 (1976).
(b) E. I. Negishi and T. Takahashi, *Aldrichimica Acta,* **18**, 31 (1985).
46. B. M. Trost and C. R. Hutchinson (Eds.), *Organic Synthesis Today and Tomorrow,* J. Schwartz and Co, pp. 55, 64 (1981).
47. U. M. Dzemilev, O. S. Vostrikova and G. A. Tolstikou, *J. Organometal. Chem.,* **304**, 17 (1986).
48. K. I. Gell, J. Schwartz, *J. Am. Chem. Soc.,* **100**, 3246 (1978).
49. S. Datta, M. B. Fisher and S. S. Wreford, *J. Organometal. Chem.,* **188**, 353 (1980).
50. J. L. Costa, A. F. Noels, A. J. Hubert and P. Teyssié, *Tetrah. Lett.*, 649 (1984).
51. J. Schwartz, M. J. Loots and H. Kosugi, *J. Am. Chem. Soc.,* **102**, 1333 (1980).

A. J. HUBERT, A. DEMONCEAU AND A. F. NOELS

APPLICATION OF TRANSITION METALS IN NATURAL PRODUCT AND HETEROCYCLE SYNTHESIS

1. Introduction

Transition metal catalysis is now at the basis of major industrial processes [1] but this concept has also led to many very interesting applications in fine chemicals and natural products synthesis.

Moreover, transition metals based organometallic derivatives are also most useful reagents both in stoichiometric and in catalytic reactions and many classical organic processes have been thus improved by appealing to such reagents. Actually the catalysis of organic reactions by soluble metal complexes has become a major synthetic tool, particularly useful not only for functionalizing organic substrates, but also for investigating the skeletons of complex molecules, sometimes in a remarkably straightforward way.

The main object of this contribution is to provide selected typical applications of transition metal chemistry directed towards the synthesis of fine chemicals (and more particularly of biologically active compounds), each topic being illustrated by some significant examples. This chapter is therefore not an exhaustive review but rather an overview of the field with references to further publications. Moreover, we propose to establish a correlation between some classical organic reactions and processes based on the application of transition metals.

1.1. INTRODUCTION OF FUNCTIONAL GROUPS

Transition metal-based reagents and catalysts are very useful for introducing efficiently and *selectively* various functions into organic substrates, the problem of selectivity being crucial where biological activity is concerned. In this regard, transition metal-based chemistry has been particularly useful for the organic chemist: very tricky problems have been solved in a very elegant way (typical examples: the synthesis of *L*-Dopa, prostaglandins, steroids, non-pollutant insecticides, and so on).

A large palette of often very selective and efficient reactions is now

A. Mortreux and F. Petit (Eds.), Industrial Applications of Homogeneous Catalysis, 93—139.

available for introducing classical organic functionalities (alcohol, ketone, epoxide, carboxylic acid, ester, nitrile, . . .) into organic substrates: chemoselective and enantioselective epoxidations, reductions, cyclopropanations, etc., are striking examples of the potentialities of these methods. Further examples of efficient processes are represented by the hydroformylation, hydrocyanation, and hydrosilylation, reactions.

1.2. IMPROVEMENT OF CLASSICAL ORGANIC REACTIONS

Many classical organic reactions can now be performed with exceptional selectivities and efficiencies when using such catalysts or reagents.

For example: the coupling reactions of aryl halides, classically based on copper metal catalysis (Ullmann coupling), can now be advantageously realized when using Ni(0) or Pd(0) reagents, transient aryl metal derivatives being formed as reactive intermediates.

Similarly, the coupling reactions of allyl systems in intermolecular or intramolecular (cyclization) ways is also best performed with nickel(0) reagents (a Corey reaction) [2].

The 1,4-addition reactions of nucleophiles onto α, β-unsaturated ketones are well-known in Michael-type processes: such reactions are run with high regioselectivities when using organocuprate reagents.

Even the classical malonic synthesis and other related nucleophilic substitution reactions based on stabilized carbanions proceeds very smoothly when using π-allyl palladium complexes as the electrophilic partner.

A particularly significant and useful contribution of transition metals in fine organic synthesis as well at the industrial level is based on their use as *catalysts*. This aspect is of course particularly important with *expensive* transition metals (Rh, Os, Pd, etc.). Indeed, there are numerous examples of selective processes which have never been developed up to the industrial stage because of catalyst costs, especially when some (even minor) loss of the catalyst could not be avoided. This was, for example, the case for palladium-catalyzed benzylic acetoxylation reactions, and several rhodium-catalyzed reactions, such as the direct ethylene glycol production from syngas (prohibitive pressures being an additional major drawback in this latter case).

1.3. CONSTRUCTION OF THE SKELETON OF ORGANIC MOLECULES

The set of reactions classically used in organic synthesis for elaborating multicyclic networks found in complex molecules (steroids, alcaloids, antibiotics, etc.) can be substantially enlarged by using transition metal chemistry. Unique access to sometimes very elaborate molecules directly from quite simple starting materials have been developed. For example, the nickel-catalyzed oligomerization of diolefins such as butadiene leads directly to cyclododecatriene derivatives in one single step (Wilke process).

Another classical example is the synthesis of cyclooctatetraene in one step from acetylene when using a nickel-based catalyst (a Reppe process), whereas the first classical synthesis required a painful, multistep approach.

Of major importance was the finding that cobalt catalysis (typically with $CpCo(CO)_3$ as catalyst) permitted the cocyclooligomerization of functionalized alkynes, the catalyst being, moreover, compatible with functionalities such as ketones, ethers, esters. In this context, Vollhardt [3] even reported a straightforward synthesis of a steroid by a cyclooligomerization reaction of acetylenic precursors (eq. 1).

[Co] (1)

$[Co] = Cp\,Co(CO)_2$

Another particularly significant breakthrough results from the direct synthesis of pyridine and of its derivatives by Co oligomerization of two molecules of acetylenes with a nitrile (Bönnemann synthesis [4]). The potentialities of the Bönnemann reaction are well illustrated by its application to the synthesis of heterocyclic systems, a particularly illustrative example being the synthesis of vitamine B6 [5g].

Today, many heterocyclic systems (including indole derivatives which are useful precursors for alkaloid synthesis) have been prepared by transition metal-mediated reactions.

However, despite the immense value of these methodologies, some drawbacks appeared to be associated to their applications. In fact, a lack of generality is apparent for many reactions.

Indeed, the most selective and efficient processes can often be successfully applied only to a limited range of reagents and substrates. This feature can of course be related to the high sensitivity of the catalysts to stereoelectronic parameters, just as is observed in many enzyme-catalyzed reactions.

The role of the transition metal in these reactions can be rationalized by the classical concepts of coordination catalysis: template effects, activation or, in some cases, stabilization of labile intermediates. Electron transfer processes and the participation of polynuclear complexes are also involved in some particular reactions.

As far as natural and biologically active compounds are concerned, it is noteworthy that two metals (copper and palladium) are particularly frequently encountered in recent applications of organometallic reagents or in catalysts used in fine chemical synthesis: the former is often used in stoichiometric reactions (cuprate reagents) whereas the second one is particularly useful in catalytic compositions owing to the high cost of this last metal. Nickel catalysts are also well represented in organic synthesis, together with some other metal-based systems (rhodium, vanadium, molybdenum, etc.) which are of interest for some important applications.

2. Stoichiometric Reactions: Organocopper Derivatives

2.1. PREPARATION OF ORGANOCOPPER REAGENTS

Organo Cu reagents are commonly formed by reacting 2 equivalents of organo Li compound to a Cu(I) salt or complex:

$$2\ RLi + CuI \rightarrow [R_2CuLi] + LiI \qquad (2)$$

$$\text{cyclohexenone} + [R_2CuLi] \rightarrow \text{3-R-cyclohexanone} \qquad (3)$$

The stereochemistry of the migrating group is retained. Moreover, vinyl cuprates are readily prepared from acetylenes [5m, 6, 7].

$$R_2CuLi + 2\ HC{\equiv}CH \xrightarrow[\text{add.}]{\text{syn.}} (R\text{–CH=CH})_2CuLi \quad \mathbf{A} \qquad (4)$$

$$\text{A} + 2\,\underline{\text{R}'}\text{X} \xrightarrow{70-95\%} 2\ \text{R-CH=CH-}\underline{\text{R}'}\ (\text{I}) \qquad (5)$$

$$(\text{A} + \text{I}_2) \longrightarrow$$

2.2. STABILITY OF CUPRATES [10]

Simple organocopper compounds are unstable even at 0 °C: their instability is related to the facile β-elimination reaction of a copper hydride moiety. The general use of cuprates instead of simple organocopper reagents is linked to their improved stability: in fact, blocking vacant coordination sites on copper increases the stability of the reagent as it is recognized that such vacancies are responsible for the copper—hydride elimination reaction. Heterocuprates therefore display quite different stabilities according to steric effects and absence of β-hydrogen. The presence of non-transferable ligands can also play a determining role (alkylcyanocuprates are particularly stable, the cyano ligand being non-transferable). Similarly, the improved stabilizing effect of tetrahydrofuran, as compared to diethylether, is explained by its more effective complexation on the copper vacant site.

Organocopper(I) reagents form the basis of two extremely important and general sets of selective and efficient reactions:

— addition reactions (particularly conjugate addition), and
— coupling reactions.

2.3. CONJUGATE ADDITIONS — ORGANOCUPRATES

Conjugate additions of organic groups to α, β-unsaturated carbonyl compounds through the intermediacy of organometallic reagents is now a commonly employed method in organic synthesis. Of crucial significance to the development of this general type of reaction was the observation that transition-metal salts could divert the course of addition of Grignard reagents to α, β-enones from 1,2 to 1,4: the implication of organo Cu species as the reactive intermediates led to the extensive development of *stoichiometric* organo Cu reagents.

Cuprates containing a non transferable chiral group leads to induction of asymmetry, the optical yields being however moderate (15—50% enantiomeric excess) but the presence of a chiral center in the substrate can also govern an induction of asymmetry as shown in a synthesis of $S(-)$ citronellic acid [8].

To date, a comprehensive mechanism of the addition process has not evolved. However, organo Cu compounds are known to be efficient electron transfer reagents and reaction with R_2CuLi produces an enolate anion as initial product.

$$[R_2CuLi] \longrightarrow [\text{charge transfer complex}] \longrightarrow R_2Cu\text{–(enolate)} \longrightarrow RCu + R\text{–(enolate)} \quad (6)$$

Examples:

(1) Application to prostaglandin synthesis:

RLi, CuI, L; OCH-R, BF_3 — THPO, O, OH, OCH$_3$, R, OH (7)

(2) Synthesis of cymopol [9]:

CH_3O, OCH_3, CH_3O, OCH_3 — a) BuLi b) Cu_2I_2DMS c) R Br — OH, R, Br, OH, CYMOPOL

$R = C_{10}H_{17}$ (8)

(3) Synthesis of a pheromone [7]:

$$[\ldots]_2CuLi + HC{\equiv}CH \xrightarrow{-30°}$$

$$[\ldots Z \ldots]_2CuLi \xrightarrow{I\ldots OAc} \quad (9)$$

OAc

2.4. SOME PARTICULAR APPLICATIONS OF ADDITION REACTIONS OF CUPRATES

2.4.1. *1,6-conjugate addition*

This has been applied in one step of the muscopyridine synthesis [11].

$CH_3MgI - CuCl$ (10)

2.4.2. *Homoallylic addition to epoxides*

Has been used in steroid synthesis [12].

2.4.3. *Ring opening reactions*

Strained ring systems (epoxides, lactones, etc.) are cleaved by cuprate reagents: these reactions have found application in prostaglandin chemistry and in pheromones synthesis.

Example (see also [13, 14] for other examples): ring opening reaction of lactones [15]: synthesis of a pheromone of *Danam chrysippys* (an African butterfly).

CuI, MgBr, OH (11)

HO, OH

2.4.4. *Substitution of acetoxy groups*

The substitution of an acetoxygroup by a cuprate reagent has been reported, for example, in (5*R*, 6*R*)-carbapenem synthesis [16]. This example is an additional illustration of the chemioselectivity of the organo-copper reagent as compared to classical organometallics (derived from Mg or Li). Moreover, the substitution takes place with retention of configuration, a result which can be explained by the participation of an internal transfer of groups within a cyclic transition state involving the copper center complexing the β-lactam substrate.

RNH, OAc, O, N, H $\xrightarrow{R'_2CuLi}$ RNH, R', O, N, H (12)

Moreover, the substitution of allylic acetates affords the product corresponding to a formal S_N2' substitution reaction [17]. The reaction is therefore remarkably regioselective and stereoselective (in favour of the *trans*-isomer) and also chemoselective, since a homoallylic acetate does not react. The stereoselectivity is an indication of the participation of a cyclic transition state involving copper, a transient π-allyl copper species can also be formulated but the existence of such π-allyl intermediates has never been clearly established (in opposition to metals of Group VIII).

OAc, OAc $\xrightarrow[\text{MgBr}]{\text{RCu, LiBr,}}$ R, OAc (13)

trans: 96 , cis 4%, no 1, 2-isomer

2.5. COUPLING REACTIONS

2.5.1. *Aromatic coupling reactions*

The most classical reaction of this type is the Ullmann coupling of aromatic halides in the presence of copper(0).

This reaction has received some improvements. For example, highly reactive metal powders are obtained from Co, Ni and Fe halides and Li metal. Copper powder in DMF is used to couple polychloroaromatics, but pentachloropyridines are reduced and do not couple (for example [18]).

Cross coupling of aryl halides involves one metallation step (e.g. with Li) followed by exchange with copper halide and reaction with Ar iodide [19].

2.5.2. *Copper mediated coupling of an organometallic reagent with an alkyl or vinyl halide*

The nucleophilic substitution of a halogenocompound by an organometallic reagent (e.g. an organolithium or -magnesium) proceeds smoothly through the formation of cuprates. This technique has been applied to the synthesis of various natural products (Equation 14) and of biologically active compounds such as antibiotics (Equation 15) (see also [13]).

Examples

(1) Synthesis of cymopol monomethylether (a green algae extract)

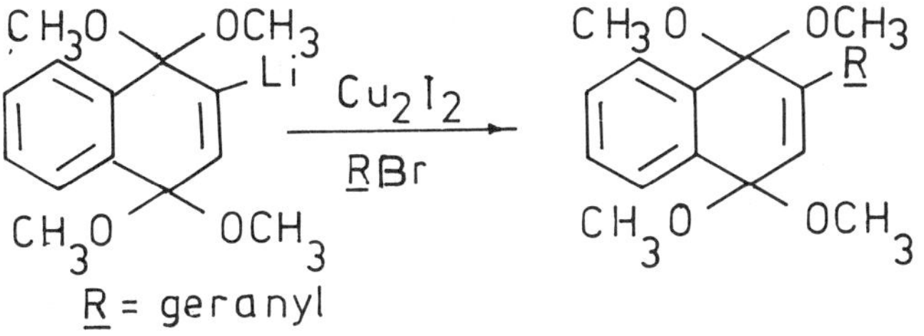
CH3O
OCH3
Li
Cu2I2
RBr
CH3O
OCH3
R = geranyl
CH3O
OCH3
R
CH3O
OCH3

(14)

(2)

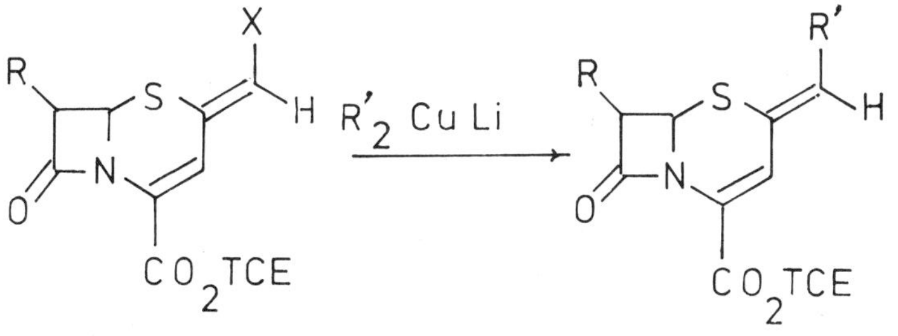
X
R
S
H
N
O
CO2TCE
R'2 CuLi
R'
R
S
H
N
O
CO2TCE

(15)

3. Catalytic Reaction: Palladium and Nickel Organometallic Reagents.

Palladium and nickel based organometallic derivatives are remarkably represented in organic chemistry. Of course, the catalytic reactions present a major interest, particularly with palladium, a rather expensive metal.

The interest in these reagents (and more particularly of palladium) arises from their compatibility with various functional groups. These reagents appear, therefore, as very chemoselective.

The mechanisms of the typical reactions of palladium (and nickel) reagents imply the classical general pathways which are now well recognized in coordination chemistry: oxidative addition, reductive elimination, insertion reactions, coupling reactions, β-eliminations, transmetallation, rearrangements (e.g. *cis*), and electron transfers (the most important oxidation states of nickel ranging from (0) to (III) and those of palladium from (0) to (II) in these processes, the low oxidation states being particularly significant as far as the early activation stages of the reactants are concerned).

3.1. THE KEY INTERMEDIATES

In most of these processes the key intermediates are π-allyl type complexes and σ-carbon bonded metal species. The detailed mechanistic features of these reactions are extensively reported in the literature (see references). π-Allyl nickel and palladium complexes offer a particularly large set of useful reactions both on the industrial and laboratory scales. When compared to those of non-transition-metal σ-allylorganometallic reagents (such as allylmagnesium halides), the most dramatic improvement is the possibility to promote catalytic cycles. Moreover their preparation is relatively easy as the techniques for working under a controlled atmosphere are now generally available (a factor of major importance in the nickel case, but notably less stringent requirements are required with palladium). The π-Allyl nickel or palladium intermediates are obtained in different ways, the oxidative addition of allylic halides or esters (trifluoroacetates, for instance) onto low oxidation state complexes (Ni(0), Pd(0)) being a very general method, together with the reaction of conjugated dienes with metal hydrides (often formed *in situ*), a particularly significant reaction in diene polymerization and telomerization processes.

π-allyl complexes are also formed in more particular reactions such as the palladium induced ring-opening of 1,3-diene monoepoxides: an example of an application of this approach is found in the steroid field [5b], a highly regio- and stereoselective introduction of a 15 β-hydroxy group and of a side chain to steroids could thus be realized (typically 20-*R*/20-*S* = 95 : 5). A significant feature of π-allyl nickel species is the dynamic equilibrium between the π-allyl and the σ-allyl forms: the occurrence of such equilibria has been well established in some cases [21] and may help us to understand different processes (for example the Corey coupling of allylic halides in the presence of nickel(0) as further vacancies are thus liberated at the metal center. The dimeric nature of the π-allyl nickel and nucleophilic character of the Ni—C bond must also be born in mind.

By contrast, the essential feature of the π-allyl palladium complexes is their particularly strong electrophilic character which allows smooth attack by various nucleophiles. This mode of activation has been extensively applied for realizing selectively various coupling reactions (see below).

3.2. ACTIVATION BY π-COMPLEX FORMATION

In fact, even a simple olefin is activated by π-complexation, the attack by nucleophiles being then possible under smooth conditions: this principle is the basis of the Wacker process. Therefore, π-complexation of olefins or the formation of π-allyl metal complexes offers a versatile and efficient alternative to the organic chemist for decreasing the electron density on unsaturated systems.

In fact, Wacker type oxidations (largely applied for aldehyde synthesis, acetoxylation reactions) can be considered as an intra or, more probably according to the recent literature, as an out-of-sphere nucleophilic attack on a palladium-olefin π-complex.

3.3. REMARK

Nucleophilic aromatic substitution can also be promoted by π-complex formation but chromium complexes are particularly useful for realizing this type of activation as palladium or nickel do not lead to stable π-complexes with arenes. For example, π-arenechromium tricarbonyl complexes react easily with nucleophiles, even when electrodonating substituents are present on the aromatic ring. This reaction has been used in two steps for the synthesis of carenone B (a sesquiterpene), one of these steps being an *intra*-molecular nucleophilic attack with formation of a spirobicyclic system [22].

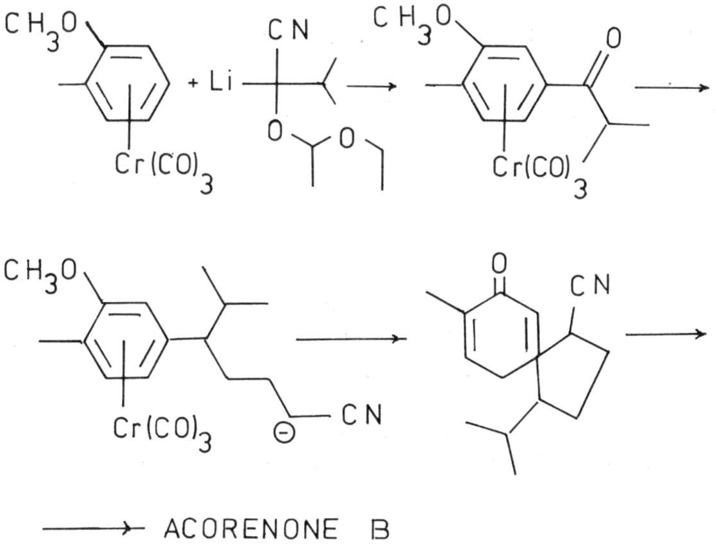

(16)

Another interesting application of arene group activation by chromium complex formation is found in the facile metallation of coordinated arene moieties with lithium, the lithiated complex being activated towards subsequent substitutions by electrophilic reagents. This principle has been applied to the synthesis of anthracyclone analogues [5e].

$(OC)_3Cr$ X BuLi $(OC)_3Cr$ Li X $E^{\oplus}$ $(OC)_3Cr$ E X (17)

Coordination of the chromium tricarbonyl group onto an arene enhances the kinetic acidity of the aryl C—H bonds. In order to avoid nucleophilic attack of the organolithium reagent onto a CO ligand, the reaction has to be run at low temperature. The reaction is regioselective as ortholithiation is observed with arene substituted by OCH_3, F, Cl.

These reactions offer therefore an alternative for modifying the usual reactivity of the aromatic ring.

4. Applications of Palladium and Nickel Complexes in Natural Product Synthesis

(Together with some significant applications of other transition metals complexes.)

4.1. COUPLING REACTIONS

Several coupling reactions, particularly those involving aryl halides, have been discussed in the chapter devoted to copper (see above). However, Pd(0) and Ni(0) complexes are also efficient reagents for realizing such coupling reactions, σ-aryl palladium entities being formed as intermediates.

Example: one key cyclization step in an alnusone synthesis involves such a nickel(0) (or palladium (0)) promoted intramolecular coupling of aryl iodide groups [23].

CH_3O ... $\xrightarrow[Pd]{Ni^0}$ CH_3O, CH_3O (18)

It is noteworthy that the use of Ni or Pd catalyst also limits homocoupling to less than 5% in many classical organometallic-organohalides nucleophilic reactions [24].

Example:

$$PhZnCl + IAr \xrightarrow[Pd]{Ni} Ph{-}Ar + ZnICl \qquad (19)$$

Palladium catalysed coupling reactions are also effectively applied to vinyl derivatives and thus constitute a particularly convenient access to conjugated diene systems [23b].

Similarly, homocoupling and coupling of two different allyl moieties (cross coupling) can be efficiently run with Ni or Pd complexes, the reaction proceeding by π-allyl type intermediates formation through the oxidative addition of the allyl halide to the metal centre. A mixture of isomers is usually obtained as a result of the allylic resonance, but the reaction is however of interest for the preparation of macrocycles: the selectivity being in such cases mediated by geometrical and conformational factors.

Example:

Br, Br $\xrightarrow{Ni(CO)_4}$ ... $\xrightarrow[10\%]{h\nu}$ HUMULENE (20)

humulene (10%) = mixture of products.

4.1.1. *Typical cross coupling reactions of allyl groups*

(A) *Cross coupling reaction of allyl halides* are efficiently run in the presence of nickel (or palladium (0)) complexes. The key intermediate is a π-allyl nickel complex resulting from an oxidative addition of one of the two allylic bromides implied in the coupling process.

Example: a synthesis of geranyl acetate.

Br OAc

Br $\xrightarrow{Ni(CO)_4}$ NiBr ⟶ OAc (21)

geranyl acetate

This reaction is another illustration of the compatibility of the organonickel reagent with different functional groups (an ester in the present case).

(B) *Cross coupling reactions of aromatic halides* can sometimes be performed with copper reagents but they also take place cleanly in the presence of nickel or palladium(0) complexes through formation of arylmetal species.

(C) *Cross coupling reactions of aryl halides with π-allyl-nickel complexes* has received several useful applications: for example, such a cross coupling reaction has been applied in one key step of a coenzyme Q_1 synthesis.

OAc, CH_3O, Br, CH_3O, CH_3, OAc + Ni Br ⟶ Ar. ⟶ O, CH_3O, CH_3O, CH_3, O (22)

REMARKS

(1) The alkylation of quinones by π-allyl nickel complexes is another interesting synthetic route to vitamin K and coenzyme Q, the selectivity of this process being remarkably improved relatively to the classical approach based on acid-catalyzed addition of phytol to quinones.

(2) Similarly, the coupling of vinyl bromides with olefins is very selectively promoted by palladium(II) acetate. Conjugated polyenes are obtained: in fact, this reaction is potentially useful for the synthesis of isoprenoids such as vitamin A [25].

(D) *Palladium catalyzed cross-coupling reactions of organometallics* [26] have been developed to a very valuable synthetic tool and have received applications in natural product chemistry the vinylalane case being a particular but useful one as such alanes are readily available by zirconium catalyzed carboalumination of acetylenes. More classical organometallics have also been successfully used for such coupling reactions.

Example (see also [27, 28, 29] for other examples): A dehydrovitamin A synthesis:

R AlMe$_3$ Cl_2ZrCp_2 R AlMe R'I Pd^0, Zn^{++} R R' (23) I_2 I Pd^0 ClZnR'

Similarly, copper and zirconium reagents lead to transmetallation reactions with organopalladium compounds. These reagents are particularly useful to introduce a vinyl group into various substrates and has received applications in steroids [30], prostaglandins (Ref. 31) and natural diterpenes [32, 33] synthesis.

4.2. ALKYLATION REACTIONS

Nucleophilic substitutions of allylic functional groups (allyl esters, ethers, alcohols, halides, fluorosilanes, etc.) are efficiently catalyzed by palladium derivatives; a very electrophilic π-allyl palladium being formed as an intermediate.

π-Allyl Pd derivatives are also readily formed by reaction of olefins with $PdCl_2$ in an organic solvent (e.g. DMF) or by coupling of 1,3-dienes (telomerization reactions).

As the yield of these reactions is generally good and as they proceed smoothly under very mild conditions, these processes have therefore received many applications in natural products synthesis. Cyclic compounds are obtained by similar but intramolecular reactions.

It is noteworthy that the attack of π-allyl Pd can be selective despite the fact that two different sites are available for nucleophilic attack. Various stabilized carbanions and nucleophiles can be used as nucleophilic reagents for attack of the π-allyl palladium center.

Allylic organometallic reagents (Mg, Sn) [5f] afford the coupled products through formation of the corresponding bis-π-allylpalladium intermediate, the yield being sometimes practically quantitative in the organomagnesium case. The reaction is regioselective and stereoselective, particularly in the tin case, since the coupling takes place by joining the least-hindered ends of the allylic ligands. However, the yield is notably higher with the organomagnesium reagent. The main advantage of the former approach (in addition to the improved regioselectivity) lies in the compatibility with functional groups such as esters.

4.2.1. *Examples of nucleophiles useful in π-allylpalladium substitution processes*

(A) *Malonates* and related stabilized carbanions.
Examples:
Preparation of cyclopentanones by Pd-catalyzed intramolecular reaction of active methylene compounds with allylic ether moieties: [34].

R, CO_2CH_3, OPh; $Pd(OAc)_2PR_3$ / acetonitrile; R, CO_2CH_3 (24)

This reaction has been applied to the elaboration of steroid skeletons and to sarkomycin (a natural product which shows inhibitory effect on Ehrlich ascites tumor) synthesis. Coronafacic acid synthesis is another interesting example of a palladium-catalyzed cyclization.

A synthesis of a Queen substance [35] and of the pheromone of Monarch butterfly [36] are additional examples of applications of this particularly useful process.

(B) *Sulfones*
Example: synthesis of vitamin A.

OAc; $PdCl_2$ / $CuCl_2$; PdCl, OAc; $R\frown SO_2Ph$ / NaH/ PPh_3 (25)
PhO_2S, R, OAc → R, OAc; R =

(C) *Nitroalkanes*
Examples: [36]: Synthesis of recifeiolide

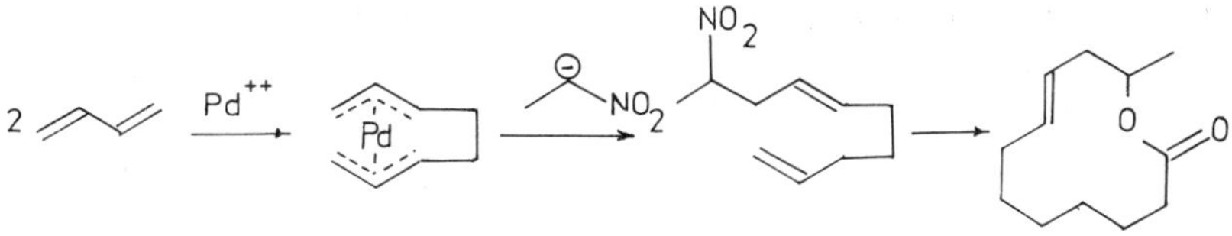

2 (butadiene) $\xrightarrow{Pd^{++}}$ Pd $\longrightarrow$ NO_2 ... $\longrightarrow$ O, O (26)

Stereoselectivity: trans-attack of a π-allyl moiety is the prefered pathway: in fact, steric factors and the complexing ability of double bonds govern the regioselectivities and the coupling occurs generally through joining the less hindered termini of the reacting groups.

The use of *chiral ligands* on the π-allylpalladium complexes leads to induction of chirality with however rather low optical yields [37].

Examples of stereoselectivity and regioselectivity:

(1) *Trans*-chrysanthemic acid (an insecticide) synthesis [38]:

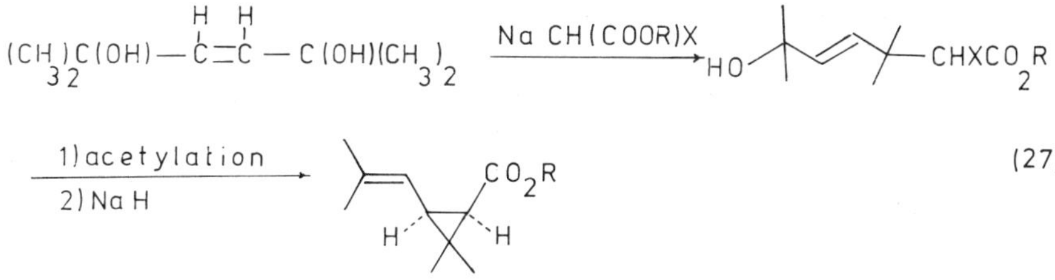

$$(CH_3)_2C(OH)—\overset{H}{C}=\overset{H}{C}—C(OH)(CH_3)_2 \xrightarrow{Na\ CH(COOR)X} HO—\ldots—CHXCO_2R$$

$$\xrightarrow[2)\ NaH]{1)\ acetylation} \ldots CO_2R \qquad (27)$$

(2) A further significant example is the stereospecific synthesis of (R*, S*)-5-hydroxy-2-methylhexanoic acid lactone (the major component of the Carpenter bee pheromone) [5c].

4.3. CYCLIZATIONS

Intramolecular attacks of π-allyl Pd complexes or of aryl palladium intermediates have been applied to the synthesis of cyclic systems: in fact, this type of reaction appears to be particularly interesting for the preparation of macrocyclic compounds.

Examples:

(1) Intramolecular reactions with aryl halides

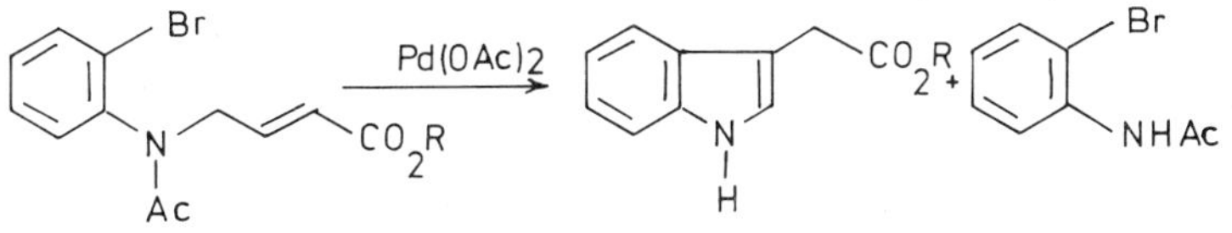

Br, N–Ac, CO_2R $\xrightarrow{Pd(OAc)_2}$ N–H, CO_2R + Br, NHAc (28)

(2) Recifeiolide synthesis (cyclization reaction based on a nucleophilic attack of a π-allyl Pd intermediate).

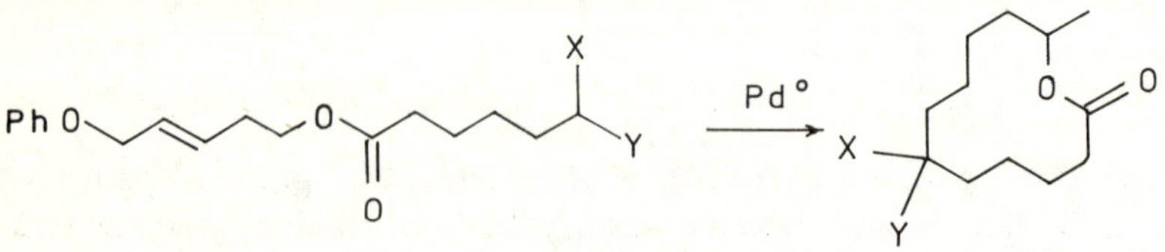

(29)

Similar reactions have been applied to humulene, macrolide [36] and steroid [57] synthesis and to the preparation of various heterocylic systems (Examples [22, 37, 38]).

4.4. 1,4-ADDITION TO CONJUGATED SYSTEMS

We have already discussed the very useful 1,4-addition of organocopper reagents to conjugated enones. In fact, various transition metal complexes catalyze such additions, even to simple conjugated dienes [44].

For example, nickel acetylacetonate reduced by Dibal H (diisobutylaluminium hydride) catalyzes the 1,4-addition of vinylzirconium reagents onto cyclic enones, a reaction which has received application in prostaglandin synthesis.

Example: 1,4-addition to cyclopentenones (of application in prostaglandin chemistry).

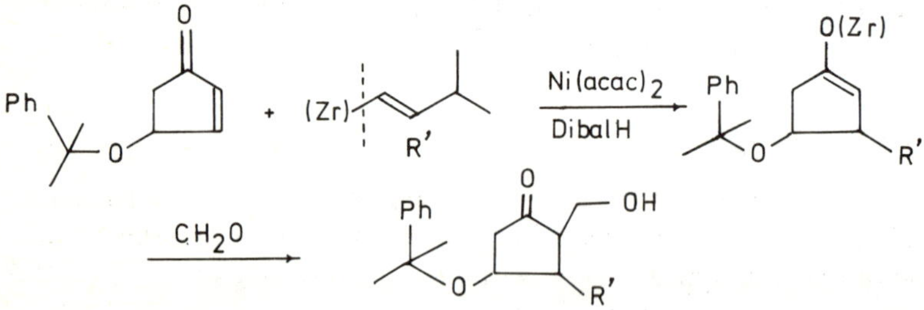

(30)

The mechanism of action of the nickel addend proceeds probably through a reduced form of nickel (e.g. nickel(I)) which is involved in a electron transfer process with the enone. Further reaction with the organozirconium reagent affords the addition product through transfer of the organic radical from zirconium to the nickel center, followed by a reductive elimination step (see [45] for a similar mechanistic proposal).

Remark: in this context, it is of interest to note that the classical Michael addition of β-carbonyl anions to a conjugated enone can be efficiently promoted by nickel acetylacetonate in the absence of a strong base [46]: this particular procedure is of particular value when base-sensitive groups are present in a molecule.

4.5. TELOMERIZATIONS AND OLIGOMERIZATIONS

Telomerization and oligomerization reactions of unsaturated substrates are certainly one of the most useful application of transition metal catalysis (together with Ziegler—Natta polymerizations). This problem is considered in other chapters of the present book: we shall therefore present only a few typical examples of applications.

An important difference between Pd and Ni catalysis is that Pd affords linear dimerization products whereas Ni catalysts are more suitable when cyclic dimers or trimers are required. However, linear oligomers can also be obtained with Ni catalysts.

4.5.1. *Preparation of linear telomers and oligomers*

Various natural products have been synthesized from telomers prepared by a Pd or Nickel catalyzed telomerization of conjugated dienes such as butadiene and isoprene.

Steroids have also been prepared from such telomers:

Example: Synthesis of homoestrenenone [47].

2 + AcOH —Pd(II)→ OAc → O

$PdCl_2$, CuCl (Wacker oxidation) / O_2,H_2O → O → → (31)

HOMOESTRENENONE

It is noteworthy that natural terpenoids have not yet been prepared by selective dimerization of isoprene with Pd or Ni catalysts. Even telomerization of isoprene with methanol does not lead to citronellol since the methoxy group is not located at the proper position: the synthesis of the natural isomer requires additional steps [35, 36].

Asymmetric synthesis of citronellol has been reported with Pd-dialkylmenthylphosphonites (Men $P(OR)_2$) as catalyst [48].

4.5.2. *Preparation of cyclic oligomers*

Example [49]: Preparation of muscone from butadiene, allene and CO.

$$\text{Ni complex} \xrightarrow{C{=}C{=}C} \text{Ni complex} \xrightarrow{CO} \text{cyclic ketone} \xrightarrow{H_2} \text{MUSCONE} \quad (32)$$

Cyclotrimerization and cyclotetramerization of acetylenes takes place also with nickel catalysts. However, cobalt derivatives appear as particularly convenient as they are compatible with some functional groups.

Example: see [50] and equation 1.

4.6. CARBONYLATION REACTIONS

Carbonylation is a particularly elegant and convenient method for introducing an aldehyde, a ketone or a carboxylic acid functional group into organic molecules. The main industrial application of the hydroformylation of olefins is the industrial scale production of alcohols or aldehydes.

In fine chemistry, such processes offer a particularly straightforward access to various natural products. Dicobaltoctacarbonyl as well as rhodium derivatives are the classical catalysts for carbonylation reactions.

Example (see also [44, 51, 52]).
Synthesis of royal jelly acids:

$$2\ \text{butadiene} + CO + ROH \xrightarrow[PPh_3]{Pd(II)} \text{CH}_2{=}\text{CH(CH}_2)_3\text{CH}{=}\text{CHCH}_2CO_2R$$

$$\xrightarrow[H_2O]{CO,\ Co_2(CO)_8} HO_2C\text{–(CH}_2)_6\text{–CH}{=}\text{CH–}CO_2H \quad (33)$$

A particularly useful aspect of these processes is illustrated by the Khand reaction [5i] in which the cyclopentenone ring is directly synthesized through cobalt(0) catalyzed cyclooligomerization of ethylene and acetylene under an atmosphere of carbon monoxide. In this way, improved synthesis of jasmone (a valuable fragrance), jasmonic acid (a plant growth regulator), cyclopentanoic antibiotics (methylenomycin A and B, sarcomycin), and prostaglandins analogs have been carried out.

The key intermediate is a binuclear alkyne hexacarbonyledicobalt complex (X) which reacts stoichiometrically with an alkene to form a cobaltacyclopentene, CO insertion followed by a reductive elimination yield the cyclopentenone.

X 33 % (34)

Another interesting extension of the above reaction consists in the intramolecular alkyne—alkene cyclocarbonylation reaction which has been used for the preparation of precursors of linearly fused 5-membered rings (such as the tricyclo-[6.3.0.0^{2,6}]-undecane system) represented by hirsutene, hirsutic acid and the antibiotic coriolin.

In fact, the intramolecular carbonylation of ene—yne systems (e.g.: 1-heptene-5-yne) affords bicyclo-[3.3.0]-1-octene-3-ones which are suitable precursors for these natural products [5j].

Example of the Khand reaction [53, 54]. An efficient synthesis of methylenomycin B from (2-butyne)-hexacarbonyldicobalt has been reported [54]: the complex (X, R = R′ = CH_3) reacts with allyloxy-TPH under milder conditions and with higher regioselectivity than with other olefinic substrates: a cyclopentenone is obtained without contamination by any isomeric adduct.

Hydrolysis of (Y) followed by dehydration of the corresponding alcohol affords methylenonomycin B.

y methylenomycin B (35)

4.7. PROTOTROPIC ISOMERIZATIONS AND REARRANGEMENTS

Metal catalysis is a general method for promoting prototropic isomerizations and to reach the thermodynamic equilibrium between positions and/or geometrical unsaturated isomers.

Example: one step of a tetrahydrofuranecarboxaldehyde synthesis involves a rhodium catalyzed isomerization of the double bond in a dioxacycloheptene ring:

R1 R2 O + HO OH → R1 R2 O O —Ru(II)→

R1 R2 O O —BF_3Et_2→ O R1 R2 (36)

Such a prototropic isomerization has also been applied in an original method for protecting hydroxy groups in carbohydrate chemistry.

Rearrangements: As π-allyl Pd species are easily formed from allyl acetates, 1,3-migration of an acetoxy group in allylic compounds is readily carried out in the presence of Pd catalysts (in the absence of added nucleophiles).

Example: Synthesis of Matsutake alcohols:

OAc —Ru / H_2→ OAc

—Pd (II)→ OAc → OH (37)

Moreover, Pd species promote sigmatropic rearrangements, mainly of the 3,3-type, particularly when heteroatoms are involved: [44].

Example:

Ac N N S —Pd(II)→ Ac N N S (38)

This reaction is formally similar to a Claisen rearrangement.

Similarly, a palladium catalyzed allyl—vinyl ether shift reaction has been applied to a prostaglandin synthesis [51, 55].

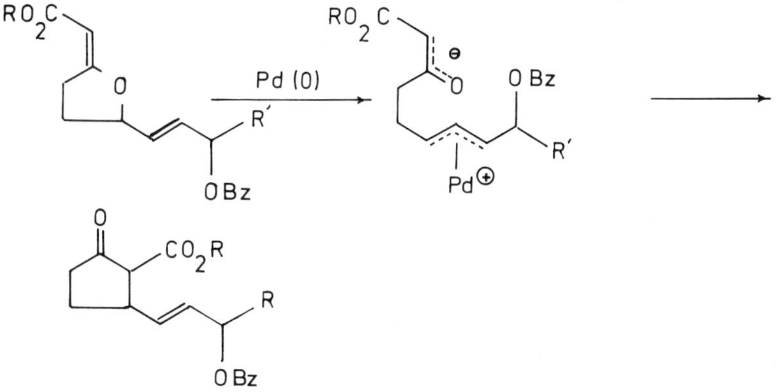

(39)

However, it has been stated that the regiospecificity of these rearrangements ruled out the intermediacy of a π-allyl complex. Some of these reactions are in fact reminiscent of Cope rearrangements and could perhaps be explained as concerted type processes.

4.8. ELIMINATION AND DECARBOXYLATION REACTIONS

Allylic acetates afford olefins by elimination of acetic acid whereas carboxylic acids are readily decarboxylated with Pd catalysts.

Application in vitamin A synthesis [56].

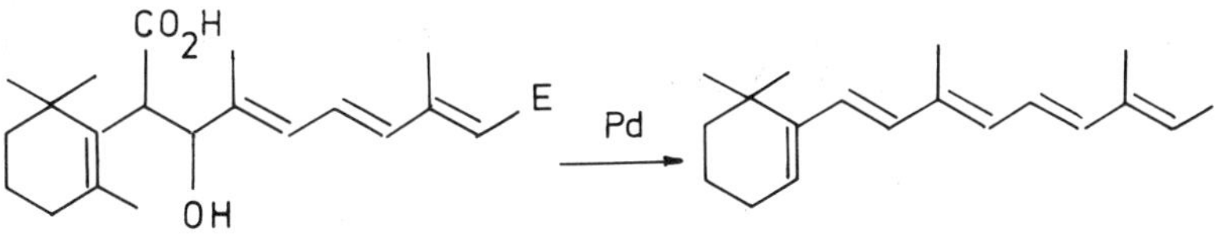

(40)

The reaction is also potentially useful for the synthesis of drugs active for treating tumors and keratizing dermatosis (Eq. 41).

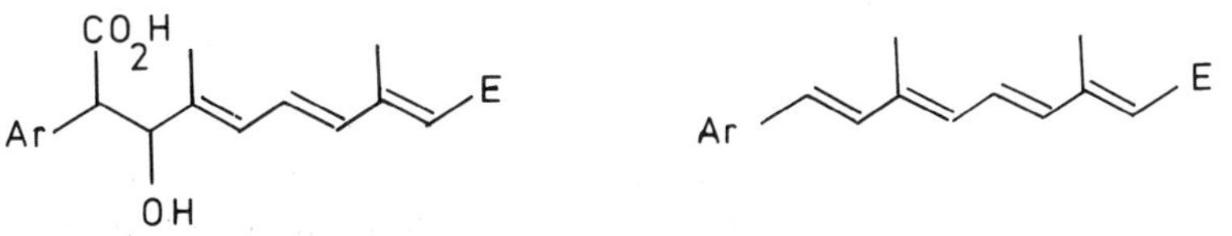

(41)

(AR = 2,3,6-trimetyl-4-methoxyphenyl, E = $CO_2C_2H_5$)

4.9. TRANSMETALLATION

Metal exchange reactions are general methods for generating transition metal intermediates, the transition metal can be used sometimes in catalytic amounts [57].

Examples:

(1) $R\text{-}HgX + PdY_2 \longrightarrow R\text{-}PdY \xrightarrow{RCH=CH_2} RCH=CHR'$ (42)

Application: Pterocarpine synthesis:

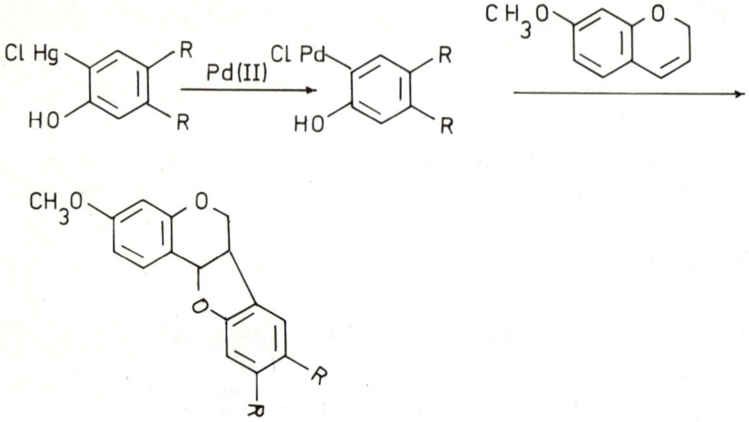

(43)

(2) Thallium—Pd exchange reactions offer a practical method for coupling aromates with unsaturated lactones [20b]. This reaction has received an application in indole alkaloid synthesis.

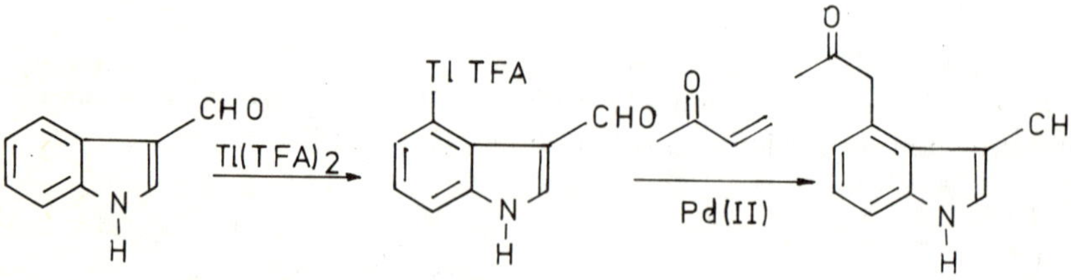

(44)

(3) Organozirconium reagents (readily available by hydrozirconation of acetylenes) are particularly convenient for generating various transition

metals intermediates by adding the corresponding salt into the reaction medium in a one pot reaction.

4.10. METALLATION

Direct metallation of aromates can be conveniently performed with palladium acetate. The reaction is also of application with heterocycles [57] (example 2):

$$ArH + Pd(OAc)_2 \longrightarrow ArPdOAc + HOAc$$

Examples

(1)

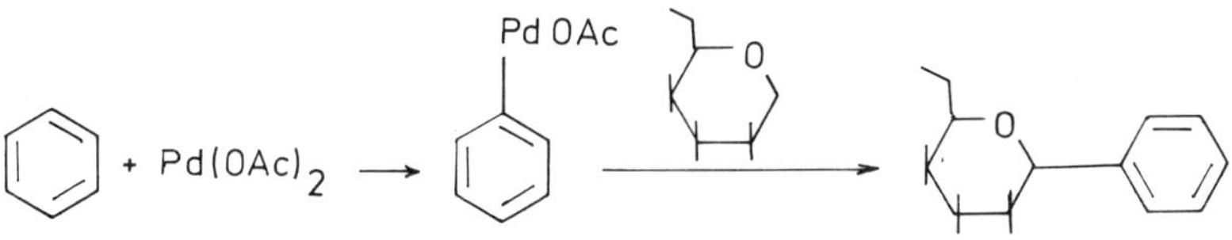

(45)

(2)

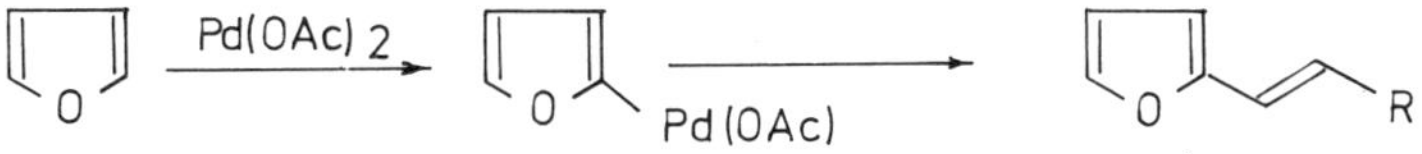

(46)

4.11. APPLICATIONS OF OXIDATION AND HYDROGENATION

Hydrogenation and oxidation certainly constitute two areas where transition metal catalysis has brought the most impressive contributions. As these fields cannot be treated in detail within the frame of the present chapter, only a few selected applications will be reported as illustrations of their tremendous potentialities and general interest (both on the huge industrial scale as well as in the synthesis of elaborate molecules).

4.11.1. *Oxidations*

4.11.1.1. The Wacker reaction is an important process at the industrial level but several applications of related reactions have been reported in natural products synthesis.

Example (see also [58].)
Application in prostaglandin synthesis [36].

$$CH_2=CH(CH_2)_8COOR \xrightarrow{Pd(II)\ Cu(II)} CH_3CO(CH_2)_8COOR \longrightarrow \text{cyclopentane-1,2,4-trione}-(CH_2)_6COOR \quad (47)$$

4.11.1.2. Oxidation of phenols. Copper catalyzed oxidation of phenols, notably of pyrocatechol, has been extensively studied as a biomimetic model, muconic acid being formed.

More useful from the preparative point of view, are the palladium catalyzed oxidations which proceed either by addition of hydroxyl groups to double bonds when an olefin is present (a) (telomerization can even occur with butadiene), or by formation of quinoid systems (b):

(a)

OH Ph —Pd Cl2→ O Ph (48)

O O

(b)

O O OH → O O O O O O → O O O O O O O CARPANONE (49)

4.11.1.3. Epoxidations. Vanadium based catalysts are largely used for olefin epoxidation by hydroperoxides (such as *t*-butyl hydroperoxide). These catalysts are very *regioselective* for epoxidation of double bonds of allylic alcohols [56].

Titanium alkoxides appear particularly useful catalysts as they can be highly stereospecific when associated to a chiral ligand such as L(+)-diethyltartrate (Sharpless reaction).

Examples:

(1) The synthesis of disparlure:

OH ⟶ OH ⟶ P (50)

(2) [58].

CH_3O — t.BuOOH / Mo(CO)$_6$ ⟶ A + B, A/B= 5 (51)

The stereoselectivities of Ti and V catalysts are very different, Ti is the more selective catalyst, the selectivity being reversed relative to the MCPA case.

4.11.1.4. Another important application of transition metals in the oxidation of organic molecules is encountered in the modified oxidation reactions of expensive oxidizing reagents such as ruthenium and osmium oxides: a cheaper oxidizing reagent (typically a periodate or a hypochloride) is used to reoxidize the expensive metal complex which acts, therefore, as a catalyst [59].

Example

$(CH_2)_n$, N, R — RuO_2 (catalytic amount) / Na metaperiodate ⟶ $(CH_2)_n$, N, R (52)

4.11.2. *Hydrogenations*

The Wilkinson catalyst is in fact the most popular and best studied hydrogenation catalyst in the homogeneous phase. The complex is soluble in benzene but the rate of the reaction is enhanced by addition of hydroxylic solvents or of nitrobenzene but the hydrogenation reaction is inhibited by coordinating solvents.

Its remarkable selectivity results from its high sensitivity to steric hindrance: the less hindered double bonds in polyenes always reacting best. Moreover, this catalyst is compatible with a variety of functional groups such as the carboxylic acid, ester, ether, cyano- and nitro groups. The thiophene ring is not an inhibitor of the Wilkinson catalyst, in contrast to many other classical transition metals systems [60, 61].

Highly stereospecific hydrogenation catalysts have thus been developed, particularly rhodium based complexes bearing a chiral ligand: the synthesis of L-Dopa with RhCl (DIOP+) catalyst is the most successful realization in this field. The mechanism of this reaction has been extensively investigated, particularly by Halpern. Bidentate coordination of the prochiral unsaturated precursor is a key factor to explain the induction of chirality.

Chiral di- and tripeptides prepared by similar hydrogenation reactions [62] have been applied to the synthesis of enkephalin analogues:

Ph
CBzNH C-N COOMe
O H
$Rh\,L^{*}/H_2$
Ph
CBzNH C-N COOMe
O H

ENKEPHALIN (53)

AcNH C-N C-N COOMe
O H O H
$Rh\,L^{*}/H2$
AcNH C-N----
O H

5. Particular Applications of Transition Metals

5.1. GROUP PROTECTION BY COMPLEX FORMATION

The protection of functional groups is of strategic importance in the synthesis of natural products. Complexation of functionalities by transition metals can modify their relative reactivities.

The protection of a conjugated diene system by complexation with the $Fe(CO)_3$ group (from $Fe(CO)_5$ or $Fe_2(CO)_9$) has received several applications in synthesis, particularly in pheromone synthesis. Deprotection is achieved by oxidation, Me_3NO being particularly recommended [63].

The protected diene group can also be formed from a cationic iron complex by nucleophilic attack (see below for examples of applications of cationic iron complexes in synthesis). This principle has been extended to other metal complexes and functional groups and applications in prostaglandin, steriod, pyrethroid, peptide chemistry [27, 60, 64] and various pheromone syntheses.

5.2. IRON COMPLEXES: CATIONIC COMPLEXES

Different types of cationic iron complexes because of their high electrophilicity are now currently reported as highly efficient and selective reagents in organic synthesis.

5.2.1. Early applications of iron cationic species resulted from the facile reaction of α, α'-dibromo-(and tetrabromo) acetone with $Fe_2(CO)_9$ with formation of the cation (A). Such cationic complexes can be efficiently trapped by olefin with formation of five-membered ketones. The reaction with conjugated dienes affords a straightforward access to tropones. Furan is a particularly good acceptor of these cationic complexes whereas their addition to enamines constitutes a potentially useful pathway to prostaglandins (cyclopentenones being thus obtained).

A = 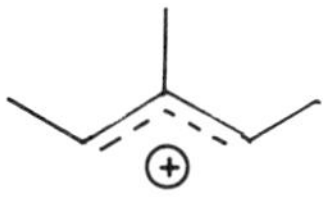 (54)

Applications: this reagent has been applied to the formation to five- or seven-membered rings for example in alkaloid [66, 67, 51] and thuyaplicin [68] syntheses.

5.2.2. A second class of cationic complexes corresponds are enones equivalents such as (B):

B = 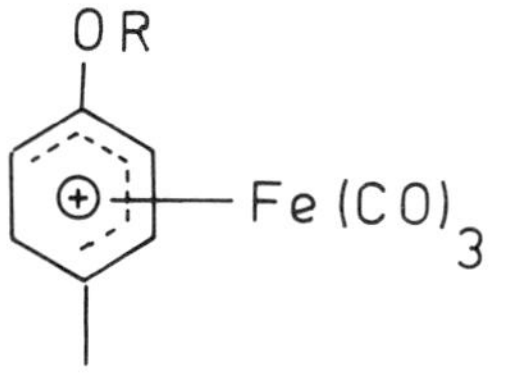 OR ⊕ $Fe(CO)_3$ C =  O ⊕ R (55)

((B) is in fact synthetic equivalent of (C)) which react readily with nucleophiles such as the malonate anion. This approach has been recommended for the synthesis of macrolide antibiotic precursors [5d].

5.2.3. The third class of cationic iron reagents is synthesized by the addition product of an electrophile E^+ onto ($^1\eta$ allyl)F_p complexes (F_p = $^5\eta$ cyclopentadienyl dicarbonyl iron): a dihapto cationic complex is thus obtained.

Fp
E
base
Fp E
(56)

In the presence of a base, the complex rearranges to the corresponding monohapto complex which can be further functionalized for example by coupling with an allylic iodide.

This reaction has been applied to a pheromone synthesis (californian red scale sex pheromone [5a]): the electrophile E^+ being the dioxolylium cation.

By contrast, strong nucleophiles add onto the olefinic ligand of the dihapto complex to afford mixtures of the positional isomeric adducts.

5.2.4. Other iron complexes: the impact of iron based complexes in organic synthesis is further illustrated by the preparation of quinones [5h] from acetylenes (complexes of Co, Mo, Mn, Rh, Pt having also been used). The mechanism of this reaction is well-established as the corresponding maleoylmetal complexes have been isolated. The reaction was also used for the preparation of naphthoquinones, the phthaloyliron intermediates are readily available from orthodiiodobenzene and ironpentacarbonyl in the presence of UV light (under CO atmosphere) or from benzobutanedione and iron pentacarbonyl. Further reaction with acetylenes affords the corresponding naphthoquinone (sometimes in quantitative yield). Nanaomycin A was successfully prepared by using this technique. Phthaloylcobalt complexes appear to be particularly convenient in this field.

5.3. ANIONIC TRANSITION METAL REAGENTS

Anionic reagents suitable for nucleophilic substitutions are formed from alcoholates and metal carbonyls.

$$Ni(CO)_4 + CH_3ONa \longrightarrow \underset{D}{CH_3ONi(CO)_4^{\ominus}} \qquad (57)$$

Such complexes are of importance in stoichiometric reactions.

I
D
COOCH$_3$
CH$_2$OH
(58)

5.4. TITANIUM AND ZIRCONIUM

Reduced titanium species are particularly useful for promoting coupling reactions of ketones [65, 69] and methylenation of carbonyl groups, the active species in the coupling reaction are low valent Ti species (See also 7.2.2.). Organo Zr reagents couple readily with π-allyl Pd species [59].

Examples

(1) a humulene synthesis [27].

$PdCl_2$ → Pd

$(CH_3O)_2CH$ ... $ZrClCp_2$ → OHC ...

(59)

(2) The coupling of retinal by the $TiCl_3$—$LiAlH_4$ reagent affords carotene in 85% yield [27].

5.5. METATHESIS

The metathesis of olefins is one of the most thoroughly investigated modern catalytic reaction mediated by transition metals. The reaction was initially applied to internal olefins at least under "pseudohomogeneous" conditions, with the exception of a nitrosylmolybdenum-based catalyst, but further developments led to the extension of the reaction to alkynes and even to functionalized olefins, particularly unsaturated esters. The main modification of the initial catalyst (WCl_6—AlR_3) consisted in the replacement of the organoaluminum partner by organotin reagents and, in some particular cases, by organotitanium derivatives.

Applications: synthesis of macrocyclic ketones and lactones [70].

$WOCl_4$ / $Cp_2Ti(CH_3)_2$

(60)

6. Applications of Transition Metals in Hydride Chemistry

(Applications to natural product synthesis)

Hydrides, essentially *mixed* hydrides and boron hydrides, are extensively used in organic synthesis as selective (both chemoselective and stereoselective) reagents for the reduction of various functionalities.

The association of transition metal complexes and of hydrides offer a convenient method for modifying the selectivities (and efficiencies) of the classical hydride reagents. Ill-defined, very active reducing species are formed (transition-metal hydrides?) that promote selectivities quite different from those of the corresponding non-catalyzed reactions.

Application: Matsutake alcohol (a fragrance compound in a Japanese mushroom) is synthesized by reduction of one terminal double bond by $LiAlH_4$—Cp_2TiCl_2 after protection of an allylic alcohol group as a diisobutylaluminium alkoxide:

OH DIBALH O Al(Bu)$_2$

OH (61)

Li Al H_4 / Cp_2 Ti Cl_2

Particular transition metal hydrides have found very promising applications in organic reactions, as is shown for example by the hydrozirconation reaction, which appears as a remarkable improvement of the classical hydroboration reaction (the high cost of the reagent, however, limits its application in large scale preparations).

6.1. ORGANOBORON CHEMISTRY

The classical hydroboration procedures offer an easy entry to the preparation of alkyl- and vinylboranes, which are valuable intermediates for further synthesis.

The *coupling reactions* of alkyl (and vinyl) boranes is particularly interesting as the latest developments of this principle have resulted in very selective procedures, some of them having already found application in biologically active compound synthesis (prostaglandins).

The application of transition metal catalysis to borane chemistry has received little attention up to now. Highly active hydrogenation catalysts have been devised by reacting diborane with a transition metal salt. More recently, very stereospecific coupling reactions have been carried out in the presence of a palladium(0) catalyst: the quantitative formation of pure *trans*-styrenes is observed by coupling a vinylcatecholborane with an aryl iodide under basic conditions [71, 72].

6.2. ALANE CHEMISTRY

Besides their application at the industrial level (e.g. in the Alfol process or in Ziegler—Natta polymerization catalysts), organoaluminums are the basis of various useful organic reactions. In fact, hydroalumination of acetylenes (a Cp_2ZrCl_2 catalyzed process when Me_3Al is the metallation reagent) is particularly useful for the stereospecific synthesis of prostaglandin precursors.

Example: Application to a prostaglandin synthesis.

$$(CH_3)_2Al-\overset{H}{C}=\underset{H}{C}-C_6H_{13} + \text{(2-(CH}_2)_n\text{COOR-cyclopent-2-enone)} \longrightarrow \text{(3-octenyl-2-(CH}_2)_n\text{COOR-cyclopentanone)} \qquad (62)$$

The *1,4-addition* of acetylenic alanes to conjugated enone systems proceeds through a cyclic transition state and is therefore limited to enones which are able to achieve a *cis*-conformation, transoid olefins giving the 1,2-adduct.

However, the 1,4-adduct can be obtained in both cases when $Ni(acac)_2$ is added as a catalyst [73].

The case of cyclopentenone presents a particularly interesting case in connection with prostaglandin synthesis. In this case, the 1,4-addition is observed when an OH group is present in the 4-position, the pure *cis*-addition product being thus obtained (the OH group presenting a directing effect) [74] but the *trans*-isomer is obtained in the presence of $Ni(acac)_2$ when the OH group is protected as a dimethylbenzylether.

$$\text{(4-HO-2-R-cyclopent-2-enone)} \xrightarrow{Al-(C\equiv C-R')_3} \text{(4-HO-2-R-3-(C}\equiv\text{C}-\text{R}')\text{-cyclopentanone)} \qquad (63)$$

6.3. TIN HYDRIDE CHEMISTRY

n-Bu_3SnH is the most popular tin hydride in organic synthesis. It is a particularly useful reagent for achieving the removal of a halogen in an organic molecule even in the presence of other functional groups. The mechanism of these reactions corresponds to a radicalar chain process as UV light and radical initiators (e.g.: AIBN) are acting as promoters.

Example

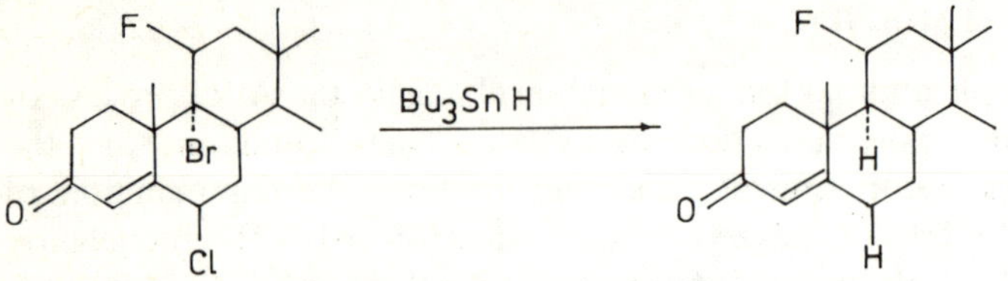

Bu_3SnH

(64)

Conjugated addition of trialkylbutyltin hydride is promoted by the presence of water, the former acts as a hydride donor whereas water participates to the reaction as a protonating partner. In this context, the role of the proton carrying reagent is further demonstrated by the optimalization of the yield when ammonium chloride is added to the reaction medium as [75].

Transition metal catalysis affords an efficient way for controlling the selectivity of these tin hydrides. For example, in the withanolide case, the selectivity of the reaction is completely modified by adding $Pd(PPh_3)_4$ to the medium.

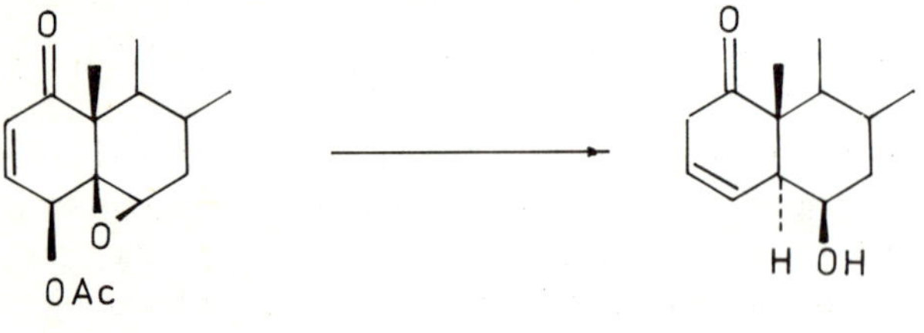

(65)

In fact, the enone system is reduced through a 1,4-addition mechanism in the absence of the transition metal catalyst.

6.4. HYDROZIRCONATION

The *hydrozirconation* reaction can lead to a series of reactions formally identical to the hydroboration processes. The addition to multiple carbon—carbon bonds is also a very stereoselective (*cis*) and regioselective (anti-Markownikov type) reaction. Moreover, the isomerization of the addition product towards the formation of the less crowded isomer occurs at a much lower temperature than in the boranes or alanes cases (25 °C instead of > 150 °C).

6.4.1. *Applications of hydrozirconation to the synthesis of biologically active compounds*

Preparation of *n*-triacontanol [76] (a growth stimulating component for plants) was prepared through a hydrozirconation based process, taking advantage of the facile isomerization reaction of the organozirconium intermediate.

$$C_{14}H_{29}-CH=CH-C_{14}H_{29} \xrightarrow[t.BuOOH]{Cp_2ZrClH} CH_3(CH_2)_{29}OH \quad (66)$$

The organic group of the organozirconium entity can be transferred to a transition metal and, therefore, hydrozirconation reaction offers a very general entry into various organometal intermediates. The latter are easily prepared in one-pot reactions by adding a transition metal salt to the reaction medium.

Example: Zirconium—palladium exchanges reaction [77]. (π-Allyl) palladium reagents afford coupled 1,4-dienes when treated with alkenylzirconium reagents.

Cp Zr(
PdCl 2
Pd
(67)

When applied in steroid chemistry, the reaction appears to be poorly selective. The regioselectivity could be controlled to some extent by adding bridge-breaking ligands; moreover, these addends strongly increase the rate of the reaction.

6.4.2. *Particular applications of organozirconium reagents*

The *carboalumination* of acetylenes and of olefins is a zirconium catalyzed reaction. Moreover, the alkenyl aluminum derivatives obtained from acetylenes can be used in cross coupling reactions with unsaturated organohalides in the presence of *Ni or Pd* catalysts [78]. This reaction has been applied to a prostaglandin synthesis [79].

6.5. HYDROSILYLATION

6.5.1. Hydrosilylation of ketones and aldehydes is a transition metal catalyzed process, typical catalysts for such reactions being rhodium (Wilkinson complex), Pd and Pt derivatives [80].

$$M + HSiX_3 \longrightarrow HMSiX_3 \xrightarrow{RR'C{=}O} R{-}\underset{R'}{\overset{M-H}{C}}{-}O{-}SiX_3 \longrightarrow \quad (68)$$

$$RR'CH{-}O{-}SiX_3 + M$$

This reaction can be preferable to catalytic hydrogenation when the resulting alcohol is prone to hydrogenolysis. Moreover, the application of this method to terpenoid ketones permits the highly stereospecific reduction of the carbonyl group to any stereoisomeric alcohols (menthone afford the less stable neomenthol under these conditions whereas dimethylphenylsilane yields the more stable menthol).

The conjugate addition of $HSiR_3$ onto α, β-unsaturated ketones is of particular interest in synthesis. Silylenolates are thus easily obtained and can be further used. **Cobalt catalysis** has been recommended as a particularly convenient method for the preparation of such silylenolates [82]. Earlier reports refer to rhodium catalysis [82], and palladium and nickel [83] catalysts.

Example

$$\text{cyclohexenone} \xrightarrow{Co_2(CO)_8} \text{1-(O–SiR}_3\text{)cyclohexene} \quad (69)$$

Induction of chirality can be observed when a chiral catalyst is used for the hydrosilylation of a ketone [84]. Similarly, optically active amines (50% e.e.) can be obtained from imines [85].

Examples: RhL^+, $L^+ = (-)(S)$-benzylmethylphenylphosphine of DIOP

$$CH_3-\overset{O}{\overset{\|}{C}}-COOC_3H_7 \xrightarrow[(-)\,DIOP]{\alpha\text{-}NpPhSiH_2} -\overset{OH}{\overset{|}{\underset{H}{\underset{|}{C^*}}}}- \quad 85\%\ ee \qquad (70)$$

Application:

$$\text{R(-) CARVONE} \xrightarrow{Rh\,L^*} \text{CARVEOL (25-58 ee)} \qquad (71)$$

L = DIOP, (+)BMPP(39—42% e.e.)
BMPP = benzylmethylphosphine
DIOP = 2,3-*O*-isopropylidene-2,3-dihydroxy-1,4-bis(diphenylphosphino)butane

6.5.2. Hydrosilylation of olefins is also a transition metal catalyzed reaction (Pt, Co, Pd, Rh, etc., derivatives being active catalysts) [86].

7. Application of Transition Metal Catalysis in Heterocyclic Synthesis (Typical Examples)

We have already mentioned the importance of transition metal catalysis in the field of heterocyclic synthesis, and numerous examples of such applications have been reported throughout the text (see also the References section for further reviews). Generally, the problem of heterocycle formation can be solved by the application of typical transition metal catalyzed processes to *intramolecular* reactions.

Also, some very specific reactions lead directly to the formation of heterocyclic nuclei. This approach offers, in some cases, the possibility of building up an elaborate heterocyclic system directly from simple molecules.

A typical example of such an application is the one step synthesis of pyridines directly from acetylenes and nitriles (Bönnemann synthesis [4]).

7.1. TYPICAL EXAMPLES OF HETEROCYCLIC SYSTEM SYNTHESIS

7.1.1. Indole synthesis is of particular importance, since the indole nucleus is the essential part of many biologically active compounds (alkaloids derived from *tryptophane*, phalloidin, and so on). Moreover, indole chemistry is particularly intricate and presents many challenging problems to the organic chemist. Actually, some key steps in indoles synthesis have been carried out by appying Pd catalysis to aminobenzene derivatives.

7.1.1.1. The intramolecular reaction of an aryl bromide with a double bond is an application of the oxidative addition—olefin insertion—β-hydride elimination process developed by Heck.

Example [87].

$Pd(OAc)_2$, PPh_3; R = $COOCH_3$ (72)

Moreover, this reaction seems to be particularly useful in the synthesis of ergot type molecules when applied to bromoindole.

7.1.2. The palladium-catalyzed cyclization of 2-ethenylaninine to indole.

Example

Pd(II) (73)

This reaction corresponds, in fact, to the application of the classical palladium catalyzed amination reaction of olefins. The above example is of particular interest as it offers a novel entry to bromoindole, a synthon of particular importance for ergot alkaloid and tryptophanes synthesis.

7.1.3. *Copper catalysis* has also been used for cyclization reactions of aryl aminoenone systems leading to the indole nucleus [57, 88].

7.2. PYRROLE SYNTHESIS

An interesting pyrrole synthesis is based on an application of a copper catalyzed 1,4-addition of an activated CH_2 group to a conjugated ene—azo system [89].

$$RSO_2-N=N-CR=CHR \; + \; RRN-\overset{O}{\overset{\|}{C}}-CH_2-\overset{O}{\overset{\|}{C}}-R$$

$$\xrightarrow{Cu\,Cl_2} \left[RSO_2-NH-N=CR-CHR-CH(C(=O)R)(C(=O)NRR) \right] \qquad (74)$$

$$\longrightarrow \text{pyrrole}$$

Another approach to pyrrole synthesis is based on transition metal catalyzed rearrangement of azirines [57].

$$\text{Ph (azirine)} \longrightarrow \text{Ph, N–H} \; + \; \text{N–H} \qquad (75)$$

Arizines also afford pyrrole through catalyzed cycloaddition reactions [57].

$$\text{N, Ph, Ph} \; + \; CH_3-O_2C-C\equiv C-CO_2-CH_3 \xrightarrow{M(CO)_6} \text{Ph, Ph, N–H}, CO_2CH_3, CO_2CH_3 \qquad (76)$$

7.3. ISOQUINOLINE AND QUINOLINE

Interestingly, palladium catalysis affords isoquinoline derivatives when applied to aromatic Shiff bases: orthoimino aryl palladium intermediates react with acrylonitriles which yield isoquinoline when heated at 160° [90]. This reaction is of particular interest for the production of CNS drugs, as the isoquinoline system is frequently encountered in active compounds, such as poppy alkaloids.

$$\text{R, C}\equiv\text{N, N} \longrightarrow \text{R, C}\equiv\text{N, N} \; + \qquad (77)$$

Quinoline derivatives have been obtained by the palladium catalyzed condensation of nitrobenzenes with aldehydes or (in the presence of rhodium catalysts) with alcohols [91, 92].

7.4. β-LACTAM CHEMISTRY

Lactams, and β-lactams in particular, are interesting owing to their occurrence in biologically active compounds such as antibiotics related to penicillines. Insertion reactions of carbenes offer useful access to poly heterocyclic systems contain a β-lactam nucleus, particularly when using rhodium and copper catalysis. Moreover, palladium catalyzed carbonylation of azirines affords β-lactam derivatives [93] in one step.

7.5. LACTONE SYNTHESIS

Numerous examples of lactone synthesis have been reported [57, 94], particularly as applications of transition metal-catalyzed carbonylation [95, 96] reactions and as applications of CO_2 [97].

Example of carbonylation

Pd, CO (78)

OH O O

Examples of CO_2 activation

R—C≡C—R Ni (0) / CO_2 (79)

R R R R O O + R R R O O R

N.B.: Cyclic anhydrides are formed by carbonylation of lactone precursors (Example [98]).

7.6. CYCLIC ETHER SYNTHESIS

Various oxygen containing heterocycles, particularly furan derivatives, have been obtained from various precursors such as diols, allyl, homoallyl and progargyl ethers. Typical catalysts are: Pd(II), $TiCl_4$, or cobaloxime(III).

7.7. MISCELLANEOUS EXAMPLES

We have already reported above the interest of Pd(0) and Ni(0) as catalysts for the synthesis of macrolides [39] and of various heterocycles through intramolecular reactions [40, 43]. Moreover, carbene chemistry also affords some interesting pathways to heterocyclic systems: for example, the insertion reaction of carbalkoxy carbenes into an enol (acetylacetone) yields furan derivatives [98] whereas the insertion reaction into diols affords lactones [99].

Isocyanates are precursors of various heterocycles such as pyrones, maleimides [100], and 1,2,3-triazolidines, [101].

Ureas react with vinyl halides in the presence of Pd(II) catalysts to produce pyrimidines [102].

Interestingly, palladium catalyzed carbonylation of amido bromoanilines affords a straightforward route to anthramycin and diazepam [57].

8. Transition Metal-Catalyzed Reactions of Carbenes

Carbene chemistry constitutes a particular but challenging field in organic synthesis. Carbenes offer a straightforward access to small rings (cyclopropanes, cyclopropenes) as well as to cycloheptatriene derivatives (the Büchner reaction) from cheap raw material (olefins, acetylenes, benzenic compounds, etc.).

However, the selectivity of these processes is difficult to mediate, as may be expected from the high reactivity of carbenes: in addition to the above reaction, they also lead to insertion reactions into OH, NH and even aliphatic CH bonds. Moreover, the formation of the formal dimerization and polymerization products of carbenes very often contributes to a decrease in the yield of the desired reaction products.

Transition metal catalysis offers, in fact, a powerful means to mediate their reactivity and selectivity: since the end of the last century, copper catalysis was largely applied in this context. More recently, the discovery of efficient catalysis by Group VIII complexes (particularly of rhodium(II) carboxylates, and (in some particular cases) of palladium(II) carboxylates)) now offers novel opportunities for preparative chemistry.

Other promising applications of transition metals in carbene based synthesis result from stoichiometric reactions of particular carbene precursors such as:

— stable isolable carbene-transition metal complexes (Fischer complexes),
— a titanium-aluminium methylene bridged complex,
— thiophenium ylides.

However, we shall report here only some typical applications of these reactions (an exhaustive review of this interesting but specialized area of chemistry being outside the frame of the present publication).

8.1. CATALYTIC REACTION

8.1.1. *Cycloaddition of carbenes to alkenes*

Various cyclopropanated molecules such as chrysanthemic acid, and pyrethrins are well known as nonpolluting insecticides, and one of the synthetic routes to these biologically active compounds appeals to the copper or rhodium catalyzed decomposition of diazo esters to promote the generation of the reactive carbenoid species [103].

$$ + N_2CH{-}CO_2R' \xrightarrow{Rh_2(OAc)_4} \quad (80)$$

R = CH_3, Cl

8.1.2. *Insertion reactions*

Carbonyl carbenes insert into various X—H bonds (X = O, N, C) in the presence of rhodium carboxylates. An application of the insertion of carbalkoxy carbene into the OH group of alcohols is illustrated by a synthesis of chorismic acid [104].

$$\xrightarrow[Rh_2(OAc)_4]{N_2C(CO_2R')_2} \quad (81)$$

CO_2R, OH; CO_2R, $OCH(CO_2R')_2$; CO_2H, O, CO_2H, OH (chorismic acid)

The intramolecular insertion reaction into the N—H bond offers an interesting entry into the 1-carbapenam ring system [105].

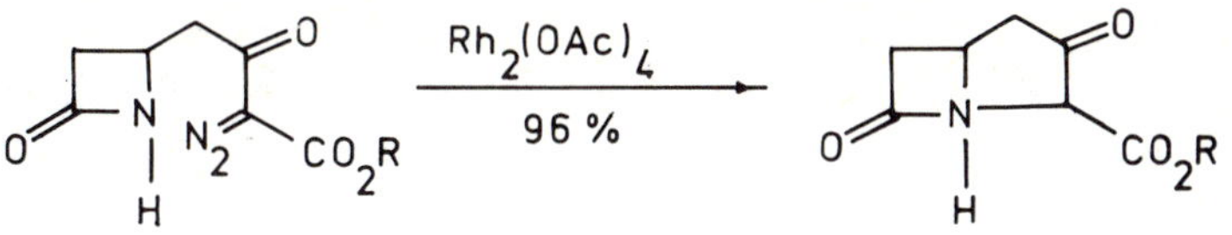

(82)

Similarly, the application of an intramolecular insertion reaction into an alkyl C—H has resulted in a straightforward (and stereoselective) access to cyclopentanone derivatives [106].

8.2. STOICHIOMETRIC REACTIONS OF CARBENOIDS AND YLIDES

8.2.1. Stable *carbenoids* (carbene—metal complexes) have been readily available since Fischer's classical work.

A key step in a synthesis of vitamin E [107] constitutes an illustration of the potentialities of these reagents in organic synthesis.

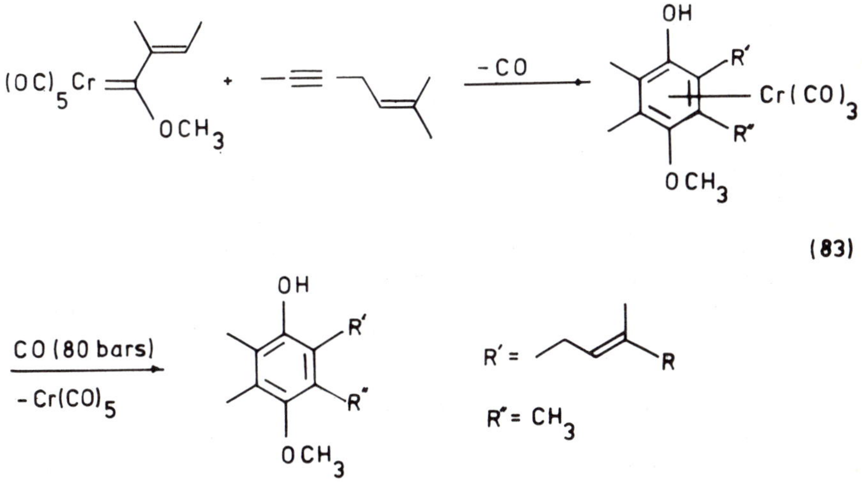

(83)

8.2.2. A reactive titanium carbene intermediate is formed from a stable bridged titanium—aluminum methylene complex.

Besides being active in cyclopropanation of olefins, this reagent is particularly useful to convert carbonyl groups to methylene compounds.

The potentialities of this reagent are illustrated by the synthesis of C-glycosides and for example in a project directed toward the synthesis of citroviridin [108].

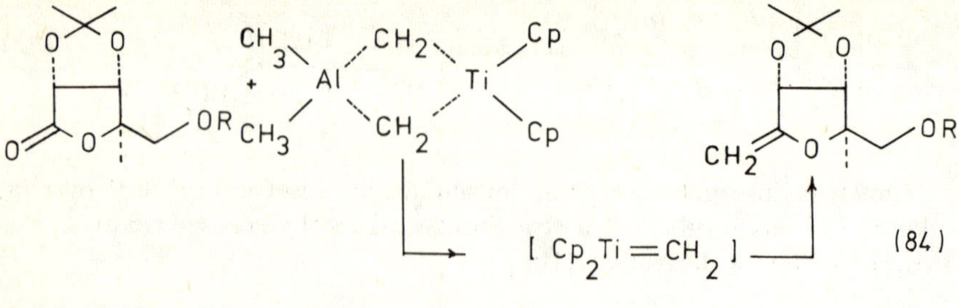

(84)

8.2.3. Sulfur ylides (such as thiophenium ylides) may act as carbene precursors in the presence of rhodium or copper catalysts. As the reported examples refer to their preparation from diazomalonates, this approach appears as a sophisticated development of the more classical carbene generation from diazoesters.

This original and promising pathway has been successfully applied to typical carbene reactions (cyclopropanation of olefins, OH insertion, C—H insertion into activated arenes) [109, 110].

University of Liège, Belgium

References

1. G. W. Parshall, *Homogeneous Catalysis*, Wiley (1980).
2. L. Crombie, G. Kneen, G. Pattenden and D. Whybrow, *J. Chem. Soc. Perkin I*, 1711 (1980).
3. K. P. C. Vollhardt, *Ann. N.Y. Acad. Sci.,* **333**, 241 (1980); S. H. Lecker, N. H. Nguyen and K. P. C. Vollhardt, *J. Amer. Chem. Soc.,* **108**, 856 (1986).
4. H. Bönnemann and R. Brinkmann, *Synthesis*, 600 (1975).
5. (a) J. Celebuski and M. Rosenblum, *Tetrahedron* **41**, 5741 (1985); (b) T. Takahashi and A. Ootake, *ibid.* 7747; (c) J. E. Bäckvall, E. S. Byström and J. E. Nyström, *ibid.* 5764; (d) A. J. Pearson and T. Ray, *ibid.* 5765; (e) M. Uemura, K. Take, K. Isobe, T. Minami and Y. Hayashi, *ibid.* 5771; (f) A. Goliaszenski, J. Schwartz, *ibid.* 5779; (g) C. A. Parnell and K. P. C. Vollhardt, *ibid.* 5791; (h) L. S. Liebeskind, S. L. Baysdon, M. S. South, S. Iyer and J. P. Leeds, *ibid.* 5839; (i) P. L. Pauson, *ibid.* 5855; (j) P. Magnus, C. Exon and P. Allbought-Robertson, *ibid.* 5861; (k) S. T. Hodgson, D. M. Hollinshead and S. V. Ley, *ibid.* 5871; (l) R. Noyori and Y. Hayakawa, *ibid.* 5879; (m) M. Gardette, A. Alexakis and J. F. Normant, *ibid.* 5887.
6. M. Gardette, A. Alexakis and J. F. Normant, *Tetrahedron Lett.*, **23**, 5155 (1982).
7. M. Gardette, N. Jabri, A. Alexakis and J. F. Normant, *Tetrahedron* **40**, 2741 (1984).
8. W. Oppolzer, R. Moretti, T. Godel, A. Meunier and H. Löher, *Tetrahedron Lett.,* **24**, 4971 (1983).
9. B. L. Chenard, M. J. Manning, P. W. Raynolds and J. S. Swenton, *J. Org. Chem.,* **45**, 378 (1980).
10. S. H. Bertz and G. Dalbagh, *J. Chem. Soc. Chem. Commun.*, 1030 (1982).

11. H. Saimoto, T. Hiyama and H. Nozaki, *Tetrahedron Lett.* 3897 (1980).
12. G. Teutsch and G. Costerousse, *J. Chem. Res.* 294 (1983).
13. S. Mubarik Ali, M. A. W. Finch and S. W. Roberts, *J. Chem. Soc. Chem. Commun.*, 74 (1980).
14. J. P. Marino and H. Abe, *Synthesis*, 872 (1980).
15. T. Fuyisawa, T. Sato, T. Kawara and A. Nada, *Tetrahedron Lett.*, **23**, 3193 (1982).
16. W. Koller, A. Linkies, H. Pietsch, H. Rehling and D. Reuschling, *Tetrahedron Lett.*, **23**, 1545 (1982).
17. A. Carpita and R. Rossi, *Synthesis* 469 (1982).
18. L. Crombie, G. Kneen, G. Pattenden and D. Whybrow, *J. Chem. Soc. Perkin I*, 1711 (1980).
19. F. K. Ziegler, I. Chliwner, K. W. Fowler, S. J. Kanfer, S. J. Kuo and N. D. Sinha, *J. Amer. Chem. Soc.*, **102**, 690 (1980).
20. C. B. Chapleo, M. A. W. Finch, S. W. Roberts, G. T. Woolley, R. F. Newton and D. W. Selby, *J. Chem. Soc. Perkin, I*, 1847 (1980).
21. R. Warin, M. Julémont and Ph. Teyssié, *J. Organometal. Chem.*, **185**, 413 (1980).
22. M. F. Semmelhack and A. Yamashita, *J. Amer. Chem. Soc.*, **102**, 5924 (1980).
23. (a) M. F. Semmelhack and L. S. Ryono, *J. Amer. Chem. Soc.*, **97**, 3875 (1975); (b) R. F. Heck, *Pure and Appl. Chem.*, **53**, 2323 (1981).
24. E. Negishi, A. O. King and N. Okukado, *J. Org. Chem.*, **42**, 1821 (1977).
25. W. Fischetti, K. T. Mak, F. G. Stakem, J. J. Kim, A. Rheingold and R. F. Heck, *J. Org. Chem.*, **48**, 948 (1983).
26. E. Negishi, *Pure and Appl. Chem.*, **53**, 2333 (1981).
27. R. Noyori: *Transition Organometallics in Organic Synthesis* (*Coupling Reactions via Transition Metal Complexes* V. 33—I, H. Alper Ed.), pp. 83—(119)—187. Academic Press (1976).
28. E. Negishi, L. F. Valente and M. Kobayashi, *J. Amer. Chem. Soc.*, **102**, 3298 (1980).
29. M. Kobayashi and E. Negishi, *J. Org. Chem.*, **45**, 5223 (1980).
30. J. S. Temple and J. Schwartz, *J. Amer. Chem. Soc.*, **102**, 7381 (1980).
31. R. C. Larock, D. R. Leach and S. M. Bjorge, *Tetrahedron Lett.*, **23**, 715 (1982).
32. J. E. McMurry, J. R. Matz, L. Kees and P. A. Bock, *Tetrahedron Lett.*, **23**, 1777 (1982).
33. J. E. McMurry, M. P. Fleming, K. L. Kees and L. R. Krepski, *J. Org. Chem.*, **43**, 3255 (1978).
34. J. Tsuji, *Pure Appl. Chem.*, **53**, 2371 (1981).
35. J. Tsuji, K. Masaoka and T. Kakahashi, *Tetrahedron Lett.*, **26**, 2267 (1977).
36. J. Tsuji, *Topics in Current Chemistry*, **91**, 29 (1979).
37. B. M. Trost, T. J. Dietsche, *J. Amer. Chem. Soc.*, **95**, 8200 (1973).
38. J. P. Genêt, F. Piau and J. Ficini, *Tetrahedron Lett.*, **21**, 3183 (1980).
39. B. M. Trost, *Pure Appl. Chem.*, **53**, 2357 (1981).
40. B. T. Khai; C. Concilio and G. Porzi, *J. Org. Chem.*, **46**, 1759 (1981).
41. Y. Watanabe, M. Yamamoto and S. C. Shim, *Chem. Lett.*, 1025 (1979).
42. F. Camps; J. Coll, A. Messeguer, M. A. Pericas and S. Ricart, *Synthesis*, 126 (1979).
43. J. M. O'Connor, B. J. Stallman, W. G. Clark, A. Y. L. Shu, R. E. Spada, T. M. Stevenson and H. A. Dieck, *J. Org. Chem.*, **48**, 807 (1983).
44. T. Takahashi, H. Ikeda and J. Tsuji, *Tetrahedron Lett.*, **21**, 3885 (1980).
45. J. Schwarz, M. J. Looks and H. Kosuji, *J. Amer. Chem. Soc.*, **102**, 1333 (1980).
46. A. Biavata, G. P. Chiusoli, M. Costa and G. Terenghi, *Transition Met. Chem.*, **6**, 398 (1979).

47. J. Tsuji, Y. Kobayashi and T. Takahashi, *Tetrahedron Lett.,* **21**, 483 (1980).
48. M. Hidai, H. Mizuta, H. Yagi, Y. Nagai, K. Hata and Y. Uchida, *J. Organometal. Chem.,* **232**, 89 (1982).
49. R. Baker, B. N. Blackett and R. C. Cookson, *J. Chem. Soc. Chem. Commun.*, 802 (1972).
50. R. L. Funk and K. P. C. Vollhardt, *J. Amer. Chem. Soc.,* **102**, 5253 (1980).
51. T. Takahashi, T. Nagashima and J. Tsuji, *Chem. Letters*, 369 (1980).
52. A. Cowell and J. K. Stille, *J. Amer. Chem. Soc.,* **102**, 4193 (1980).
53. R. F. Newton, P. L. Pauson and R. G. Taylor, *J. Chem. Res.* (S), 277 (1980).
54. D. C. Billington and P. L. Pauson, *Organometallics,* **1**, 1560 (1982).
55. B. M. Trost, T. A. Runge and L. N. Jungheim, *J. Amer. Chem. Soc.,* **102**, 2840 (1980).
56. (a) B. M. Trost and J. M. D. Fortunak, *Tetrahedron Lett.,* **22**, 3459 (1981); (b) T. Katsuki and K. B. Sharpless, *J. Amer. Chem. Soc.,* **102**, 5974 (1980).
57. L. S. Hegedus, *J. Organometal. Chem.,* **237**, 231 (1981); *ibid,* **207**, 185 (1982); **283**, 1 (1985); **298**, 207 (1986).
58. M. B. Groen and F. J. Zeelen, *Tetrahedron Lett.,* **23**, 3611 (1982).
59. Y. Takai, S. Yoshifuji and Y. Nitta, Heterocycles, **20**, 141 (1983).
60. J. Tsuji: *Organic Synthesis by means of metal complexes*, V. I, Springer (1985).
61. S. Nihimura, T. Ichino, A. Akimoto and K. Tsuneda, *Bull. Chem. Soc. Japan,* **46**, 279 (1973).
62. I. Ojima and N. Yoda, *Tetrahedron Lett.,* **23**, 3913 (1982).
63. G. R. Knox and I. G. Thom, *J. Chem. Soc. Chem. Commun.* 373 (1981).
64. M. Franck-Neuman, D. Martina and M. P. Heitz, *Tetrahedron Lett.,* **23**, 3493 (1982).
65. J. E. McMurry and J. R. Matz, *Tetrahedron Lett.,* **23**, 2723 (1982).
66. P. Heimbach, P. W. Jolly and G. Wilke, *Advances in Organometallic Chemistry,* **8**, 29 (1970).
67. A. J. Pearson and D. C. Rees, *Tetrahedron Lett.,* **21**, 3937 (1980).
68. R. Noyori, S. Makino, T. Okita and Y. Hayakawa, *J. Org. Chem.,* **40**, 807 (1975).
69. J. E. McMurry, M. P. Fleming, K. L. Kees and L. R. Krepski, *J. Org. Chem.,* **43**, 3255 (1978).
70. J. Tsuji and S. Hashiguchi, *Tetrahedron Lett.,* **21**, 2955 (1980).
71. H. C. Brown: *Organic Synthesis Today and Tomorrow* (*The Rich Chemistry of Vinylic Organoboranes*, B. M. Trost and C. R. Hutchinson, Eds.), pp. 121—137. Pergamon (1981).
72. H. C. Brown, *J. Organometal. Chem.,* **239**, 23 (1982).
73. J. Schwartz, D. B. Carr, R. T. Hansen and F. M. Dayrit, *J. Org. Chem.,* **45**, 3053 (1980).
74. R. T. Hansen, D. B. Carr and J. Schwartz, *J. Amer. Chem. Soc.,* **100**, 2244 (1978).
75. E. Keinan and P. A. Gleize, *Tetrahedron Lett.,* **23**, 477 (1982).
76. T. Gibson, *Tetrahedron Lett.,* **23**, 157 (1982).
77. J. Schwartz, F. T. Dayrit and J. S. Temple: *Organic Synthesis Today and Tomorrow.* (*Zirconium Reagents in Organic Synthesis*, B. M. Trost and C. R. Hutchinson Eds.), pp. 55—69. Pergamon (1981).
78. E. C. Ashby and G. Heinsohn, *J. Org. Chem.,* **39**, 3297 (1974).
79. L. Bagnell, E. A. Jeffery, A. Meister and T. Mole, *Austr. J. Chem.,* **28**, 801 (1975).
80. L. Ogima, Y. Yamamoto and M. Kumada: *Aspects of Homogeneous Catalysis.* (Asymmetric Hydrosilylation by means of Homogeneous Catalysts with Chiral Ligands. V. 3, R. Ugo, Ed.), pp. 185—228. Reidel (1977).

81. H. Sakurai, K. Miyoshi and K. Nakadaira, *Tetrahedron Lett.*, 2671 (1977).
82. R. J. P. Corriu and J. J. E. Moreau, J. Chem. Soc. Chem. Commun., 38 (1973).
83. E. Frainnet, V. Martel-Siegfried, E. Brousse and J. Dedier, *J. Organometal. Chem.*, **85**, 297 (1975).
84. T. Kogure and I. Ojima, *J. Organometal. Chem.*, **234**, 249 (1982).
85. I. Ojima: *Fundamental Research in Homogeneous Catalysis.* (Synthetic Reactions by Copper Complex Catalysts. V. 2, Y. Ishii and M. Tsutsui, Eds.), pp. 275—284. Plenum (1978).
86. (a) C. Eaborn and R. W. Bott: *Organometallics Compounds of the Group IV Elements.* (Synthesis and Reactions of the Silicon-Carbon Bond. V. I., A. G. MacDiarmed, Ed.), pp. 105—(213)—536. Marcel Dekker (1968). (b) H. Sakurai, K. Miyoshi and Y. Nakadaira, *Tetrahedron Lett.*, **23**, 2671 (1977).
87. P. J. Harrington and L. S. Hegedus, *J. Org. Chem.*, **49**, 2657 (1984).
88. T. Kametani; T. Ohsawa and M. Ihara, *Heterocycles*, **14**, 277 (1980).
89. O. Attanasi and F. R. Perrulli, *Synthesis*, 874 (1984).
90. I. R. Girling and D. A. Widdowson, *Tetrahedron Lett.*, **23**, 4281 (1982).
91. Y. Watanabe, N. Suzuki, Y. Tsuji, S. C. Shim and T. Mitsuko, *Bull. Chem. Soc., Japan*, **55**, 1116 (1982).
92. W. J. Boyle and F. Mares, *Organometallics*, **1**, 1003 (1982).
93. H. Alper and C. P. Mahatantila, *Organometallics*, **1**, 70 (1982).
94. H. Horino and N. Inoue, *Heterocycles*, **11**, 281 (1979).
95. T. Mise, P. Hong and H. Yamazaki, *J. Org. Chem.*, **48**, 238 (1983).
96. J. Tsuji, K. Sato and H. Okumoto, *Tetrahedron Lett.*, **23**, 5189 (1982).
97. (a) H. Hoberg, D. Schaefer and G. Burkhart, *J. Organometal. Chem.*, **228**, C 21 (1982); (b) G. Burkhart and H. Hoberg, *Angew. Chem. Int. Ed. Engl.*, **21**, 76 (1982).
98. R. Paulissen, E. Hayez, A. J. Hubert and Ph. Teyssié, *Tetrahedron Lett.*, 607 (1974).
99. A. F. Noels, A. Demonceau, A. J. Hubert and Ph. Teyssié, unpublished results.
100. H. Hoberg and B. W. Oster, *J. Organometal. Chem.*, **234**, C 35 (1982).
101. J. Drapier, A. J. Hubert and Ph. Teyssié, *Synthesis*, 649 (1975).
102. T. Fuchikami and I. Ojima, *Tetrahedron Lett.*, **23**, 4099 (1982).
103. Milner D. J. and Holland D. (Imperial Chemical Industries, Ltd): *German Offen.* 2.810.098 (1977); *C.A.* **90**, 38578 (1979).
104. F. Bohlmann and W. Rotard, Liebigs Ann. Chem., 1211 (1982).
105. R. J. Ponsford and R. Southgate, *J. Chem. Soc. Chem. Commun.*, 9 (1984).
106. M. J. Strauss and Z. Rapoport, *J. Org. Chem.*, **47**, 4809 (1982).
107. K. H. Dötz and W. Kuhn, *Angew. Chem. Int. Ed. Engl.*, **22**, 732 (1983).
108. C. S. Wilcox, G. W. Long and H. Suh, *Tetrahedron Lett.*, **25**, 395 (1984).
109. J. Cuffe, R. J. Gillepsie and A. E. A. Porter, *J. Chem. Soc. Chem. Commun.*, 641 (1978).
110. R. J. Gillepsie and A. E. A. Porter, *J. Chem. Soc. Chem. Commun.*, 50 (1979).

ARNO BEHR

APPLICATION OF TELOMERIZATION AND DIMERIZATION TO THE SYNTHESIS OF FINE CHEMICALS

Both the telomerization of 1,3-dienes with nucleophiles [1] and the dimerization of substituted unsaturated compounds [2] are excellent methods to synthesize functionalized products, which can be used as building blocks for fine chemicals. There is a great number of examples given in the literature and in the present review only a small selection can be demonstrated.

1. Telomerization Reactions

In homogeneous catalysis telomerization is defined as the oligomerization of dienes with incorporation of a nucleophile. For example, two molecules of butadiene react with one nucleophile HY to form telomers. This reaction is catalyzed by various organometallic compounds of the transition metals, especially by palladium and nickel complexes.

2 [butadiene] + HY —catalyst→ [1-Y-octa-2,7-diene] + [3-Y-octa-1,7-diene] (1)

In what follows, the telomerization of butadiene with acetic acid, alcohols, phenol, C—H-acidic compounds and nitroalkanes will be considered. Also some examples of carboxytelomerization and the telomerization of substituted dienes will be given. In all reactions trifunctional compounds are formed which contain two double bonds and one functional group.

These telomers are extremely useful starting materials for simple syntheses of various natural products.

1.1. TELOMERIZATION OF BUTADIENE WITH ACETIC ACID

Acetic acid and butadiene yield acetoxyoctadienes in excellent conversions and selectivities. In the presence of palladium acetylacetonate and *o*-alkyl or *o*-aryl substituted triarylphosphites in 1 : 1 molar ratio an

A. Mortreux and F. Petit (Eds.), Industrial Applications of Homogeneous Catalysis, 141—175.

almost quantitative telomerization occurs [3]. Also the catalyst system $Pd(OAc)_2/PPh_3$ can be used (Equation 2).

2 + HOAc —[Pd]→ OAc / OAc (2)

The primary telomer, the 2(*E*)-7-octadien-1-yl-acetate, can be transformed into *N*-isobutyldeca-2(*E*), 4(*E*)-dienamide, the so-called *pellitorine* [4]. Also the 2,15-hexadecanedione, a precursor of *muscone*, was synthesized by the primary acetate in eight steps [5]. Another example is the synthesis of *dihydrojasmone* [6]. The acetoxyoctadiene is reacted first to the allyl-2-octadienyl ether, which is heated with $RuCl_2(PPh_3)_3$ causing the migration of the terminal double bond to form the 1-propenyl-2-octenyl ether. The [3, 3] sigmatropic rearrangment gives 2-methyl-3-vinyloctanal which is oxidized with $PdCl_2/CuCl$ to give a ketoaldehyde. After aldol condensation and further treatment with bases the dihydrojasmone is formed (Equation 3).

OAc → O

[Ru] → rearr. → OHC → O_2 [Pd/Cu] → $OH^{\ominus}$ → O (3)

Dihydrojasmone

A similar sequence of reactions was carried out to synthesize *methyl dihydrojasmonate* (Equation 4).

OAc → O

Δ → OHC → O_2 [Pd/Cu] → $OH^{\ominus}$ → O (4)

→ O, $CH(CO_2Me)_2$ → O, CO_2Me

Methyl-dihydrojasmonate

Besides amides and ketones, also lactones with higher-membered rings are made available by telomerization. Tsuji and Mandai reported a simple synthesis of *diplodialide B*, a naturally occuring ten-membered lactone which shows interesting biological activity [7] (Equation 5).

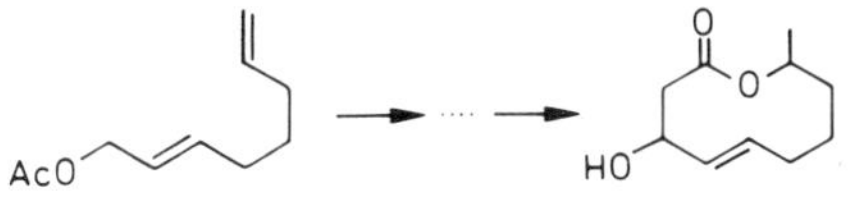

AcO ⟶ ···· ⟶ HO (5)

Diplodialide B

Also, other ten-membered lactones can be conveniently prepared starting from the primary acetates, as outlined in Equation 6. Again the terminal double bond of the telomer is oxidized in a Wacker—Hoechst-type reaction yielding a methyl ketone. After conversion into an allylic chloride and treatment with phenylthioacetylchloride an intramolecular alkylation yields the lactone-ring [8].

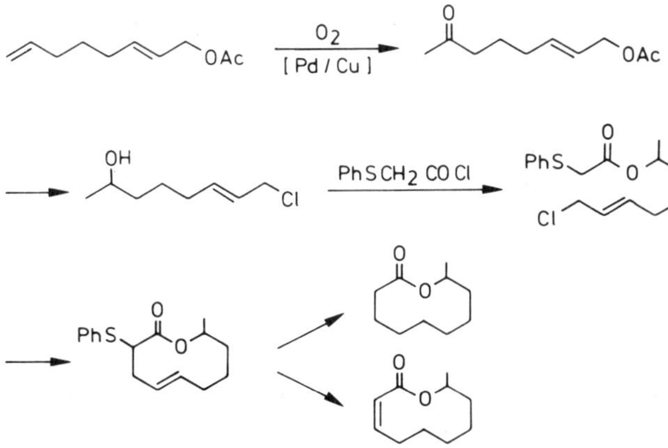

OAc $\xrightarrow[\text{[Pd / Cu]}]{O_2}$ OAc

⟶ OH Cl $\xrightarrow{PhSCH_2COCl}$ PhS Cl (6)

⟶ PhS

Also the secondary telomer, the 3-acetoxy-1,7-octadiene can be a useful starting material. The simplest natural product prepared from this telomer is *Matsutake alcohol*, the 1-octen-3-ol, a fragrant compound of a Japanese mushroom [9] (Equation 7).

OH ⟶ ···· ⟶ OH (7)

Matsutake Alcohol

Another example is the synthesis of α-lipoic acid, a naturally occuring, sulfur containing vitamin [10]. The starting compound is once again the alcohol 3-hydroxy-1,7-octadiene, obtained by hydrolysis of the acetate (Equation 8).

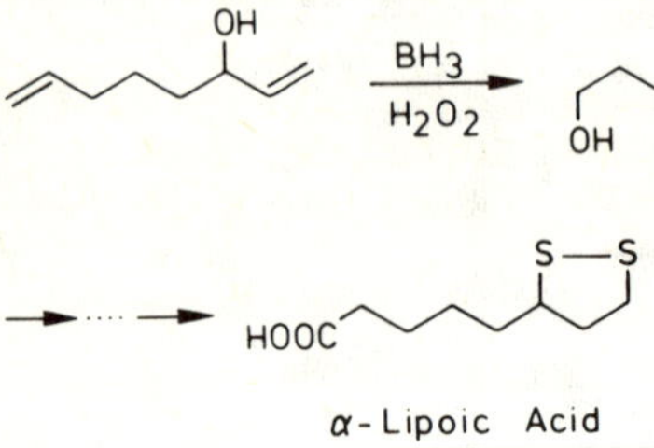

$$\xrightarrow[H_2O_2]{BH_3} \quad \longrightarrow \cdots \longrightarrow \quad \text{HOOC} \ldots \quad \alpha\text{-Lipoic Acid} \tag{8}$$

This alcohol can also be dehydrogenated, and the reaction is catalyzed by Cu/Zn alloy, yielding the 1,7-octadien-3-one, a very useful bis-annelation reagent. This compound enables the simple synthesis of important intermediates in steroid chemistry, as outlined in Equation 9.

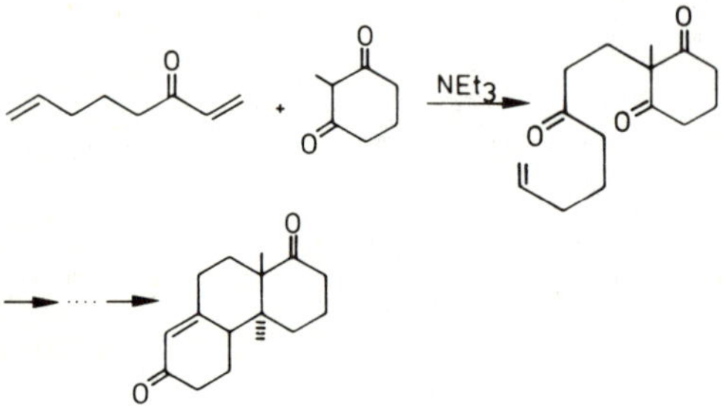

$$\xrightarrow{NEt_3} \quad \longrightarrow \cdots \longrightarrow \tag{9}$$

1.2. TELOMERIZATION OF BUTADIENE WITH ALCOHOLS AND PHENOL

Primary and short-chained *alcohols* react very easily with butadiene to form ethers. Hagihara showed that the telomerization of methanol proceeds smoothly even at low temperatures (40—100 °C) and with a short reaction time (1—3 h). The primary telomer is always the main product which is accompanied by the secondary telomer and small amounts of the by-product 1,3,7-octatriene.

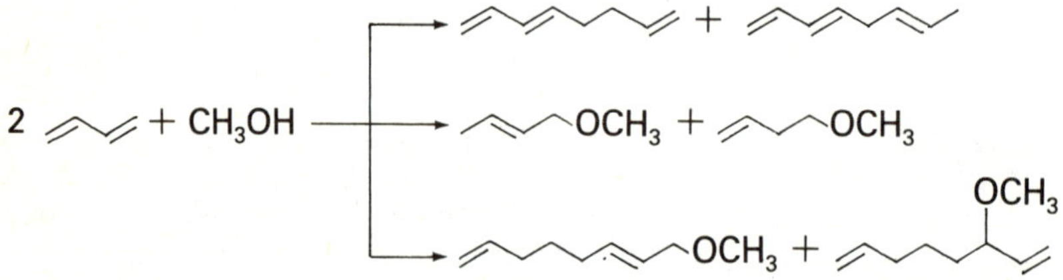

$$2 \text{ butadiene} + CH_3OH \longrightarrow \ldots + \ldots; \quad \ldots OCH_3 + \ldots OCH_3; \quad \ldots OCH_3 + \ldots (OCH_3) \tag{10}$$

A new type of telomerization was found recently by Tkatchenko *et al.*, when they used cationic palladium complexes as catalysts [11]. Not only methoxyoctadienes were formed by butadiene and methanol, but also C_{12}-, C_{16}-, C_{20}- and C_{24}-telomers. They presume, that the reaction preferentially occurs via the condensation of C_8-units.

The telomerization of various higher alcohols has also been carried out. The primary alcohols react most easily with butadiene, whereas secondary and tertiary alcohols yield only small amounts of telomers. Steric hindrance alone cannot be the only factor responsible for the reactivity because the voluminous 2,2-bis(trifluoromethyl)benzylalcohol reacts smoothly to produce the 2,7-octadienyl ether.

Phenol reacts with butadiene to produce the octadienyl phenyl ethers in high yield [12]. Phenol and butadiene which were heated together at 100 °C in the presence of $PdCl_2$ and PhONa produced *trans*-1-phenoxyoctadiene in 91% yield which was accompanied by 4% of the *cis*-isomer and 5% of 3-phenoxy-1,7-octadiene (Equation 11, route A). The condensation of butadiene and phenol, which is carried out in the presence of a tertiary aromatic phosphine or of excess phenol, takes a different course and produces the *ortho*- and *para*-octadienylphenols (route B).

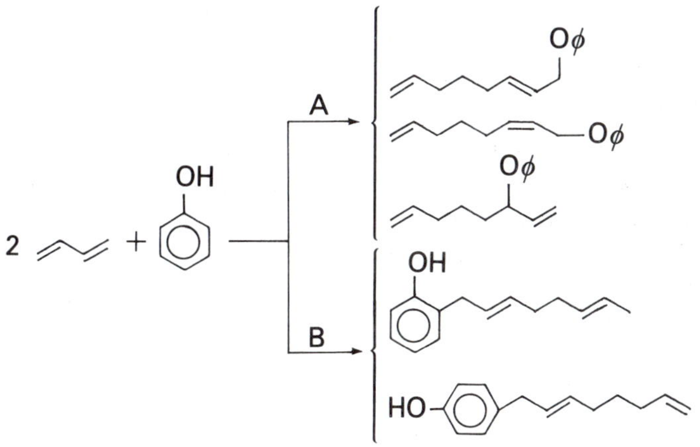

(11)

The 1-phenoxyoctadiene can be used for the synthesis of *12-acetoxy-1,3-dodecadiene* (Equation 12), a pheromone of *Diparopsis castanea* [13]. In this synthesis, the elimination of the phenol yields the required conjugated diene system.

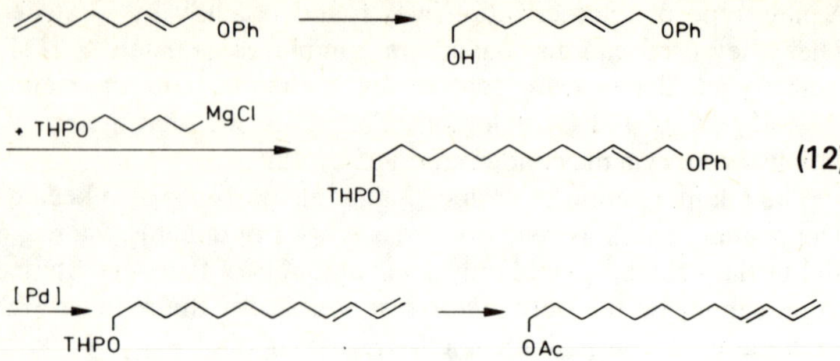

(12)

The same method of diene formation was applied to the synthesis of *pyrethrolone* (Equation 13), an alcoholic component of naturally occuring pyrethroids [14].

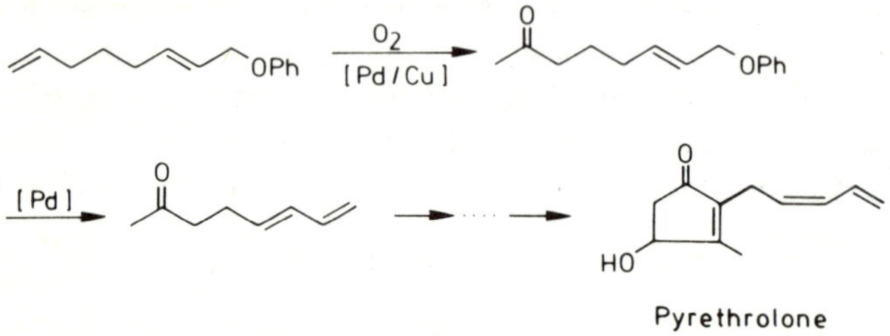

(13)

Pyrethrolone

1.3. TELOMERIZATION OF BUTADIENE WITH C—H-ACIDIC COMPOUNDS

Compounds with methylene and methyne groups in the neighbourhood of two electronegative groups X and Y undergo telomerizations with butadiene very smoothly: one or two acidic hydrogens can be replaced with the octa-2,7-dienyl group to produce the mono- and disubstituted compounds, respectively (Equation 14).

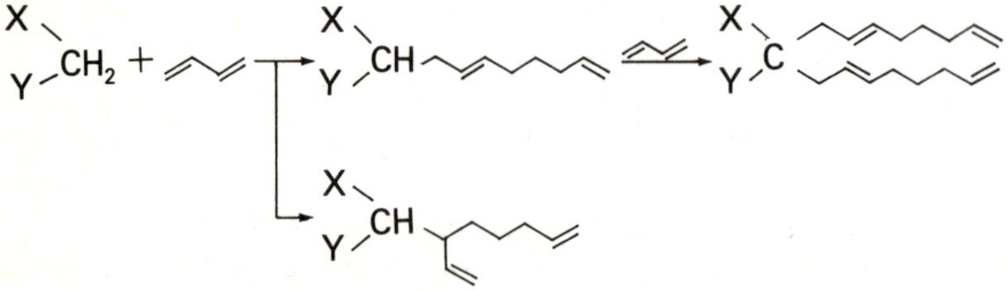

(14)

In general, the methylene and methyne groups must be activated by two carbonyl groups. The carbonyl, formyl and carboxyl group form highly active nucleophiles such as β-diketones, malonates, β-ketoesters or α-formyl ketones.

The telomer of malonate and butadiene is a particularly useful starting material. An example is given in Equation (15), which shows another synthesis of the naturally occuring pesticide *pellitorine* (compare Section 1.1) [15].

CO_2R / CO_2R → ⋯ → Pellitorine (15)

The same telomer was used by Baker *et al.* to synthesize the *dimethylpentadecan-2-ol*, the pheromone of the pine sawfly [16]. Also the *queen substance*, a well-known honey bee pheromone, is formed starting from the malonate telomer, which has exactly the right carbon number and a suitable functionality for a facile synthesis of this natural product (Equation 16) [17].

O_2 [Pd / Cu]; H_2; PhSeSePh; $NaIO_4$ — Queen Substance (16)

Zakharkin published the stereospecific synthesis of *6E, 11Z-hexadecadien-1-yl acetate*, the Antheraca polyphemus pheromone, accomplished on the basis of either the malonate or the acetoacetate telomer [18]. The acetoacetate telomer is also the precursor of one of the *royal jelly acids*, the 10-hydroxy-2-decenoic acid (Equation 17) [19].

CO_2R; OH; OR; SePh; CO_2H — Royal Jelly Acid (17)

Another synthesis of *methyl dihydrojasmonate* (compare Section 1.1), an important fragrance compound, was developed by Tsuji *et al.*, again using the acetoacetate telomer [20].

1.4. TELOMERIZATION OF BUTADIENE WITH NITROALKANES

Methylene compounds which contain only one electronegative group are normally inactive in telomerization. One exception is the nitroalkanes, which react very smoothly with butadiene [21, 22]. The telomerization of nitromethane in the presence of [$PdCl_2(PPh_3)_2$] and sodium hydroxide at room temperature produces three nitro-compounds which are accompanied by a small amount of branched products (Equation 18).

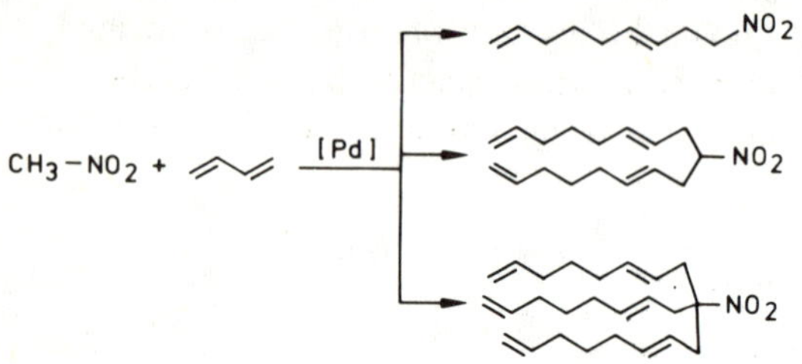

CH_3-NO_2 + butadiene —[Pd]→ (three nitro-telomers, each bearing NO_2) (18)

The telomer with two octadienyl chains fixed on the nitromethane, the 9-nitro-1,6,11,16-heptadecatetraene, can be used to synthesize civetonedicarboxylic acid, a precursor of *cis-civetone* (Equation 19) [23].

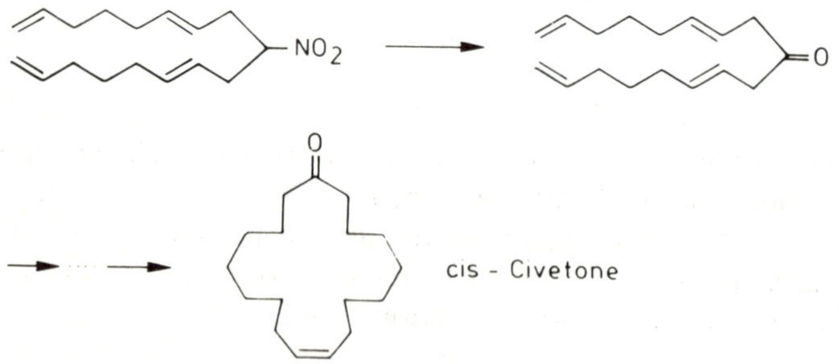

(19)

cis - Civetone

Besides nitromethane, nitroethane, 1- and 2-nitropropane and nitrocyclohexane can also be telomerized. The nitroethane telomer was used for the synthesis of *recifeiolide*, a naturally occuring 12-membered lactone (Equation 20) [24].

(20)

1.5. CARBOXY-TELOMERIZATION OF BUTADIENE

Telomerization combined with the additional incorporation of carbon monoxide is called carboxy-telomerization. The fundamental reaction which yields products containing a C_9-chain is shown in Equation 21.

(21)

For instance, with aliphatic alcohols the alkyl nona-3,8-dienoates are formed [25, 26]. This nonadienoate can be used for the synthesis of another royal jelly acid, the 2-decenedioic acid (Equation 22). Carbonylation of the nonadienoate in alcohol using $Co_2(CO)_8$/pyridine as the catalyst yields a linear diester which can be transformed into the royal jelly acid by hydrolysis and concomitant double bond migration [27].

(22)

Another synthesis of muscone (compare Section 1.1) also starts with the nonadienoate (Equation 23). Wacker/Hoechst-oxidation of the terminal double bond and hydrogenation of the internal one yields a ketocarboxylic acid, which reacts in a Kolbe electrolysis to 2,15-hexadecanedione, the precursor of muscone [28].

(23)

Muscone

A stereospecific synthesis of *endo*-brevicomin (24), a bark beetle pheromone, was published by Grigg [29]. By reduction, epoxidation and hydrolysis of the nonadienoate the 1-nonene-6,7-diol is formed, which can be cyclised directly to *endo*-brevicomin using palladium chloride/copper(II)chloride as catalyst.

(24)

1.6. TELOMERIZATION OF ISOPRENE

In contrast to butadiene, isoprene is an unsymmetrical molecule. The connection of two isoprene molecules can therefore occur on the "head" *h* or on the "tail" *t* of the molecule. Consequently, in the telomerization of isoprene the tail-to-tail (*tt*)-, tail-to-head (*th*)-, head-to-tail (*ht*)- and head-to-head (hh)-products are possible. Because the nucleophile HY can attack the chain at positions 1 and 3, eight telomers occur. The fact that the primary telomers possess an inner-standing double bond means that these molecules exist as *cis*- and *trans*-isomers and that, on the whole, twelve different molecules can be formed in isoprene telomerization.

The selective telomerization exclusively to natural type terpenoids with a head-to-tail connection has not been achieved so far, but some interesting initial steps have been done. In 1975 Hidai *et al.* had investigated the synthesis of the terpenoid *citronellol* starting from isoprene [30]. They reacted isoprene and methanol, yielding the telomer 1-methoxy-

2,6-dimethyl-2,7-octadiene. This telomer was further treated with $[NiCl_2(PBu^n_3)_2]$ and sodium methanolate as catalyst, yielding 2,6-dimethyl-1,3,7-octatriene. After hydrogenation using a carbonylchromium catalyst such as $[Cr(CO)_3(PhCO_2Me)]$ the 2,6-dimethyl-2,7-octadiene is formed which can be reacted to citronellol by a hydroboration step (Equation 25).

2 + MeOH —[Pd]→ OMe —[Ni]→

—[Cr]→ ⟶ OH (25)

This reaction is very remarkable because, in the first step of the reaction sequence, the telomerization, a palladium catalyst with an optically active phosphine can be used, yielding the telomer in an optically active form. This is one of the few examples of the asymmetric formation of a new carbon—carbon bond. These results encouraged Hidai's group to do further work, which was published recently [31]. In the telomerization of isoprene and methanol, they use a catalyst system comprising bis(π-allylpalladium chloride) and an optically active phosphorous compound with a menthyl group ('Men'). Menthyldialkylphosphines, $MenPR_2$, and bis(dialkylamino) menthylphosphines $MenP(NR_2)_2$ gave the (+)-telomer preferentially, whereas dialkylmenthylphosphonites $MenP(OR)_2$ produced the (−)-telomer. The bulky menthyldiisopropylphosphine gave the best result with respect to the optical yield (35%).

Another way to get citronellol is by the reductive dimerization of isoprene with formic acid and triethylamine using a 1% palladium phosphine catalyst [32]. The two head-to-tail dimers are formed in up to 79% yields, which can easily be separated from the head-to-head and tail-to-tail dimers by conversion with aqueous hydrochloric acid, yielding 7-chloro-3,7-dimethyl-1-octene. Hydroboration and pyrolysis of this chloro derivative produces a 1 : 3-mixture of α- and β-citronellol. The mono-chloro compound can also be oxidized with *tert*-butyl peracetate and a cuprous bromide catalyst to the chloroacetate, which is reduced with $LiAlH_4$ and pyrolyzed to *linalool* in 64% overall yield.

Tanaka and Hata [33] reported a synthesis of *dihydrocitral* from isoprene. They started with isoprene and a dialkylamine yielding the head-to-tail linked *N*-(3,7-dimethylocta-2,6-dienyl)dialkylamine. The base-catalysed isomerization of this terpene amine proceeds smoothly to give, in high yields (~80%), the *N*-(3,7-dimethylocta-1,3-dienyl)dialkylamine, which is stabilised by the conjugation of the two double bonds and the nitrogen atom (Equation 26). As basic catalyst an alkali metal 2-aminoethylamide was used which can be prepared *in situ* by adding KH, BuLi or Li to ethylenediamine. The activity of these catalysts is much higher than that of *t*-BuOK in DMSO.

NR_2 → NR_2 (26)

The hydrolysis of the isomerized amine gives dihydrocitral in yields of about 70% (Equation 27).

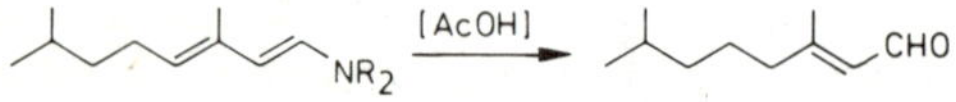

NR_2 —[AcOH]→ CHO (27)

Also isoprene and water undergo telomerization when a palladium or a platinum catalyst and carbon dioxide as the cocatalyst are used [34]. The *monoterpenic alcohols*, which can be used as fragrances and perfumes, are formed in one reaction step. Seven terpenols (28) have been synthesized in yields up to 50% in which the isoprene units are combined in tail-to-tail, tail-to-head and head-to-tail manners.

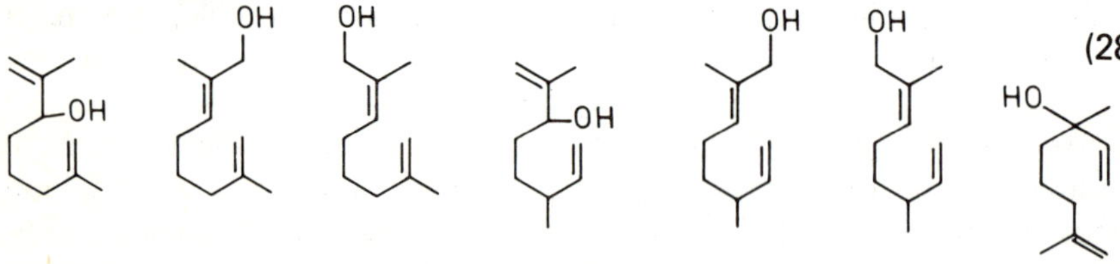

OH OH OH OH OH OH HO (28)

The highest conversions of isoprene are obtained with palladium catalysts such as $[Pd(PPh_3)_4]$, but the highest amount of terpenols was found with platinum catalysts as $[Pt(PPh_3)_4]$ or $[Pt(PPh_3)_2(CH_2{=}CH_2)]$. The head-to-tail linked terpenoid α-linalool could be synthesized in selectivities up to 20%.

The terpene alcohols are important precursors of *vitamins*. As is shown in Equation (29) vitamin E can be synthesized by starting from β-linalool. Vitamin K can also be produced with β-linalool as starting material.

OH

(a) + AcAcOET
(b) + HC≡CH
(c) + H_2

OH (a) + AcAcOEt
(b) + 3 H_2
(c) + HC≡CH
(d) + H_2

OH

(29)

+ trimethyl-hydroquinone

HO

O

AcAcOEt = ethylacetoacetate

vitamin E

1.7. TELOMERIZATION OF PIPERYLENE

1,3-Pentadiene (piperylene) is, like isoprene, a major component in the C_5-cut of naphtha pyrolysis. It can be expected that in the near future the C_5-cut will be used more extensively by the chemical industry and that a greater number of C_5-separation plants will be built. In this case 1,3-pentadiene could become a useful starting material for technical synthesis.

Also piperylene can be used in telomerization; however, so far only poor yields have been obtained [35]. Therefore, our investigations were focused on finding reactive nucleophiles and active catalyst systems which allow the telomerization of piperylene in high yields and selectivities [36]. We found that the three alcohols benzyl alcohol, furfuryl alcohol and 2,2,2-trifluoroethanol proved to be very reactive nucleophiles in the telomerization of piperylene.

It can be assumed that the electron-withdrawing effect of the aromatic, the furanoic and the fluorine-containing substituents influences the polarity of the OH-bond, thus facilitating the attack of the alcohol.

Besides smaller amounts of 1 : 1-adducts of piperylene and alcohol, the head-tail-linked telomer *2-alkoxy-6-methyl-3,8-nonadiene* was formed in good selectivities (Equation 30).

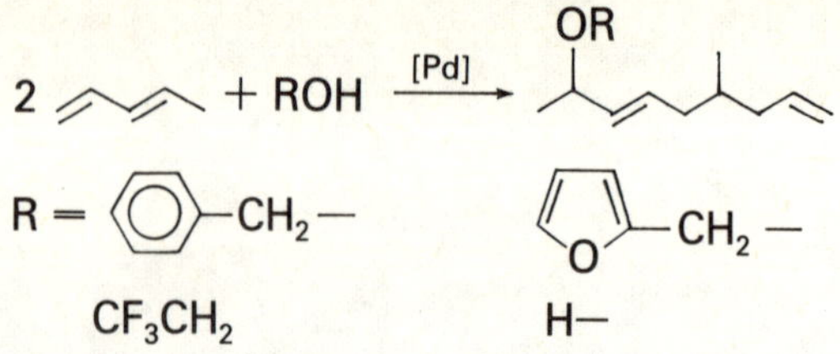

(30)

The most favourable catalyst proved to be an *in situ* system of palladium-bis(acetylacetonate) modified by a phosphorus ligand. With regard to the benzyl alcohol, best results (~60% yield of telomer) have been obtained with 1,1,1-trimethylolpropanephosphite, a ligand of low basicity and of low steric hindrance. A difficulty with the phosphite ligands consists in the fact that their coordination to the metal is low and that they do not stabilize the palladium sufficiently at higher reaction temperatures. To avoid the precipitation of palladium metal, temperatures lower than 60 °C are favourable.

Furfurylalcohol and 2,2,2-trifluoroethanol behave very similar to benzyl alcohol; other ligands must be applied however: with furfurylalcohol, best yields (37%) were obtained using triphenylphosphine, with trifluoroethanol the optimum ligands were trimethylolpropanephosphite (24%) and triphenylphosphite (22%).

Also the rather inert nucleophile water reacted with piperylene yielding the telomer *6-methyl-3,8-nonadien-2-ol*, only in low yields, however.

The application of the new piperylene telomers, for instance as pharmaceuticals or insecticides, is still under investigation.

1.8. TELOMERIZATION OF 2,3-DIMETHYLBUTADIENE

2,3-Dimethylbutadiene is a symmetrical molecule like butadiene and, therefore, only three isomeric telomers can be expected. However, this diene is a very inactive one and only in one case has telomerization been reported. The reaction of 2,3-dimethyl-1,3-butadiene with ethylacetoacetate in the presence of a $[PdCl_2(PPh_3)_2]$/NaOPh catalyst results in the formation of ethyl-2-acetyl-2,3,6,7-tetramethyl-2,7-decadienoate (31) with a yield of 7% [37, 38].

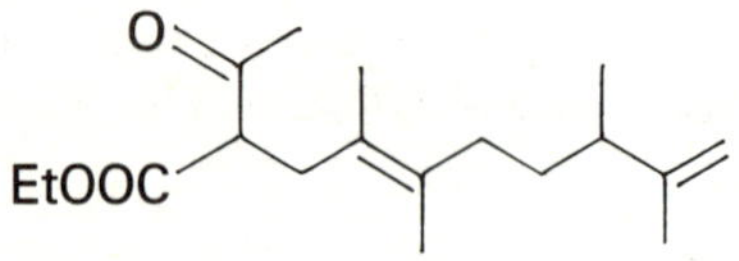

(31)

In other experiments the formation of 1 : 1 adducts was observed. For instance, when the reaction of 2,3-dimethylbutadiene with methyl acetoacetate is carried out in the presence of a catalyst composed of palladium dichloride and 3-methyl-1-phenyl-Δ^3-phospholene the 1 : 1 adduct 3-carbomethoxy-5,6-dimethyl-5-hepten-2-one is formed (Equation 32). Alkali decomposition of this ketone gives 5,6-dimethyl-5-hepten-2-one, a key intermediate for the synthesis of *α-irone* [39].

[Pd] CO_2Me CO_2Me (32)

α-Irone

2. Dimerization Reactions

There are different possibilities to synthesize functionalized fine products using dimerization reactions. A list of important examples is given in the following:

(a) Two molecules of the same type of a functionalized monoolefin dimerize yielding bifunctional monoenes:

$$2\,C{=}C{-}X \rightarrow X{-}C{=}C{-}C{-}C{-}X \qquad (33)$$

(b) Two different functionalized monoolefins dimerize:

$$X{-}C{=}C + C{=}C{-}Y \rightarrow X{-}C{=}C{-}C{-}C{-}Y \qquad (34)$$

(c) One diene co-dimerizes with a functionalized monoolefin:

$$C{=}C{-}C{=}C + C{=}C{-}X \rightarrow C{=}C{-}C{=}C{-}C{-}C{-}X \qquad (35)$$

This reaction is often accompanied by the formation of longer-chained products, e.g.:

$$2\,C{=}C{-}C{=}C + C{=}C{-}X \rightarrow C{=}C{-}C{-}C{-}C{-}C{=}C{-}C{-}C{=}C{-}X \qquad (36)$$

(d) Two dienes dimerize yielding an unsaturated product, which is functionalized in a subsequent reaction:

$$2\,C{=}C{-}C{=}C \longrightarrow C{=}C{-}C{=}C{-}C{-}C{-}C{=}C$$

$$\downarrow +HX \qquad (37)$$

$$\overset{\displaystyle X}{\overset{|}{C}}{-}C{-}C{=}C{-}C{-}C{-}C{=}C$$

The products formed by reaction (37) often contain a similar structure to the telomers described in Section 1.

2.1. DIMERIZATION OF FUNCTIONALIZED OLEFINS

The dimerization of functionalized olefins by transition metal catalysts is a well studied reaction. A first review was given by Lefebvre and Chauvin in 1970 [40]. In 1974 Hidai and Misono described in detail the dimerization of acrylic compounds [41]. These starting olefins have found considerable attention from the industrial point of view, because they can be produced at high quantities and at low costs by the petrochemical industry. For instance, acrylonitrile, acrylates and allyl esters are favorable starting molecules.

Acrylonitrile is an olefin of special interest, because its linear dimerization yields 1,4-dicyanobutenes, which are possible precursors of nylon-6,6 (Equation 38).

$$2\,CH_2{=}CH{-}CN \longrightarrow NC{-}CH{=}CH{-}CH_2{-}CH_2{-}CN \qquad (38)$$

As early as 1964, Rhône-Poulenc had investigated the hydrodimerization of acrylonitrile yielding adipodinitrile (Equation 39). In a patent they claimed a catalyst system composed of iron pentacarbonyl and an aqueous alkaline solution [42].

$$2\,CH_2{=}CH{-}CN + H_2 \longrightarrow NC{-}(CH_2)_4{-}CN \qquad (39)$$

Adipodinitrile was also produced using the catalyst $(CH_2{=}CHCN)_2Co(CO)_2$ prepared by reacting $Co_2(CO)_8$ with acrylonitrile [43]. Almost at the same time, Du Pont claimed, in a patent, the hydrodimerization of acrylonitrile to 2-methylglutarodinitrile (Equation 40). The catalyst was prepared *in situ* from reduced cobalt, carbon monoxide and water [44].

$$2\,CH_2{=}CH{-}CN + H_2 \longrightarrow NC{-}\overset{\displaystyle CH_3}{\overset{|}{C}H}{-}CH_2{-}CH_2{-}CN \qquad (40)$$

In a more recent work, acrylonitrile was dimerized using a $[Co(salen)]^-$

catalyst [45]. As a major product, 2,4-dicyano-1-butene was formed (Equation 41).

$$2\ CH_2{=}CH{-}CN \longrightarrow CH_2{=}\overset{\overset{\textstyle CN}{|}}{C}{-}CH_2{-}CH_2{-}CN \qquad (41)$$

Besides iron and cobalt, ruthenium is an important catalyst metal to dimerize acrylonitrile [41, 46, 47]. Using ruthenium trichloride the reaction occurs at 100—160 °C and moderate hydrogen pressure (5—20 atm) to give a 50—65% yield of dimers. The reaction is usually carried out in pure acrylonitrile or sometimes in the presence of alcohols, which act as promoters. Besides $RuCl_3 \cdot 3\ H_2O$, also other ruthenium complexes such as $RuCl_2(C_{12}H_{18})$ and $Ru(acac)_3$ are effective catalysts for the dimerization. The details of the mechanism are not well understood, although the hydrogen requirement suggests that ruthenium hydride complexes are involved as intermediates.

Also the dimerization of *acrylates* has found great interest in numerous academic and industrial research groups. Alderson and co-workers of Du Pont [48] investigated the dimerization of methyl acrylate in the presence of rhodium chloride. By heating a solution of methyl acrylate in methanol to 140 °C some dimethyl 2-hexenedioate (= dimethyl-α-dihydromuconate) was formed (Equation 42).

$$2\ CH_2{=}CHCOOMe \rightarrow MeOOC{-}CH{=}CH{-}CH_2{-}CH_2{-}COOMe \qquad (42)$$

For example, starting from 129 g of methylacrylate after a reaction time of 10 h, about 9 g of dimer (7%) and only small amounts (0.6 g) of nonvolatile higher molecular weight products were formed. The dimerization was also accomplished with ruthenium chloride as catalyst. At temperatures of 210 °C a remarkably higher yield of dimers (44%) was observed; however, the yield of the nonvolatile residue (47%) increased also. Addition of a small amount of ethylene to this latter system permitted use of a much lower temperature (150 °C) and gave a 56% yield of dimer. Nevertheless, the results of this experiment, too, did not permit a technical application, because the amount of higher boiling products and the required amount of catalyst (1 g $RuCl_3$ producing 17 g of dimers) were both too high.

A more selective catalyst system was found by Barlow *et al.* [49]. Using dichlorobis(benzonitrile)palladium a yield of 93% dimers was obtained. The product contained approximately 90% linear dimers with dimethyl *trans*-2-hexenedioate (67%) as the major isomer, while the next most prevalent isomer was dimethyl *trans*-3-hexenedioate. The surprising selectivity of the palladium catalyst encouraged Pracejus and Oehme to further studies. They found that the addition of *p*-benzoquinone gives an increase in conversion and yield, however, the *p*-benzoquinone could not hinder the precipitation of palladium metal, which shows no further catalytic activity [50]. The same problem was observed when palladium chloro complexes were activated by silver tetrafluoroborate. During the reaction the metal precipitates and maximum turnovers of about 100 mol substrate per mol catalyst were achieved [51].

Recently, Nugent and McKinney of Du Pont studied the combination of a Group VIII metal halide and Lewis acids, which are less powerful than $AgBF_4$ [52]. The Lewis acids $ZnCl_2$ and $CdCl_2$ enhanced the rate of acrylate dimerization by $(PhCN)_2PdCl_2$ (Table I). However, catalyst life remained comparable to that in the $AgBF_4$ promoted reaction. Yields were relatively insensitive to Lewis acid concentration, but with either $ZnCl_2$ or $CdCl_2$ as promoter, a Pd/Lewis acid ratio of 1 : 1 appeared optimal.

TABLE I
Effect of Lewis Acids on Dimerization of Methyl Acrylate by $PdCl_2(NCPh)_2$[a]

Additive	Equivalents	Dimers (mol/mol Pd)
none	—	6
$AgBF_4$	2	26
$ZnCl_2$[b]	2	18
$CdCl_2$[b]	2	15
$AlCl_3$[c]	4	20
$FeCl_3$[c]	4	8
$CoCl_2$[c]	4	1
$NiCl_2$[c]	4	5

[a] 1.0 mmol $PdCl_2(NCPh)_2$, 50 ml methyl acrylate, T = 60 °C, t = 17 h.
[b] Run also containing 3.0 mmol methanol.
[c] Run also containing 0.5 mmol methanol.

Table II shows that in the presence of Lewis acids, acrylate dimerization by $[(C_2H_4)_2RhCl]_2$ occurred at 90 °C. No dimer was formed at this temperature in the absence of Lewis acid. The reactions listed in Table II also include a small amount of methanol as a proton source (see below). It can be seen from Table II that the yield of dimer is extremely dependent on the nature of the Lewis acid, with $FeCl_3$ being the best of those tested. A ratio of $FeCl_3 : Rh$ of 5 : 1 was found to be optimal at 90 °C. One can conceive of three distinct roles that $FeCl_3$ might be playing in this system:

(1) reversible removal of chloride ion to afford a cationic rhodium species,
(2) reoxidation of reduced rhodium metal to the +1 state,
(3) providing a source of HCl by the protonolysis of some Fe—Cl bonds.

It is not possible at present to identify which of these functions are important.

TABLE II
Effect of Lewis Acids on the Rhodium-Catalyzed Dimerization of Ethyl Acrylate[a]

Lewis Acid	Equivalents	Time[b] (h)	Catalyst turnovers
$AlCl_3$	5	2	0.1
Ph_3SnCl	5	5	2.7
Cp_2TiCl_2	5	15	3.1
$AgBF_4$	1	3	6.4
$PhSnCl_3$	5	72	12
HCl	10	15	16
$RuCl_3 \cdot 3\,H_2O$	5	17	36
$ZnCl_2$	4	15	37
$HgCl_2$	5	17	60
$SnCl_4$	5	16	68
$FeCl_3$	5	17	150

[a] 0.25 mmol $[RhCl(C_2H_4)_2]_2$, 2.5 mmol methanol, 25 ml ethyl acrylate, T = 90 °C.
[b] No further reaction detected after indicated time.

Dimer yields using the $ZnCl_2$-promoted $(PhCN)_2PdCl_2$ catalyst system were initially found to vary erratically from run to run. This irreproducibility was traced to low levels of methanol in commercial methyl acrylate. As shown in Table III, the presence of a proton source such as methanol significantly increased the yield in Rh-catalyzed acrylate dimerization.

Nugent and McKinney found that the distribution of linear product isomers from acrylate dimerization is markedly dependent on the combination of catalyst and promoter used (Table IV). For instance, changing the promoter from $FeCl_3$ to $SnCl_4$ in the Rh-catalyzed dimerization

TABLE III
Effect of Protic Acids on Rh-Catalyzed Dimerization using $FeCl_3$ Promoter[a]

Acid	Dimers (mmol/mol Rh)
none	25
methanol	150
HCl	38
trifluoroethanol	50
$Me_2O \cdot HBF_4$	15

[a] 0.25 mmol $[RhCl(C_2H_4)_2]_2$, 1.25 mmol $FeCl_3$, 2.5 mmol protic acid, 25 ml ethyl acrylate, T = 90 °C, t = 17 h.

changed the ratio of *trans*- to *cis*-Δ^2 hexenedioate from about 14 : 1 to 1 : 2. Pd catalysis with the $AgBF_4$ promoter produced overwhelmingly the *trans*-Δ^2 dimer. Remarkably, the same catalyst with a $CdCl_2$ promoter produced 66% of the Δ^3 isomers.

TABLE IV
Effect of Catalyst and Promoter on Isomer Distribution from Dimerization of Methyl Acrylate.

Catalyst	Promoter	Isomeric hexenedioates as % of total[a]			
		cis-Δ^2	*trans*-Δ^2	*cis*-Δ^3	*trans*-Δ^3
$PdCl_2(NCPh)_2$[b]	$ZnCl_2$(1)	2	37	12	48
$PdCl_2(NCPh)_2$[b]	$CdCl_2$(1)	3	31	15	51
$PdCl_2(NCPh)_2$[b]	$AgBF_4$(2)	4	91	1	4
$[RhCl(C_2H_4)_2]_2$[c]	$FeCl_3$(5)	6	85	3	6
$[RhCl(C_2H_4)_2]_2$[c]	$SnCl_4$(5)	62	33	1	4

[a] Linear dimers only.
[b] 1.0 mmol catalyst, 50 ml methyl acrylate, T = 60 °C, t = 17 h.
[c] 0.25 mmol catalyst, 25 ml methyl acrylate, T = 80 °C, t = 17 h.

Tkatchenko and coworkers reported that cationic palladium complexes are very effective in the dimerization of methyl acrylate [53], as it has been already observed for other oligomerization and telomerization reactions

[11, 54—56]. An active and selective catalyst proved to be (η^3-methallyl)(η^4-cycloocta-1,5-diene)palladium tetrafluoroborate activated by a phosphorus ligand. It can be seen from Table V that the nature of the phosphorus ligand and the P : Pd ratio strongly affect catalytic activity and product selectivity. The best results are observed for tributylphosphine and a P : Pd ratio of 0.5 : 1 to 1 : 1. Noteworthy is the inhibition of the reaction for P : Pd ratio of 2 or more. The main dimer obtained is dimethyl *trans*-hex-2-enedioate when tributylphosphine was used. In this instance, practically no change in selectivity is observed by varying the P : Pd ratio. Lower selectivities are observed for less basic phosphorus ligands such as triphenylphosphine or -phosphite.

TABLE V
Dimerization of Methyl Acrylate with $[(C_4H_7)Pd(1,5\text{-cod})]BF_4$[a]

Ligand (mmol)	P/Pd ratio	Methyl Acrylate Conversion (%)	Selectivity (%)	
			1[b]	Others
none		<1	—	—
PBu_3 (0.2)	1	48	89	11
PBu_3 (0.4)	2	<1	—	—
PBu_3 (0.1)	0.5	58	92	8
PPh_3 (0.2)	1	11	74	26
$P(OPh)_3$ (0.2)	1	4.5	57	43

[a] 0.2 mmol catalyst, 120—140 mmol methyl acrylate, $T = 80$ °C, $t = 20$ h.
[b] Compound *1* = dimethyl *trans*-hex-2-enedioate.

Concerning the mechanism, model reactions suggested that a cationic hydridopalladium species is formed prior to dimerization of the methyl acrylate.

Also cationic nickel complexes have been studied in acrylate dimerization [57]. The catalyst $[NiYL(\eta^3\text{-}C_3H_5)]$ was investigated with a broad variation of the anion Y and the phosphorus ligand L. As is shown in Table VI, the trimethylphosphine is by far the best ligand, followed by other phosphines with linear alkyl substituents such as PBu^n_3 or $POct^n_3$. The sterical property of the ligand seems to have a great influence on the activity of the catalyst, whereas the selectivity is not influenced remarkably. In all experiments about 85—90% linear dimers were formed with dimethyl *trans*-hex-2-enedioate as the main product.

TABLE VI
Dimerization of Methyl Acrylate with $[(\eta^3\text{-}C_3H_5)Ni(BF_4)(PR_3)]$[a]

Ligand	Methyl Acrylate Conversion (%)	Turnover	Selectivity (%)	
			1[b]	Others
PMe_3	84	42	82	18
PBu^n_3	71	31	74	26
$POct^n_3$	50	26	74	26
PPr^i_3	—	—	—	—
PPh_3	—	—	—	—
Me_2PPh	48	24	78	22
$P(OMe)_3$	13	8	82	18

[a] catalyst/methyl acrylate = 1 : 100, T = 0 °C, t = 10 h.
[b] Compound 1 = dimethyl *trans*-hex-2-enedioate.

The influence of the anion Y was also investigated. The highest conversion was obtained using BF_4^- (84%), whereas PF_6^- and SbF_6^- gave only conversions of 46%. Other anions such as $CF_3SO_3^-$ or $EtAlCl_3^-$ proved to be inactive.

Another important parameter is the starting nickel compound. As seen from Table VII a nearly quantitative conversion of methyl acrylate was observed using $Ni(C_3H_5)_2/HBF_4/PMe_3$. However, this very labile system is not very suitable for an industrial synthesis. The more stabilized nickel

TABLE VII
Dimerization of Methyl Acrylate with Different Ionic Nickel Catalysts[a]

Catalyst System	Methyl Acrylate Convers. (%)	Turnover	Selectivity (%)	
			1[b]	Others
$(\eta^3\text{-}C_3H_5)Ni(PMe_3)Cl + AgBF_4$	54	53	85	15
$Ni(cod)_2 + HBF_4 + PMe_3$	44	43	77	23
$Ni(\eta^3\text{-}C_3H_5)(\eta^5\text{-}C_5H_5) + HBF_4 + PMe_3$	19	19	87	13
$Ni(\eta^3\text{-}C_3H_5)_2 + HBF_4 + PMe_3$	98	93	83	17
$Ni(\eta^3, \eta^3, \eta^2\text{-}C_{12}H_{18}) + HBF_4 + PMe_3$	90	98	78	22

[a] catalyst/methyl acrylate = 1 : 200, 50 ml CH_2Cl_2, T = 0 °C, t = 10 h.
[b] Compound 1 = dimethyl *trans*-hex-2-enedioate.

complex [$Ni(\eta^3, \eta^3, \eta^2\text{-}C_{12}H_{18})$] which can be prepared by the reaction of $NiCl_2$ or $Ni(acac)_2$ with butadiene and $AlEt_3$ seems to be more attractive to transfer the reaction into an industrial scale.

In spite of the novel encouraging results in acrylate dimerization, the technical exploitation is still a great challenge. The best turnover in the nickel catalyzed reaction are 400 cycles per nickel atom, obtained in a batch reactor after a reaction time of 72 h. All experiments showed that the catalyst decomposes during the reaction.

Besides acrylonitrile and acrylates other functional olefins such as vinyl or allyl compounds can also be dimerized. An important example for vinyl compounds is *styrene*, which can be dimerized to 1,3-diphenyl-1-butene. Owing to its high tendency to polymerize spontaneously, the reaction conditions must be chosen carefully. Thus the dimerization catalyzed by $PdCl_2$ at 100 °C yielded only 33% of a dimer fraction with more than 60% of a dark polymeric residue [49]. Using $Ni(\eta^3\text{-}C_3H_5)_2$ as the catalyst, styrene is converted to 1,3-diphenyl-*trans*-1-butene [58, 59]. With $[PdCl(\eta^3\text{-}C_3H_5)]_2$, dimers and trimers are obtained (Equation 43) [60].

$$n\ CH_2{=}CH{-}Ph \rightarrow \underset{Ph}{CH}{=}CH{-}\underset{Ph}{CH}{-}CH_3 + \underset{Ph}{CH}{=}CH{-}\underset{Ph}{CH}{-}CH_2{-}\underset{Ph}{CH}{-}CH_3 \qquad (43)$$

An oxidative coupling of *vinyl acetate* with the catalyst palladium acetate was reported by Kohll and van Helden of Shell research [61]. They observed the formation of 1,4-diacetoxy-1,3-butadiene which was hitherto available only from cumbersome multistep syntheses (Equation 44).

$$2\ CH_2{=}CH{-}OAc + Pd(OAc)_2 \rightarrow Pd^0 + \underset{OAc}{CH}{=}CH{-}CH{=}\underset{OAc}{CH} + 2\ HOAc \qquad (44)$$

However, this reaction seems to be only stoichiometric and the palladium acetate is converted into palladium metal.

Vinyl silanes too can be dimerized using Ziegler-type catalyst systems. For instance, trimethylvinylsilane yielded the linear dimer with the catalyst $Ni(acac)_2/Et_3Al_2Cl_6/PPh_3$ [62] and vinylsilanes containing methoxy groups reacted with a catalyst composed of tetrabutoxytitanium, PPh_3 and $AlEt_3$ in a 1 : 1 : 6 ratio [63] (Equation 45).

$$2\ CH_2{=}CH{-}SiR_3 \longrightarrow R_3Si{-}CH{=}CH{-}CH_2{-}CH_2{-}SiR_3 \qquad (45)$$

$R = Me, OMe$

Some examples for the dimerization of allyl compounds are also known. An intramolecular connection was reported by Bogdanović [64]. He describes the cyclization of *diallyl ether* yielding 3-methylene-4-methyltetrahydrofuran (Equation 46).

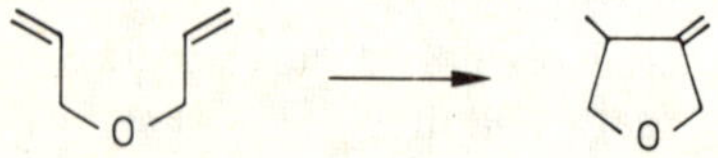

(46)

When the catalyst $Ni(1:4\text{-}5\text{-}\eta\text{-}C_8H_{13})(PR_3)(SO_3CF_3)$ was modified with a methylphosphine which contains chiral centers, an asymmetric cyclization occured.

2.2. CODIMERIZATION OF DIFFERENT OLEFINS

The codimerization of *acrylonitrile and methyl acrylate* is catalyzed by ruthenium complexes under an atmosphere of hydrogen. Besides the hydrodimerization products of each monomer, up to 50% of the head-to-head codimerization products are formed [65] (Equation 47).

$$CH_2{=}CH{-}CN + CH_2{=}CH{-}COOMe \rightarrow NC{-}(CH_2)_4{-}CN + NC{-}CH{=}CH{-}(CH_2)_2{-}CN \quad (47)$$

Barlow reported some examples of codimerization using the catalyst dichlorobis(benzonitrile)palladium [49]. The reaction of *styrene and methyl acrylate* yielded the straight-chain isomer methyl *trans*-5-phenyl-4-pentenoate (Equation 48).

$$Ph{-}CH{=}CH_2 + CH_2{=}CH{-}COOMe \rightarrow Ph{-}CH{=}CH{-}CH_2CH_2COOMe \quad (48)$$

The codimerization of a functional olefin with a non-functional olefin is an interesting possibility for the synthesis of longer-chain monofunctional products. One example is the rhodium or ruthenium chloride catalyzed codimerization of *methyl acrylate with ethylene* yielding linear monounsaturated acids [48]. The main product is methyl-3-pentenoate (47%), but also esters of acids containing seven and nine carbon atoms were isolated in yields of 12 and 9%, respectively (Equation 49).

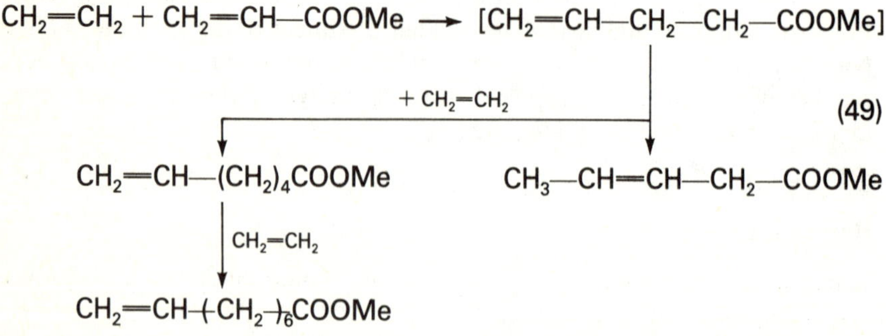

$$CH_2{=}CH_2 + CH_2{=}CH{-}COOMe \rightarrow [CH_2{=}CH{-}CH_2{-}CH_2{-}COOMe]$$

$$[CH_2{=}CH{-}CH_2{-}CH_2{-}COOMe] \rightarrow CH_3{-}CH{=}CH{-}CH_2{-}COOMe$$

$$[CH_2{=}CH{-}CH_2{-}CH_2{-}COOMe] \xrightarrow{+\,CH_2=CH_2} CH_2{=}CH{-}(CH_2)_4COOMe \xrightarrow{CH_2=CH_2} CH_2{=}CH{-}(CH_2)_6COOMe \quad (49)$$

Dichlorobis(benzonitrile)palladium is also an effective catalyst, and the codimers are formed in 47% yield [49]. The non-conjugated esters predominate in the mixture of products (60% *trans*- and 10% *cis*-3-pentenoate), but methyl *trans*-2-pentenoate is also present in substantial amounts of about 30% (Equation 50).

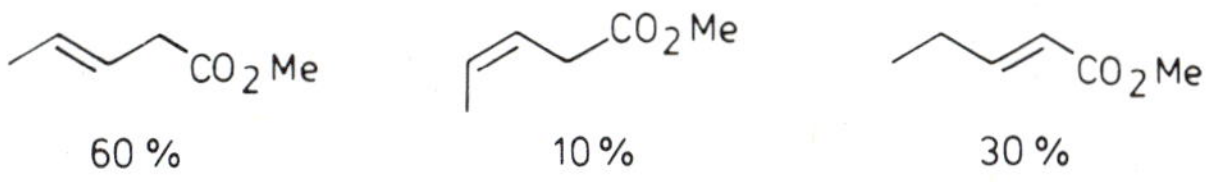

CO_2Me 60 % CO_2Me 10 % CO_2Me 30 %   (50)

The codimerization of methylacrylate and ethylene can be obtained with nickel catalysts, too [57]. A suitable catalyst is Ni(1 : 4-5-η-C_8H_{13})$PR_3(SO_3CF_3)$.

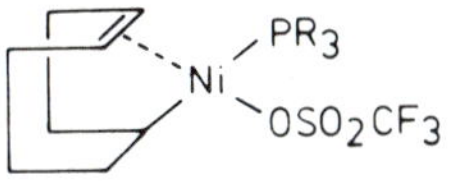

Ni PR_3 OSO_2CF_3 (51)

With tricyclohexylphosphine as PR_3 the reaction in chlorobenzene or methylenechloride gave conversions of 48% or 35%, respectively. Only codimers, and no higher boiling fractions or polymers were formed. The only side-products were the oligomers (C_4, C_6) of ethylene. The codimers identified are shown in Equation 52.

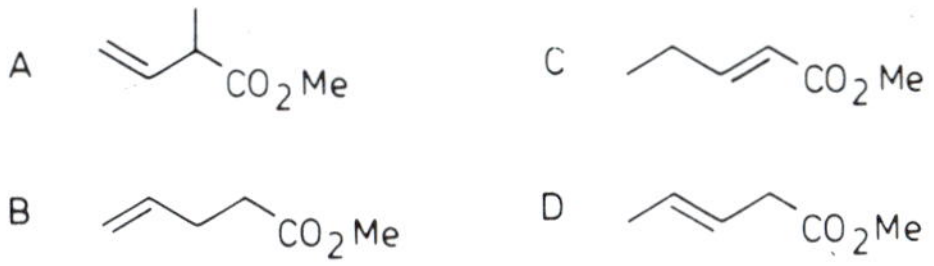

A CO_2Me C CO_2Me
B CO_2Me D CO_2Me (52)

Typical results are summarized in Table VIII. These data show that chlorobenzene is the solvent with the best conversions. The mild reaction

TABLE VIII
Codimerization of Methyl Acrylate and Ethylene with $[Ni(C_8H_{13})PR_3(SO_3CF_3)]$[a]

Solvent	Ligand	Conversion (%)	Isomer-Distribution (%) A	B	C	D
PhCl	PCy_3	48	41	32	8	2
CH_2Cl_2	PCy_3	35	45	31	19	—
PhCl	PPh_3	55	29	32	4	33
CH_2Cl_2	PMe_3	35	14	17	16	46

[a] $T = 20\ ^\circ C$, $t = 24$ h, 100 bar ethylene.

conditions (room temperature), the conversions up to 55% and the high selectivity with regard to the codimers are important advantages of this catalyst system.

The isomer distribution depends to a remarkable extent on the ligand used. The ratio of the linear (B, C, D) to the branched (A) products seems to be influenced by the steric qualities of the phosphine. Tricyclohexylphosphine yields predominantly the branched codimer 2-methyl-3-butenoate (A), whereas PMe_3, which has no significant steric influence, yields methyl-3-pentenoate (D) as the main product.

The nickel catalyst $[(\eta^3\text{-}C_3H_5)Ni(BF_4)(PR_3)]$, which proved to be highly active in the dimerization of methyl acrylate (compare Section 2.1), yielded a broad product mixture in the codimerization with ethylene. Besides codimers (45%), also dimers of methyl acrylate and cotrimers of two molecules of methyl acrylate and one molecule of ethylene were observed. The codimerization of *methyl acrylate with propene* using $[(\eta\text{-}C_3H_5)Ni(BF_4)(PCy_3)]$ gave a conversion of 96% with regard to the methyl acrylate [57]. 45% codimers and 55% dimers and cotrimers were formed. The codimers identified are shown in Equation 53.

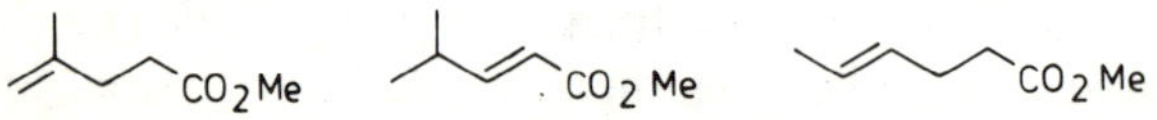

CO_2Me CO_2Me CO_2Me (53)

The main product was the branched methyl 4-methyl-2-pentenoate. The codimerization of *methylacrylate with 1-butene* using the same catalyst yielded methyl 4-methylene-hexanoate (54).

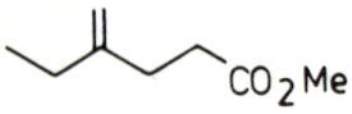

CO_2Me (54)

Styrene and ethylene were successfully codimerized by palladium chloride [49]. The major product was *trans*-1-phenyl-1-butene (90%), but also 2-phenyl-2-butene (5%) and 1-phenyl-2-butene (5%) were present (Equation 55). These codimers were obtained in yields up to 45%.

$$Ph{-}CH{=}CH_2 + CH_2{=}CH_2 \longrightarrow PhCH{=}CH{-}CH_2{-}CH_3 + \overset{Ph}{\underset{CH_3}{>}}C{=}CH{-}CH_3 + PhCH_2CH{=}CHCH_3 \quad (55)$$

The corresponding rhodium chloride catalyzed codimerization has been reported to yield the branched product, 2-phenyl-2-butene, exclusively [48]. The formation of this codimer can only be explained by an attack of the ethylene residue to the α-carbon of styrene. It is presumed that the primary adduct is 3-phenyl-1-butene which rearranges to the more stable conjugated isomer (Equation 56).

$$CH_2{=}CH_2 + \underset{\displaystyle Ph}{\underset{|}{CH}}{=}CH_2 \longrightarrow [CH_2{=}CH{-}\underset{\displaystyle Ph}{\underset{|}{CH}}{-}CH_3] \longrightarrow CH_3{-}CH{=}\underset{\displaystyle Ph}{\underset{|}{C}}{-}CH_3 \qquad (56)$$

2.3. CODIMERIZATION OF DIENES WITH FUNCTIONAL OLEFINS

The codimerization of *acrylates and butadiene* with cobalt catalysts can be used to produce the 4,6-heptadienic-1-ester (Equation 57) [66, 67]. This reaction proceeds under very mild conditions, for instance at 40—60 °C.

+ CO_2R [Co] → CO_2R (57)

By varying the catalyst, the 2,5- and 3,5-heptadienic-1-esters could be obtained. Beger *et al.* [68] reported that a catalyst prepared by $AlEt_3$ and $Co(acac)_3$ in a ratio of 5 : 1 yielded a mixture of esters in about 34% yield. The analogous iron system gave similar results; however, higher ratios of Al : Fe (10 : 1 to 20 : 1) were required.

The mechanism of this reaction was studied by Tolstikov [69] using deuterated butadiene. With nickel catalysts, butadiene and acrylic ester form 2 : 1 adducts (Equation 58) [70].

2 + CO_2Me [Ni] → CO_2Me ; MeO_2C (58)

Instead of acrylic esters, *methacrylic esters* can be used [71, 72]. As catalyst, $Ni(acac)_2/AlEt_3$/phosphine yields C_{12} and C_{20} esters (Equation 59).

2 + CO_2R [Ni / Al] → CO_2R (59)

2 → CO_2R

At a temperature of 0 °C and in the presence of triphenylarsine or triphenylstibine as ligands, the undecatriene-2-carboxylic ester is formed in yields of about 25%. Using temperatures of 30—80 °C the C_{20}-esters turn out to be the main products (~ 90%).

Unsaturated dicarboxylic diesters also give 2 : 1 cooligomerization products [73]. The reaction of *butadiene and methacrylic nitrile* forms both 2 : 1 and 4 : 1 cooligomers in yields of about 20 or 6%, respectively (Equation 60).

CO2Me CO2Me 2 CO2Me CO2Me

CO2Me CO2Me 2 CO2Me CO2Me (60)

CN 2 CN 2 $C_{20}H_{29}N$ (mixture)

Another interesting example is the reaction of *isoprene with methacrylic ester* yielding C_{14} esters [74, 75] (Equation 61).

2 + CO2Me [Ni] CO2Me CO2Me CO2Me (61)

Chiusoli *et al.* reported the codimerization of *butadiene with 3-butenoic acid*, chosen as an example for β, γ-unsaturated acids [76]. In the presence of catalytic amounts of rhodium(I) complexes two 3,6-octadienoic acids were formed (Equation 62).

+ COOH [Rh] COOH COOH (62)

When $[Rh(cod)(PPh_3)_2]PF_6$ was used as catalyst, the yield was 66% based on the butenoic acid. This corresponds to the unusual value of 20 000 mol of product per mol of complex. Higher homologues of butadiene such as *isoprene* also react *with 3-butenoic acid* (Equation 63).

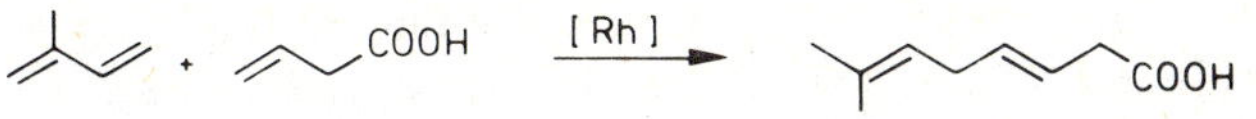

(63)

Bochmann and Thomas of ICI [77] investigated the codimerization of *telomers* (containing two olefinic bonds) *with butadiene*. In the reaction of 1-acetoxy-2,7-octadiene with excess butadiene in the presence of $RhCl_3 \cdot n\,H_2O$ a mixture of linear and branched C_{12}-products was formed (Equation 64).

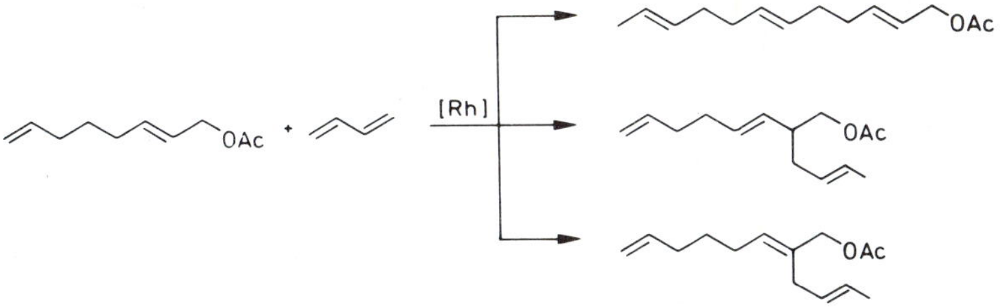

(64)

The catalyst is poisoned by donor molecules such as phosphines, amines and even molecular nitrogen. By contrast, non-phosphine promoters greatly increase the catalyst activity. Allylic chlorides and hydrated chromium(III) chloride are particularly effective. The linear C_{12} products obtained by codimerization of telomers and butadiene remind on products which are formed by cotelomerization of butadiene, 1,3,7-octatriene and a nucleophile [78, 79] (Equation 65).

(65)

A further group of functional olefins which is codimerized with *butadiene* are the *vinylsilanes* [80—84]. Some typical 1 : 1- and 2 : 1-products are shown in Equation 66.

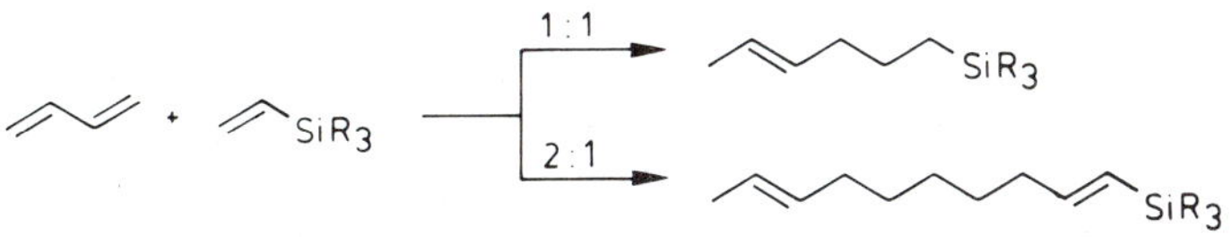

(66)

2.4. DIMERIZATION OF DIENES FOLLOWED BY FUNCTIONALIZATION

The selective dimerization of isoprene followed by functionalization is of particularly great interest, since this reaction sequence enables the synthesis of many terpenoid fine products. In the present review only some typical examples of this far-reaching chemistry can be given.

2.4.1. *Dimerization of Isoprene*

A great number of catalysts active in dimerization of isoprene is known. Ziegler-type catalysts prepared by an aluminum compound and a titanium-, cobalt-, zirkonium-, iron- or vanadium compound yield, besides cyclic products [85—87], mostly the linear tail-to-head dimer 2,6-dimethyl-1,3,6-octatriene [88—95]. Only in a few cases, is the tail-to-tail product 2,7-dimethyl-1,3,6-octatriene also formed [86]. Using nickel catalysts, the tail-to-tail dimer 2,7-dimethyl-1,3,7-octatriene is generally the main product [96—98], but also the head-to-tail dimer allocymene can be obtained [99].

Palladium compounds seem to be the most favourable and versatile catalysts. They enable the selective tail-to-tail, tail-to-head and head-to-head dimerization of isoprene. The catalyst $Pd(PPh_3)_2$(maleic anhydride) yields the 2,7-dimethyl-1,3,7-octatriene in selectivities up to 98% [100, 101], and also $Pd(PPh_3)_4$ [102, 103], and $Pd(acac)_2/PPh_3$ [104, 105] favour this type of linkage. Small amounts of the *tail-to-head* dimer 2,6-dimethyl-1,3,7-octatriene are formed in the presence of $Pd(PPh_3)_2(OAc)_2$ [106] or $Pd[P(OPh)_3]_4$ [107]. To get *head-to-head* dimerization the catalyst $PdBr_2$(dppe)/NaOPh/HOPh can be used [108, 109]. Also the formation of *head-to-tail* dimers was reported: the naturally occuring monoterpenes myrcene and ocimene could be prepared with the catalyst $Pd(NO_3)_2$/KOPh/PPh_3 in allylic alcohol as solvent [110, 111]. Besides dimethyloctatrienes, also dimethyloctadienes were synthesized by reaction of isoprene with formic acid [112, 113].

2.4.2. *Functionalization of Isoprene Dimers*

The dimers of isoprene can be transformed into a great number of terpenoids. For instance, terpene alcohols can be formed by hydroboration [30, 114, 115] and epoxidation [116] reactions. Also the reaction with sulphur dioxide yielding sulfones [117, 118] has been studied. Ploner described the reaction of the 2,7-dimethyl-1,3,7-octatriene with alcohols such as methanol or ethanol [119] yielding terpene ethers which can be used as frangrancies. As catalyst, $RhCl_3 \cdot n\,H_2O$ or H_2IrCl_6 proved to be active.

Further interesting possibilities to functionalize isoprene dimers are mentioned by Morikawa and Kitazume [120]. They started with 2,6-dimethyl-1,3,6-octatriene and reacted with maleic acid anhydride. The Diels—Alder-product obtained is useful for nonvolatile types of alkyd coatings. The termal addition of the isoprene dimer with methacrolein gives a product which possesses a linalool-like odor and which can be used as perfume.

Many other ideas on the functionalization of terpenoid compounds are given by Hoffmann [121], McQuillin [122] and Boelens [123].

3. Conclusions

As is shown in a simplified manner in Equation 67, a great number of unsaturated and/or functional fine products can be produced by dimerization reactions of dienes and functional monoenes.

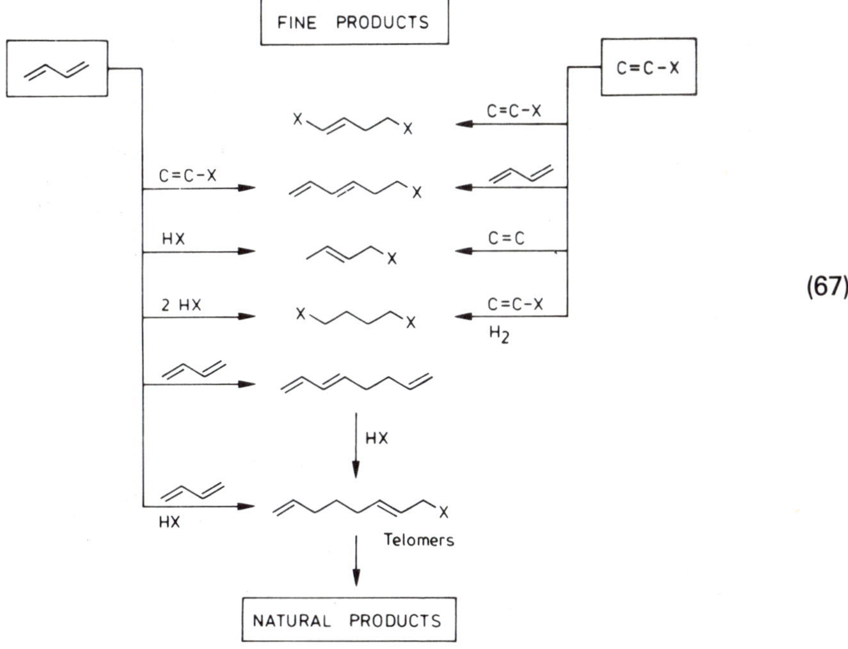

(67)

Technical University, Aachen, F.R. Germany.
New address: *Henkel KGaA, Düsseldorf, F.R. Germany*

References

1. A. Behr, in: *Aspects of Homogeneous Catalysis* (Ed.: R. Ugo), Reidel Publ. Co. (Dordrecht), Vol. 5, p. 3—73 (1984).

2. W. Keim, A. Behr and M. Röper, in: *Comprehensive Organometallic Chemistry* (Ed.: G. Wilkinson), Pergamon Press (Oxford), Vol. 8, p. 371—462 (1982).
3. D. Rose and H. Lepper, *J. Organomet. Chem.* **49**, 473 (1973).
4. L. I. Zakharkin and E. A. Petrushkina, *Izv. Akad. Nauk. SSSR, Ser, Khim.* **11**, 2589—90 (1982).
5. J. Tsuji, M. Kaito and T. Takahashi, *Bull. Chem. Soc. Jpn.* **51**, 547—9 (1978).
6. J. Tsuji, Y. Kobayashi and I. Shimizu, *Tetrahedr. Lett.* 39—40 (1979).
7. J. Tsuji and T. Mandai, *Tetrahedr. Lett.* 1817—20 (1978).
8. T. Takahashi, S. Hashiguchi, K. Kasuga and J. Tsuji, *J. Am. Chem. Soc.* **100**, 7424 (1978).
9. J. Tsuji and T. Mandai, *Chem. Lett.* 975 (1977).
10. J. Tsuji, H. Yasuda and T. Mandai, *J. Org. Chem.* **43**, 3606 (1978).
11. P. Grenouillet, D. Neibecker, J. Poirier and I. Tkatchenko, *Angew. Chem.* **94**, 796 (1982).
12. E. J. Smutny, *J. Am. Chem. Soc.* **89**, 6793 (1967).
13. T. Mandai, H. Yasuda, M. Kaito, J. Tsuji, R. Yamaoka and H. Fukami, *Tetrahedron* **35**, 309 (1979).
14. J. Tsuji, T. Yamakawa and T. Mandai, *Tetrahedr. Lett.* 3741—4 (1979).
15. J. Tsuji, H. Nagashima, T. Takahashi and K. Masaoka, *Tetrahedr. Lett.* 1917—8 (1977).
16. R. Baker, P. M. Winton and R. W. Turner, *Tetrahedr. Lett.* 1175—8 (1980).
17. J. Tsuji, K. Masaoka and T. Takahashi. *Tetrahedr. Lett.* 2267—8 (1977).
18. L. I. Zakharkin and E. A. Petrushkina, *Izv. Akad. Nauk SSSR, Ser. Khim.* 1892—3 (1981).
19. J. Tsuji, K. Masaoka, T. Takahashi, A. Suzuki and N. Miyaura, *Bull. Chem. Soc. Jpn.* **50**, 2507—8 (1977).
20. J. Tsuji, K. Kasuga and T. Takahashi, *Bull. Chem. Soc. Jpn.* **52**, 216—7 (1979).
21. T. Mitsuyasu, M. Hara and J. Tsuji, *J.C.S. Chem. Commun.* 345 (1971).
22. T. Mitsuyasu and J. Tsuji, *Tetrahedron* **30**, 831 (1974).
23. J. Tsuji and T. Mandai, *Tetrahedr. Lett.* 3285—6 (1977).
24. J. Tsuji, T. Yamakawa and T. Mandai *Tetrahedr. Lett.* 565—8 (1978).
25. W. E. Billups, W. E. Walker and T. C. Shields, *J.C.S., Chem. Commun.* 1067 (1971).
26. J. F. Knifton, *Ann. N.Y. Acad. Sciences,* **333**, 264 (1980).
27. J. Tsuji and H. Yasuda, *J. Organomet. Chem.* **131**, 133 (1977).
28. J. Tsuji, T. Yamada, M. Kaito and T. Mandai, *Tetrahedr. Lett.* 2257 (1979).
29. N. T. Byrom, R. Grigg and B. Kongkathip, *J.C.S., Chem. Commun.* 216—7 (1976).
30. M. Hidai, H. Ishiwatari, H. Yagi, E. Tanaka, K. Onozawa and Y. Uchida, *J.C.S. Chem. Commun.* 170 (1975).
31. M. Hidai, H. Mizuta, H. Yagi, Y. Nagai, K. Hata and Y. Uchida, *J. Organomet. Chem.* **232**, 89—98 (1982).
32. J. P. Neilan, R. M. Laine, N. Cortese and R. F. Heck, *J. Org. Chem.* **41**, 3455 (1976).
33. M. Tanaka and G. Hata, *Chem. Ind.* (London) 202 (1977).
34. W. Keim, A. Behr and H. Rzehak, *Tenside Detergents,* **16**, 113 (1979).
35. J. Beger, C. Duschek and H. Reichel, *J. prakt. Chem.,* **315**, 1077 (1973).
36. A. Behr, V. Falbe and W. Keim, *Proceedings* of the 4th. International Symposium on Homogeneous Catalysis, 24.—28. Sept. 1984, Leningrad.
37. G. Hata, K. Takahashi and A. Miyake, *Chem. Ind.* (London), 1836 (1969).

38. G. Hata, K. Takahashi and A. Miyake, *J. Org. Chem.* **36**, 2116 (1971).
39. S. Watanabe, K. Suga and T. Fujita, *Can. J. Chem.* **51**, 848 (1973).
40. G. Lefebvre and Y. Chauvin, in: *Aspects of Homogeneous Catalysis* (Ed.: R. Ugo), Carlo Manfredi (Milano), Vol. 1, 107 (1970).
41. M. Hidai and A. Misono, in: *Aspects of Homogeneous Catalysis* (Ed.: R. Ugo), Reidel Publ. Co. (Dordrecht), Vol. 2, 159 (1974).
42. Société des Usines Chemiques Rhône-Poulenc, *French. Pat.* 1.377.425 (1964).
43. Société des Usines Chimiques Rhône-Poulenc, *French. Pat.* 1.381.511 (1964).
44. E. I. Du Pont de Nemours and Co., *U.S. Pat.* 3.206.498 (1965).
45. D. A. White, *Synth. React. Inorg. Metal-Org. Chem.* **7**, 433 (1977).
46. A. Misono, Y. Uchida, M. Hidai, H. Shinohara and Y. Watanabe, *Bull. Chem. Soc. Jpn.* **41**, 396 (1968).
47. A. Misono, Y. Uchida, M. Hidai and I. Inomata, *J.C.S. Chem. Commun.* 704 (1968).
48. T. Alderson, E. L. Jenner and R. V. Lindsey, Jr., *J. Am. Chem. Soc.* **87**, 5638 (1965).
49. M. G. Barlow, M. J. Bryant, R. N. Hazeldine and A. G. Mackie, *J. Organomet. Chem.* **21**, 215 (1970).
50. H. Pracejus, H.-J. Krause and G. Oehme, *Z. Chem.* **20**, 24 (1980).
51. G. Oehme and H. Pracejus, *Tetrahedr. Lett.* 343 (1979).
52. W. A. Nugent and R. J. McKinney, *J. Mol. Catal.* **29**, 65—76 (1985).
53. P. Grenouillet, D. Neibecker and I. Tkatchenko, *Organometallics,* **3**, 1130 (1984).
54. P. Grenouillet, D. Neibecker and I. Tkatchenko, *J.C.S. Chem. Commun.* 542 (1983).
55. A. Sen and T.-W. Lai, *Organometallics,* **2**, 1059 (1983).
56. M. Röper, R. He and M. Schieren, *J. Mol. Catal.* **31**, 335 (1985).
57. K. Sperling, *Ph.D.-Thesis*, Ruhr-Universität Bochum, F.R.G., 1983.
58. G. Henrici-Olivé, S. Olivé and E. Schmidt, *J. Organomet. Chem.* **39**, 201 (1972).
59. F. Dawans, *Tetrahedr. Lett.* 1943 (1971).
60. Mitsubishi Chemical Ind. Co., Ltd., *Ger. Pat.* 2.234.922 (1974).
61. C. F. Kohll and R. van Helden, *Recl. Trav. Chim. Pays-Bas,* **86**, 193 (1967).
62. V. P. Yur'ev, G. A. Gailyunas, F. G. Yusupova, G. V. Nurtdinova, E. S. Monakhova and G. A. Tolstikov, *J. Organomet. Chem.* **169**, 19 (1979).
63. G. V. Nurtdinova, F. G. Yasupova, E. S. Monakhova, G. A. Gailyunas V. I. Khvostenko, S. R. Rafikov and V. P. Yur'ev, *Dokl. Akad. Nauk SSSR,* **249**, 114 (1979).
64. B. Bogdanović, *Adv. Organomet. Chem.* **17**, 105 (1979).
65. Imperial Chemical Industries, Ltd., *French Pat.* 1.519.113 (1966).
66. D. Wittenberg, *Angew. Chem.* **75**, 1124 (1963).
67. H. Müller, D. Wittenberg, H. Seibt and E. Scharf, *Angew. Chem.* **77**, 318 (1965).
68. J. Beger, N. X. Dung, C. Duschek, W. Höbold, W. Pritzkow and H. Schmidt, *J. prakt. Chem.* **314**, 863 (1972).
69. G. A. Tolstikov, U. M. Dzhemilev, R. I. Khusmutdinov and S. R. Rafikov, *Dokl. Akad. Nauk SSSR,* **224**, 609 (1975).
70. O. S. Vostrikova, U. M. Dzhemilev, G. A. Tolstikov and L. M. Zelenova, *Izv. Akad. Nauk SSSR, Ser. Khim.* 2018—23 (1975).
71. P. Heimbach, P. W. Jolly and G. Wilke, *Adv. Organomet. Chem.* **8**, 29 (1970).
72. H. Singer, W. Umbach and M. Dohr, *Synthesis* 42 (1972).
73. H. Singer, *Synthesis* 189 (1974).

74. E. Klein, F. Thömel, H. Struse, P. Heimbach and H. Schenkluhn, *Ann.* 352 (1976).
75. K.-J. Plöner and P. Heimbach, *Ann.* 54 (1976).
76. G. P. Chiusoli, L. Pallini and G. Salerno, *J. Organomet. Chem.* **238**, C85 (1982).
77. M. Bochmann and M. Thomas, *J. Mol. Catal.* **26**, 79 (1984).
78. E. J. Smutny, *Ann. N.Y. Acad. Sci.,* **214**, 125 (1973).
79. W. Keim, in: *Transition Metals in Homogeneous Catalysis* (Ed.: G. N. Schrauzer), Marcel Dekker (New York), 59 (1971).
80. G. V. Nurtdinova, F. G. Vasupova, E. S. Monakhova, V. G. Lakhtin, G. A. Gailyunas and V. P. Yur'ev, *Zh. Obshch. Khim.* **50**, 704 (1980).
81. F. G. Yusupova, G. V. Nurtdinova, G. A. Gailyunas, G. A. Tolstikov, S. R. Rafikov and V. P. Yur'ev, *Dokl. Akad. Nauk SSSR,* **239**, 1385 (1978).
82. V. P. Yur'ev, F. G. Yusupova, J. Gailyunas, G. V. Nurtdinova and E. G. Galkin, *Zh. Obshch. Khim.* **48**, 1174 (1978).
83. G. A. Gailyunas, G. V. Nurtdinova, F. G. Yusupova, L. M. Khalilov, V. K. Mavrodiev, S. R. Rafikov and V. P. Yur'ev, *J. Organomet. Chem.* **209**, 139 (1981).
84. T. Bartik, P. Heimbach and T. Himmler, *J. Organomet. Chem.* **276**, 399 (1984).
85. G. Wilke, *J. Polym. Science* **38**, 45 (1959).
86. J. Itakura and H. Tanaka, *Makromol. Chem.* **123**, 274 (1969).
87. H. tom Dieck and H. Bruder, *J.C.S. Chem. Commun.* 24 (1977).
88. F. Imaizumi, H. Hirayanagi and K. Mori, *Nippon Kagaku Kaishi* 1771—75 (1975).
89. H. J. Kaminski, *Ph.D.-Thesis*, RWTH Aachen, 1962.
90. H. Morikawa and T. Sato, *Ger. Offen.* 2.440.954 (10/4/1975).
91. Y. Uchida, K. Furuhata and S. Yoshida, *Bull. Chem. Soc. Jpn.* **44**, 1966 (1971).
92. Y. Uchida, F. Furuhata and H. Ishiwatari, *Bull. Chem. Soc. Jpn.* **44**, 1118 (1971).
93. A. Misono, Y. Uchida, K. Furuhata and S. Yoshida, *Bull. Chem. Soc. Jpn.* **42**, 1383 (1969).
94. L. I. Zakharkin, *Dokl. Akad. Nauk SSSR* **131**, 1069 (1960).
95. A. Misono, Y. Uchida, K. Furuhata and S. Yoshida, *Bull. Chem. Soc. Jpn.* **42**, 2303 (1969).
96. W. Gaube, H. Füllbier and E. Alder, *Ger.* (*East*) 124.184 (9/2/1977).
97. Takasago Perfumery Co. Ltd., *Fr. Demande* 2.230.607 (20/12/1974).
98. T. Someya, T. Sakaguchi, S. Akutagawa and A. Komatsu, *Ger. Offen.* 2.326.866 (12/12/1974).
99. M. Yagi, S. Akutagawa and A. Komatsu, *Japan.* 75 24.924 (20/8/1975).
100. A. D. Josey, *J. Org. Chem.* **39**, 139 (1974).
101. A. D. Josey and J. R. Kirchner, *Ger. Offen.* 2.150.355 (13/4/1972).
102. H. Morikawa and S. Kitazume, *Japan. Kokai* 77 23.006 (21/2/1977).
103. H. Yamazaki, *Japan. Kokai* 74 48.613 (11/5/1974).
104. L. I. Zakharkin and S. A. Babich, *Izv. Akad. Nauk SSSR, Ser. Khim.* 2099—100 (1976).
105. K. J. Ploner, *Ger. Offen.* 2.359.100 (27/11/1973).
106. S. Hattori, H. Munekata, T. Suzuki, Y. Nishikawa and T. Imaki, *Japan.* 75 07.045 (20/3/1975).
107. E. L. De Young, *U.S.* 3.691.249 (12/9/1972).
108. K. Takahashi, G. Hata and A. Miyake, *Bull. Chem. Soc. Jpn.* **46**, 600 (1973).
109. K. Takahashi, G. Hata and A. Miyake, *Japan.* 73 26.727 (15/8/1973).
100. T. Sometani, I. Sato, T. Moriya, S. Akutagawa and A. Komatsu, *Japan.* 75 24.925 (20/8/1975).
111. S. Akutagawa, *Japan. Kokai* 76 70.705 (18/6/1976).

112. D. Wright, *Ger. Offen.* 2.050.774 (29/4/1971).
113. J. P. Neilan, R. M. Laine, N. Cortese and R. F. Heck, *J. Org. Chem.* **41**, 3455 (1976).
114. G. Zweifel and H. C. Brown, *Org. Syn.* **52**, 59—62 (1972).
115. I. Uzarewicz and A. Uzarewicz, *Rocz. Chem.* **49**, 1113—18 (1975).
116. S. Watanabe, K. Suga, T. Fujita and N. Takasaka, *J. Appl. Chem. Biotechnol.* **24**, 639—43 (1974).
117. P. A. Ochsner, (a) *Ger. Offen.* 2.363.535 (11/7/1974); (b) *U.S.* 4.066.710 (3/1/1978).
118. S. Akutagawa, T. Someya and A. Komatsu, *Ger. Offen.* 2.403.630 (8/8/1974).
119. K. J. Ploner, *Ger. Offen.* 2.407.898 (12/9/1974).
120. H. Morikawa and S. Kitazume, *Ind. Eng. Chem. Prod. Res. Dev.,* **18**, 254 (1979).
121. W. Hoffmann, *Seifen-Öle-Fette-Wachse,* **101**, 89 (1975).
122. F. J. McQuillin, *Chem. Ind.* (London), 941 (1976).
123. H. Boelens, in: *Fragrance Chemistry* (Ed.: E. T. Theimer), Academic Press, Inc., Chapt. 4, p. 123 (1982).

Y. CHAUVIN

OLIGOMERIZATION OF MONOOLEFINS

Oligomerization reactions are widely used on an industrial scale either to provide high added value chemicals or to upgrade by-product olefinic streams coming from various hydrocarbon cracking or hydrocarbon forming processes (steam or catalytic cracking, Fischer—Tropsch Synthesis, methanol condensation).

Some years ago oligomerization was essentially achieved, and is still currently performed, by means of acidic catalysts, sometimes as a liquid, but mainly in a solid form. However, in spite of its economical interest owing to its low price and comparatively low sensitivity to impurities, cationic oligomerization is limited, the main drawbacks being its poor selectivity and low activity towards linear olefins. Organometallics of highly electropositive metals (aluminum, potassium) afford better selectivity but their specificity and their poor activity restrict their use to some specialized syntheses e.g. selective dimerization of propylene into 2-methyl-1-pentene ($AlPr_3$) or 4-methyl-1-pentene (allyl-K), oligomerization of ethylene into α-olefins ($AlEt_3$).

Coordination catalysts present a broader spectrum of activity (which is often the opposite of what is observed in cationic reactions) and more diversified selectivities: their practical use is expected to grow rapidly in the near future.

Some preliminary remarks have to be made concerning the selectivities of oligomerization reactions. Products resulting from oligomerization have to be defined by means of four essential features:

(1) the degree of polymerization or number of monomer units forming the hydrocarbon chain and the distribution of oligomers, e.g. for ethylene:

$$C_2H_4 \longrightarrow \begin{array}{l} x\ CH_2{=}CHCH_2CH_3 \\ y\ CH_2{=}CHCH_2CH_2CH_2CH_3 \\ \cdots\cdots\cdots\cdots\cdots\cdots \\ z\ CH_2{=}CH(CH_2CH_2)_nCH_2CH_3 \end{array}$$

A. Mortreux and F. Petit (Eds.), Industrial Applications of Homogeneous Catalysis, 177—191.

(2) the regiospecificity or mode of linking of nonsymmetrical monomer units, e.g. for propylene:

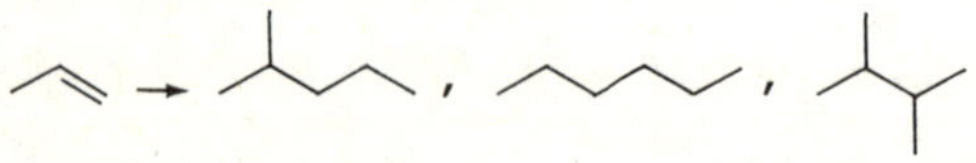

(3) the position of the double bond inside the product and its geometric isomerism, e.g.:

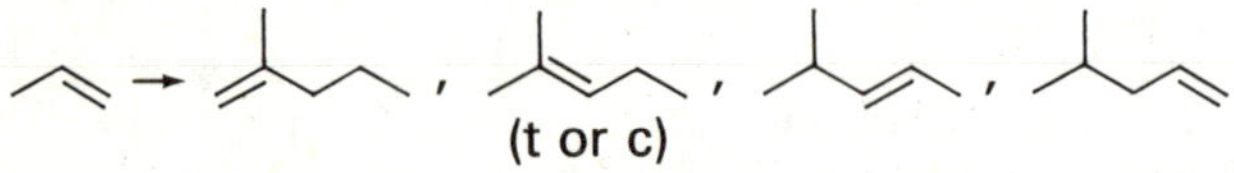

(t or c)

(4) the enantioselectivity if an asymmetric carbon atom is formed, e.g.:

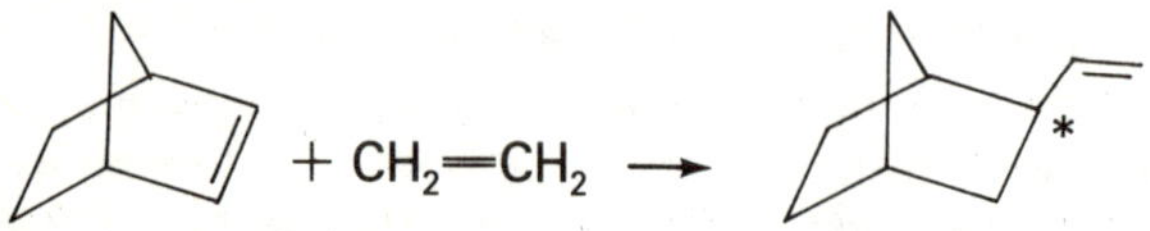

$+ CH_2{=}CH_2 \rightarrow$

From a fundamental and a practical point of view it is quite important to give an accurate definition of the conditions in which given selectivities have been obtained, because chemical selectivities may be partly, or even entirely masked by kinetics artifacts caused by consecutive reactions. This can be illustrated by the curves of Figure 1, giving the yield of dimers versus conversion in the oligomerization of propylene by a nickel based catalyst.

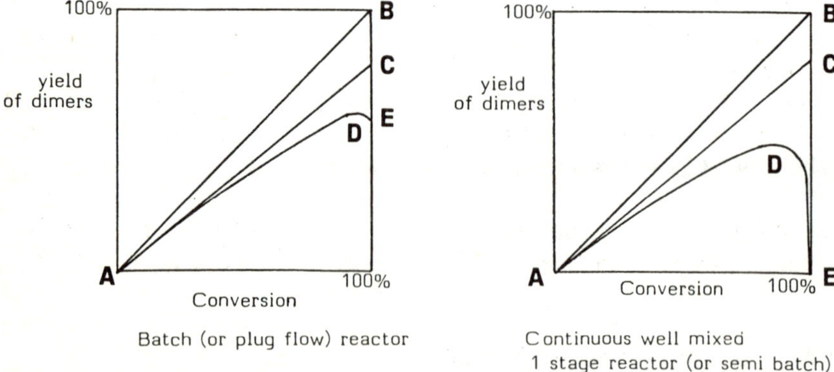

Fig. 1. AB: hypothetical 100% selective dimerization. AC: hypothetical oligomerization without any consecutive reaction (BC: % tri-, tetramers). ADE: the actual curve (DE: zone of prevailing consecutive reactions)

1. The Main Catalysts for Oligomerization

If it is difficult at the present time to classify all the known coordination catalysts according to their mechanism, from a practical point of view, they can be arranged in three main categories according to the nature of the products.

(1) The first one includes the catalysts which oligomerize ethylene, propylene, higher olefins with internal or terminal double bond (e.g. 1- or 2-butene) into a mixture of dimers, trimers, tetramers . . . of various structures (head-to-head, head-to-tail, tail-to-tail). They readily catalyse the shift of the double bond in reactants and products. Tri-, tetra-, pentamers form at the same time by parallel (growing chain) and consecutive reactions.
(2) The catalysts belonging to the second class are especially reactive towards ethylene, and afford mixtures of nearly pure linear α-olefins ranging from C_4 to C_{30} (chain length distribution of the Schulz—Flory type). These do not catalyze a double-bond shift.
(3) In the third category are classed catalysts which are highly selective for dimerization, often affording a single structure with a terminal double bond (with no isomerizing activity). Trimers form exclusively by consecutive codimerization of dimer with monomer (no parallel reaction).

1.1. CATALYSTS WITH SOME ISOMERIZING ACTIVITY

These catalysts are mainly based on Group VIII derivatives. (See Tables I and II).

Noble metal derivatives adapt to various solvents (hydrocarbons, polar, even protic) whereas Co and Ni, especially under the form of cationic complexes are only active in hydrocarbons or halogenated hydrocarbons.

Ni catalysts have given rise to the most numerous studies, and are used on an industrial scale owing to their:

— very high activity,
— low cost,
— possibility of modifying the regioselectivity of the reaction.

They can be divided into two types: cationic and non ionic complexes. In *cationic complexes* the common active species is probably a nickel hydride $[Ni—H]^+ A^-$. There are many ways to get such a species by the well-known organometallic reactions. Thus starting from η^3 allyl Ni the following scheme was proposed:

TABLE I
Some examples of complexes for the dimerization and isomerization of olefins (excluding Ni)

Complexes	Solvents	References	Complex
$RhCl_3$, 3 H_2O	Methanol	[6]	
$RuCl_3$, n H_2O	Methanol	[3]	
H_2PtCl_6	Methanol	[3]	
$PdCl_2$, L_2	Dioxane, esters	[3]	
$HCo(N_2)(PPh_3)_3$	Aromatics		
$ClCo(PPh_3)_3 + BF_3$	Chlorobenzene		
$(\eta 3\ C_3H_5)Pd(CH_3CN)^+BF_4^-$	Nitromethane	[7]	A

TABLE II
Some examples of Ni based catalysts for the dimerization and isomerization of olefins

Cationic type complexes	Solvents	References
$[(\eta 3\ C_3H_5)NiCl]_2$—PR_3—$AlCl_3$	Hydrocarbons	[1] [2]
$Ni(RCOO)_2 + AlEtCl_2$	or preferably	[1] [2]
$NiCl_2 \cdot 2PR_3 + AlEtCl_2$	halogenated	
$Ni(C_6H_5)Cl + BF_3 \cdot Et_2O$	hydrocarbons	[1] [2]
$NiL_4 + H_2SO_4 + BF_3 \cdot Et_2O$		[1] [2]
Non ionic type complexes		
$(\eta^1, \eta^2$ cyclooctenyl)Ni(RCOCH=CR) (with O→ on CR)	″	[4]
R = miscellaneous, CF_3		

$$\text{(allyl)}Ni^{(+)} + \text{cyclooctene} \rightarrow \text{(bicyclic)}Ni^{(+)} \rightarrow \text{bicyclo-octene-pentane} + NiH^{(+)}$$

From Ni^{++} derivatives and aluminium alkyls NiH may be formed by an alkylation followed by a hydride elimination:

$$NiX_2 + AlEtX_2 \longrightarrow NiEt^+AlX_4^- \longrightarrow NiH^+\ AlX_4^-$$

On the other hand Ni(0) complexes are easily protonated by a Lewis-acid/ Brønsted acid couple:

$$NiL_4 + AH \longrightarrow L_xNiH^+A^-$$

Adding phosphines to a cationic complex causes a considerable effect on the regiospecificity of the dimerization. This is particularly important in the case of propylene (Table III).

TABLE III
The influence of phosphines on the regioselectivity of propylene dimerization by cationic nickel catalysts (B. Bogdanovic *et al.*)

Phosphine	*n*-hexenes	2-methylpentenes	2,3-dimethylbutenes
Without	21	73	6
PPh_3	22	74	4
PMe_3	10	80	10
PBu_3	7	70	23
$P(C_6H_{11})_3$	3	48	59
$PiPr_3$	2	30	68

Similar effects were observed in the dimerization of *n*-butenes (Table IV). Due to the high rate of isomerization the same distribution of dimers is obtained starting from either 1-butene or 2-butene (thermodynamic equilibrium is attained soon after the dimerization starts).

TABLE IV
The influence of phosphines on the regioselectivity of 1-butene or 2-butene dimerization by cationic nickel catalysts

Phosphine	*n*-octenes	3-methylheptenes	3,4-dimethylhexenes
Without	8	68	24
PPh_3	15	76	9
PBu_3	10	81	9
$PCyhex_3$	2	64	34

It is worth underlining that, with ionic complexes whatever the ligands used, linear structures are disfavored, especially in higher oligomers: thus the trimers of propylene are highly branched and contain only 3% linear nonenes.

Remarkably *nonionic complexes* of nickel containing chelating ligands of the acetylacetonate type are able to convert olefins such as butene, hexene, or octene into predominantly linear structures. The acidity of the ligand is essential for a catalytic activity and best results are obtained with hexafluoro (or trifluoro) acetylacetone (Table V). The oligomers contain all

TABLE V
Oligomerization of olefins catalyzed by ($\eta^1\eta^3$ cyclooctenyl) Ni(CF_3COCH $COCF_3$).
W. Keim *et al.* [8]

Olefin	Conversion %	Oligomers %			Linearity %		
		di	tri	tetra	di	tri	tetra
Propene	73	59	32	8	75	76	34
1-butene	40	90	10		79	50	
2-butene	15	100			36		
1-hexene	31	92	8		62	35	
1-octene	15	98	2		85		

the possible olefinic isomers not corresponding to that of the thermodynamic equilibrium. However, compared to that of ionic complexes, the isomerizing activity is moderate and then the reactivities of 1-and-2 butene are nonequivalent.

1.2. CATALYSTS FORMING LINEAR OLIGOMERS FROM ETHYLENE (TABLE VI)

Titanium and zirconium based catalysts oligomerize ethylene into a mixture of α-olefin ranging from C_4 to C_{30} with a Schulz—Flory type distribution. The content of linear α-olefin is nearly 100% provided conversion is low enough (a high concentration of ethylene has to be maintained during the whole reaction).

TABLE VI
Linear oligomerization of ethylene

Complexes	Solvents	References	Complexes
$TiCl_4$—AEt_2Cl—*t*-butanol $Zr(OR)_4$—$Al_2Et_3Cl_3$ ZrR_4—$Al_2Et_3Cl_3$	Hydrocarbons or preferably halogenated hydrocarbons	[9] [10]	
$\overline{PPh_2CH{=}C(Ph)ON}i(PPh_3)$	Aromatic hydrocarbons. Various basic and proton active solvents	[11]	B
$Ta(CHCMe_3)(H)(PMe_3)_3I_2$	Toluene	[12]	C

Besides their comparatively low activity, the main drawback of these catalysts is the difficulty to maximize the sought-after C_{12-14} olefins. This is partly overcome by using nickel containing chelating ligands of $P\frown O$ combination. The addition of triphenylphosphine modifies the distribution as reported in Table VII. Linearity is 99% and α-olefin content higher than 98% (the catalyst does not practically isomerize α-olefins).

TABLE VII
C-number distribution of ethylene oligomerization (W. Keim *et al.*)

Catalyst	C_4	C_6	C_8	C_{10}	C_{12}	C_{14}	C_{16}	(%)
Complex "B"	6.5	8	8.5	8.5	8	7.5	7	
Complex "B" + PPh_3	52	26.5	12	5	2			

A considerable number of various types of chelating ligands has been synthesized, the complexes of which afford high flexibility catalysts giving either low α-olefins or high molecular weight polyethylene with a more or less high branching ratio.

Alkylidene tantalum complex 'C' affords with low yield α-olefins ranging from C_{40} to C_{200}. Its importance lies in its mechanistic implications (*vide infra*).

1.3. CATALYSTS WITHOUT ANY ISOMERIZING ACTIVITY

See Table VIII. Titanium based catalysts dimerize ethylene to 1-butene and codimerize ethylene and 1-butene into mixed hexenes (probably for steric reasons titanium is unable to homodimerize 1-butene or other

TABLE VIII
Complexes for the selective dimerization of α-olefins

Catalysts	Solvents	References	Complex
$Ti(OR)_4$—$AlEt_3$	Hydrocarbons, ethers	13	
$Zr(C_4H_6)_2(Me_2PCH_2CH_2PMe_2)$	Toluene	14	
$(C_5Me_5)TaCl_2(RCH{=}CH_2)$	Hydrocarbons	15	D
$Ta(CHCMe_3)(Et)(PMe_3)_2$	Hydrocarbons	16	
$(C_5Me_5)TaH_4(PMe_3)$	Hydrocarbons	17	
$Ni(PPh_3)_2$(ethylene)	Toluene or chlorobenzene	18	

α-olefins). The composition of hexenes resulting from codimerization is as follows:

- 1-hexene 9%
- 2-ethyl-1-butene 65%
- 3-methyl-1-pentene 26%

Tantalum complex 'D' not only converts ethylene to 1-butene but selectively transforms all the α-olefins into a mixture of *t,t* or *h,t* dimers, depending on the size of the substituent (Table IX).

TABLE IX
The Tantalum dimerization of α-olefins (R. R. Schrock *et al.*)

Olefin CH_2=CHR R	 R-substituted dimer (t,t)	 R-substituted dimer (h,t)
CH_3	98	2
n-Pr	88	12
neopentyl	0	100

Depending on the solvent $Ni(PPR_3)_2$(ethylene) catalyzes the dimerization of ethylene to 1-butene (chlorobenzene) or to cyclobutane (toluene). It is the only case of cyclodimer formation from an unstrained olefin (R. H. Grubbs).

2. Mechanistic Considerations

For a long time two types of mechanisms have been postulated for oligomerization, i.e.

- an "insertion mechanism" or "degenerate polymerization" which involves the three main steps of a polyaddition i.e. initiation, propagation, transfer. The same elemental steps explain the double bond shift inside olefins (Scheme 1).
- a "metallacycle mechanism" or "concerted coupling" which, like the ene synthesis, involves an intramolecular hydrogen transfer (Scheme 2).

Most of the catalysts we classified in the first two categories probably act according to Scheme 1. Substantial indirect or even direct evidence supports this hypothesis. However it seems that the essential active

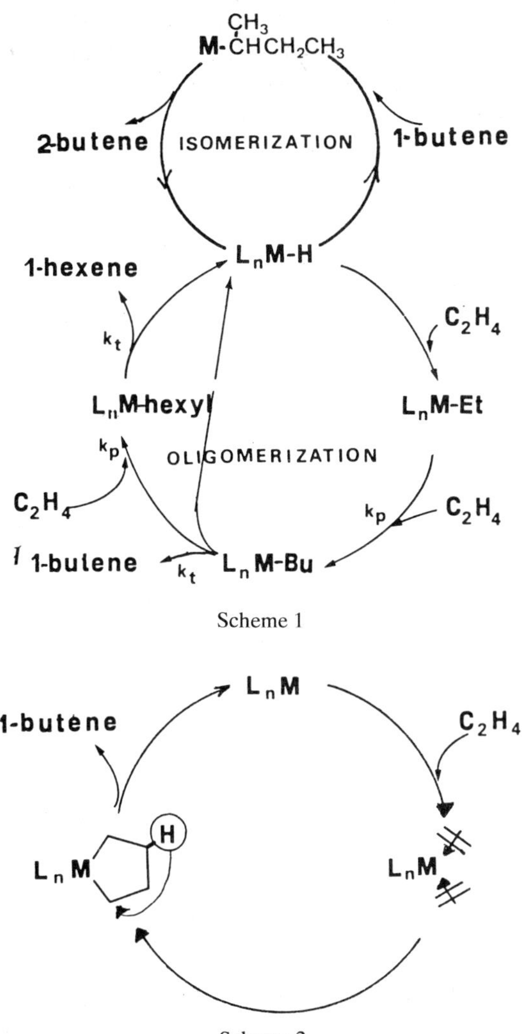

Scheme 1

Scheme 2

species, i.e. M—H, has never been isolated or characterized in significant concentration (this has been ascribed to its high reactivity) although the existence of all the other intermediates such as M-alkyl, M(η^2 olefin) . . . has been proved (see, for example the results of R. Cramer on the dimerization of ethylene catalyzed by Rh(III) in [6] and the conflicting hypothesis of R. R. Schrock in [16]).

The difference between the catalysts of the first and second category arises from the nature of the hydride: more labile in the first category, thus favoring the insertion/β-elimination equilibrium (and then the dimer formation and the isomerizing activity). More stable in the second category, thus favoring the growing chain reaction. In the case of $k_p \gg k_x$ higher olefins (or high molecular weight polymer) are formed; dimers and trimers predominate when $k_p \ll k_x$. Depending on the k_p/k_x ratio (β) the distribution of the products varies according to:

$$x_p = \frac{\beta}{(l + \beta)^p}$$

where x_p is the share of olefins of polymerization for a given amount p.

A catalyst containing a chelating ligand may complicate the simplified Scheme 1 due to a possible hydrogen transfer to the anion, thus modifying the equilibrium:

$$\text{Bu—Ni}(\text{X}\frown\text{O}) \rightleftarrows \text{CH}_2\text{=CHCH}_2\text{CH}_3 + \text{H—Ni}(\text{X}\frown\text{O}) \rightleftarrows \text{Ni}(\text{X}\frown\text{HO})$$

Formation of *linear* trimers and tetramers starting from an α-olefin (Table V) is difficult to vizualize and suggests an intramolecular rearrangement which may be ascribed to a most favored primary carbon—metal bond (or to its higher reactivity), e.g. in the case of propylene:

$$\text{M—H} + C_3H_6 \rightarrow \text{M—CH}_2\text{CH}_2\text{CH}_3 \xrightarrow{C_3H_6} \left[\text{M—CH}_2\text{CH(CH}_3\text{)CH}_2\text{CH}_2\text{CH}_3 \rightleftarrows \text{M—CH}_2\text{CH}_2\text{CH}_2\text{CH}_2\text{CH}_2\text{CH}_3 \right] \xrightarrow{C_3H_6} \text{M—CH}_2\text{CH(CH}_3\text{)}n\text{ hexyl} \rightarrow \text{MH} + n\text{-nonenes}$$

Detailed studies on tantalum complex 'C' (Table VI) demonstrate that, at least in some cases, a much more sophisticated mechanism has to be taken into account, and that metallocarbenes even play a role in polymerization reactions, i.e.:

$$\text{H—Ta=CH}t\text{-Bu} + C_2H_4 \longrightarrow \text{H—Ta(—CH}_2\text{—CH}_2\text{—)CH—}t\text{-Bu} \longrightarrow \text{Ta CH}_2\text{CH}_2\text{CH}_2t\text{-Bu} \longrightarrow \text{H—Ta=CHCH}_2\text{CH}_2t\text{-Bu} \quad \text{and so on}$$

(the transfer paths are more speculative)

The most surprising fact of this mechanism is the preference for α-elimination over β-elimination.

Lastly, the complete retention, after catalytic dimerization, of the allyl group in the Pd complex 'A' (Table I) suggests a pure cationic mechanism in which the transition metal, or more precisely the metal—carbon bond, is no longer directly responsible for the dimerization, thus:

$$(\eta^3\text{-allyl})Pd^{(+)} + CH_2{=}CHR \rightarrow (\eta^3\text{-allyl})Pd\,CH_2\overset{R}{C}H^{(+)} \xrightarrow{CH_2=CHR}$$

$$(\eta^3\text{-allyl})Pd\,CH_2\overset{R}{C}HCH_2\overset{R}{C}H^{(+)} \rightarrow (\eta^3\text{-allyl})Pd^{(+)} + CH_3\overset{R}{C}HCH{:}CHR$$

Such a concept may be proposed for some other cationic complexes which react preferentially with branched olefins. Thus $Pt(MeCN)_4[BF_4]_2$ in nitromethane [19] and $Rh(NO)(NCMe)_4[BF_4]_2$ [20] induce dimerization and oligomerization of isobutene or 2-methyl-2-butene. This is typical of a cationic mechanism.

The "metallacycle mechanism" accounts for the catalysts of the third family.

It is well documented that various metallacyclopentane (Pt, Pd, Ni, Ti, Zr . . .) decompose to produce cyclobutane, ethylene and/or 1-butene depending on the metal and the coordination number of the compound, and the fact that a bis-ethylene complex may be, in some cases, in equilibrium with the metallacycle. The intermediacy of a metallacyclopentane in a catalytic cycle of dimerization was first demonstrated by R. H. Grubbs. Reacting ethylene with $(PPh_3)_2Ni(C_2H_4)$ gives not only $(PPh_3)_2Ni(CH_2)_4$ but catalytic amounts of 1-butene and cyclobutane. Propylene reacts with the same complex to produce new substituted metallacycles and 2-hexene, but the reaction is no longer catalytic.

$$Ni^{(0)} + 2\,CH_2{=}CHCH_3 \rightarrow \text{Ni(3,4-dimethylmetallacyclopentane)} + \text{Ni(2,4-dimethylmetallacyclopentane)} + \text{2-hexene}$$

In the same way complex 'D' (Table VIII) reacts with α-olefins to reversibly form a tantalacyclopentane e.g. with propylene.

$$(C_5Me_5)Ta(CH_3CH{=}CH_2)Cl_2 + C_3H_6 \rightleftarrows (C_5Me_5)Cl_2Ta\text{(3,4-dimethylmetallacyclopentane, } CH_3, CH_3)$$

The equilibrium constant of this reaction depends on the temperature. The tantalacycle is exclusively *trans*-β, β'-disubstituted. This decomposes to give 2,3-dimethyl-1-butene. In the presence of an excess of propylene the dimer forms catalytically. Detailed studies with deuterium labelled olefins show that the dimer is not directly formed by reductive elimination from an alkenyl hydride intermediate. The most satisfactory explanation is that the tantalum hydride adds back to the alkenyl double bond to give a tantalacyclobutane which then rearranges to one of two possible olefins. This unexpected ring contraction supposes a rapid and irreversible decomposition of the less favored ring strained species.

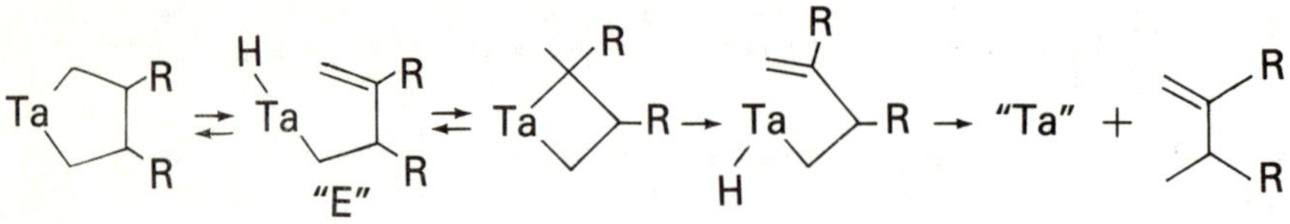

The role of metallacycles and homoallylic hydride intermediates (e.g. "E") has also been demonstrated in the case of the titanium based catalysts. In the decomposition of monosubstituted titanacyclopentane, there is a prevalence for the β-elimination from the most substituted carbon atom over the less substituted one.

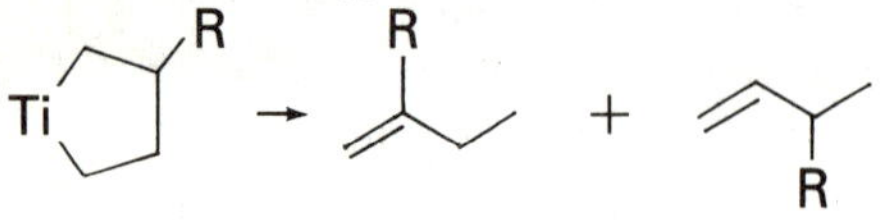

This is to be compared with the products of codimerization of ethylene with 1-butene (R = Et: *vide supra*). As for tantalum, the β-substituted metallacycle preferentially forms over the α-derivative.

In conclusion, beyond the two classical mechanisms usually accepted, many important variations on elementary steps are quite often observed which explain the differences in reactivity and selectivity. This accounts for the richness of coordination catalysis of oligomerization.

3. Heterogeneous and Supported Catalysts

Many attempts have been carried out to graft most of the previously described catalysts onto organic or inorganic supports, in order to combine the outstanding selectivity of these systems with the easy recovery of heterogeneous catalysts. This is all the more interesting when the metal or the ligand are expensive.

Generally, the performances (turnover number) of such catalysts are good. The main drawbacks concern poorer selectivities due to diffusion limitations with regard to the products, and sometimes deactivation. This latter is ascribed to elution either of a ligand or a Lewis acid (for cationic complexes) due to the dissociation equilibria:

$$ML \rightleftharpoons M + L$$

$$M^+AX_4^- \rightleftharpoons MX + AX_3 \qquad (A = Al, B;\ X = F, Cl)$$

This can be avoided by using crystalline oxides such as alumina.

4. Industrial Developments

Three main industrial processes for monoolefin oligomerization have been developed. Each of them represents one of the three abovementioned families of catalysts.

4.1. THE SHELL HIGHER OLEFIN PROCESS

The Shell Higher Olefin Process [21] converts ethylene into linear olefins. The key of the process is a Ni complex of the type B (table VI) dissolved in a glycol (e.g. butanediol) in which the α-olefin product is nearly immiscible. Reaction takes place at roughly 100 °C in a series of three phase stirred autoclaves. To insure good linearity of the product a high ethylene partial pressure is maintained ($\sim$ 10 MPa). The reactor effluent is fed to a separator, the α-olefin layer is decanted, then washed with fresh glycol to remove any catalyst residues, and distilled to recover the individual α-olefins. The glycol phase containing the Ni complex is returned to the reactors. Only small amounts of catalyst are lost.

(To cope with the unsatisfactory distribution of olefins, the SHOP includes an isomerization/disproportionation section which converts light and heavy olefins into more valuable internal $C_{12}C_{15}$ olefins).

4.2. THE DIMERSOL PROCESS

In the Dimersol process [22] a cationic nickel complex ($Ni^{++}/AlEtCl_2$ and Table II) transforms propylene and/or n-butenes into a mixture of di, tri-, and tetramers of various structures (Tables III and IV), consisting of all possible olefinic isomers. The process works in a well mixed liquid phase, without solvent, at 50 °C and 1 to 2 MPa. By contrast to the SHOP there is no recycling of the catalyst, which is deactivated by an ammonia treatment of the effluent of the last reactor, then washed out with aqueous caustic soda and water. Due to the high conversion rates (80 to 93%) there is no recycling of the unreacted monomers.

Propylene dimers are mainly used as motor gasoline, whereas heptenes, octenes and higher olefins are converted to plasticizer alcohols by an hydroformylation reaction.

4.3. THE ALPHABUTOL PROCESS

The Alphabutol process [23] selectively dimerizes ethylene into 1-butene by means of a titanium based catalyst (Table VIII). Hexenes, the only by-products of the reaction, represent 5 to 8% of the converted ethylene.

The reaction (50 °C; 1 to 3 MPa) takes place without solvent in a one stage well stirred reactor. A gaseous phase insures a steady concentration of ethylene in the liquid phase. There is no catalyst recycling and the effluent of the reactor, after deactivation of the catalyst by an amine, is directly fractionated into ethylene (recycled to the reactor) and 1-butene. The bottom fraction contains the catalyst residues.

The common usage of 1-butene is as comonomer in linear low density polyethylene.

Of these three processes the first is the most sophisticated and affords high added value chemicals. The two other processes are simpler and particularly convenient for more common olefins. It can be expected that other processes will appear in the near future.

Institut Français du Pétrole
Rueil-Malmaison

References

REVIEWS

1. P. W. Jolly, G. Wilke: *The Organic Chemistry of Nickel* v. 2, 1. Academic Press (1975).
2. B. Bogdanovic: *Advances Organomet. Chem.* **17**, Academic Press (1979).
3. G. Lefebvre, Y. Chauvin: in *Aspects of Homogeneous Catalysis* v. 1 (ed. R. Ugo) p. 107. Carlo Manfredi (1970).
4. W. Keim, A. Behr, M. Röper: in *Comprehensive Organometal. Chem.* **8** (ed. G. Wilkinson) p. 371. Pergamon Press (1982).
5. G. Henrici-Olivé, S. Olivé: *Adv. Polym. Sc.* **15**, 1. Springer-Verlag (1974).

REGULAR PAPERS

6. R. Cramer: *J. Am. Chem. Soc.* **87**, 4717 (1965).
7. A. Sen, Ta-Wang Zai: *Organometallics,* **2**, 1054 (1983).
8. W. Keim, B. Hoffmann, R. Lodewick, M. Peuckert, G. Schmitt. *J. Mol. Cat.* **6**, 79 (1979).
9. A. W. Langer: *J. Macromol. Sci.* A, **4**, 775 (1970).
10. C. J. Attridge, R. J. Jackson, S. J. Maddock, D. T. Thomson: *J. Chem. Soc. Chem. Commun.* 132 (1973).
11. W. Keim: *Ann. N.Y. Acad. Sc.* **425**, 191 (1983).
12. H. W. Turner, R. R. Schrock, J. D. Fellman, S. J. Holmes: *J. Am. Chem. Soc.* **105**, 4942 (1983).
13. K. Ziegler and H. Martin: *U.S. Patent* 2,943,125 (1960). *CA.* **54**, 257460 (1960). G. G. Belov *et al.: U.S. Patent* 3,879,485 (1974); *CA.* **83**, 27520 (1975).
14. S. Datta, M. B. Fischer, S. S. Wreford: *J. Organomet. Chem.* **188**, 353 (1980).
15. R. R. Schrock, S. Mc Lain, J. Sancho: *J. Am. Chem. Soc.* **102**, 5610 (1980).
16. J. D. Fellmann, R. R. Schrock and G. A. Rupprecht: *J. Amer. Chem. Soc.* **103**, 5752 (1981).
17. J. M. Mayer, J. E. Bercaw: *J. Am. Chem. Soc.* **104**, 2157 (1982).
18. R. H. Grubbs, A. Miyashita: *J. Am. Chem. Soc.* **100**, 7416 (1978).
19. A. de Renzi, A. Panunzi, A. Vitagliano, G. Paiaro: *J. Chem. Soc. Chem. Commun.* 47 (1976).
20. N. Connely, P. Draggett, M. Green: *J. Organomet. Chem.* **140**, C 10 (1977).
21. E. R. Freitas, C. R. Gum: *Chem. Eng. Progress* 73 (1974).
22. Y. Chauvin *et al.: Chem. Ind.* **375**, (1974).
23. D. Commereuc *et al.: Hydrocarb. Process.* **63**(11), 118 (1984).

PH. TEYSSIE

COORDINATION POLYMERIZATION OF MONOOLEFINS AND DIOLEFINS

Introduction: The Discovery

In the early fifties, K. Ziegler was working at Mülheim's Max Planck Institut für Kohlenforschung (where, ironically, he was going to develop some of the main bases of the more competitive petrochemistry), developing a very interesting catalytic process called the *aufbau* reaction:

AlH_3 $\xrightarrow[70\,^{\circ}C]{C_2H_4}$ [Al···H, C=C complex] $\xrightarrow{\text{insersion}}$ $\text{Al}-C_2H_5$ $\xrightarrow[110\,^{\circ}C]{n\,C_2H_4}$ $\text{Al}\!-\!(C_2H_4)_{n+1}H$ $\xrightarrow{>120\,^{\circ}C}$ [Al, CH_2, C, H transition state] $\xrightarrow{\beta\text{-elimination}}$ $\text{Al}-H$ + $H_2C=CH\sim\sim$

Scheme A

That process, yielding a distribution of desirable low molecular weight α-olefins, is still of great industrial importance.

At that time, it happened coincidentally that the presence of traces of Ni in the autoclave drastically changed the rate (low P and T) and the course (selective production of butene) of the reaction: it was the genius of Ziegler to recognize the significance of that phenomenon, and to investigate the influence of other transition metals. Group IV metals, i.e. Ti and Zr, yielded the so-called low pressure, high density (very crystalline) linear polyethlyene (m.p. 138 °C), the properties of which sharply contrasted with those of high pressure, low density (low crystallinity) highly branched PE (m.p. *ca.* 100°) produced up to then by radical polymerization under

A. Mortreux and F. Petit (Eds.), Industrial Applications of Homogeneous Catalysis, 193—228.

extremely drastic conditions. For obvious reasons, that was a major industrial achievement.

Another man of genius, G. Natta at Milano Polytechnic, soon extended that type of catalysis to the stereoselective polymerization of propylene into isotactic (and also syndiotactic) stereoregular polypropylene chains (highly cristalline fiber-forming material, m.p. close to 170 °C).

By the end of the 'fifties, most of that *terra incognita* had been explored in a frantic world-wide research effort, using so-called Ziegler—Natta (Z.N.) catalysts of the general composition: $Mr_aX_b + M_TY_c + L$ (see Fig. 1).

ZIEGLER – NATTA CATALYSTS

$$MR_aX_b + M_TY_c + L$$

M: Groups I to III — M_T: Groups IV to VIII

R: H, alkyl, aryl... — Y: Halogens, OR...

X: Halogen, alkoxy... — L: Electron donor

Fig. 1.

A number of new and exciting possibilities of structural control emerged, as:

— formation of ditactic (*erythro* or *threo*) iso-regular polymers from disubstituted monomers (Fig. 2);
— regioselectivity (1,2 or 1,4) in diolefins polymerizations together with stereoselectivity (*cis-* or *trans-*1,4, *iso-* or *syndio-*1,2);
— stereoselectivity, or preferential polymerization of one enantiomer of a monomer racemate, in a typical asymmetric induction by optically active catalysts (Fig. 4);
— and more recently "chronoselectivity" or the possibility to build a "code" into the catalyst, controlling the statistics of placement of the monomeric units [6] in the growing chain.

It also has to be recognized that this unique research venture had an extremely broad impact on chemistry, well beyond polymer synthesis; and similar catalytic systems have been applied to key petrochemical or fine organic synthesis processes, as illustrated by thousands of patents and articles. An extremely well-deserved Nobel Prize, granted in 1964, rapidly recognized the outstanding contributions of Professors Natta and Ziegler.

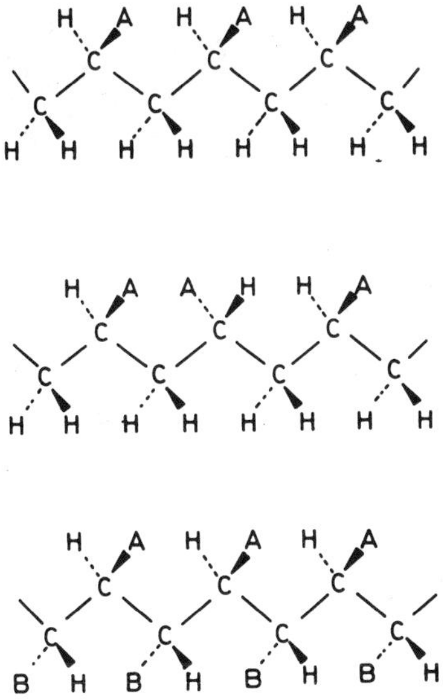

Fig. 2.

PART I: POLYMERIZATION OF MONOOLEFINS

1. Phenomenological Aspects of the Reaction

1.1. IMPORTANCE

Z.N. catalysts are the only ones able to polymerize C_2H_4 and C_3H_6, two basic products of petrochemistry, into (stereo) regular high M.W. polymers. That is an extremely important industrial achievement, as illustrated by the (approximate) data below which mean that, taking into account polydienes (see Part II) and a few experimental polymers, Z.N. catalysis alone is not responsible for a production of *ca.* 10 million tons a year of excellent products (obtained in one step from basic petrochemicals): that is a unique case in industrial organic chemistry, particularly if one considers the extreme diversification achieved in their applications.

1.2. HETEROGENEOUS SYSTEMS

Insoluble (cristalline) Z.N. catalysts are the only ones to control properly propylene conversion into the desirable isotactic polymer. They are essentially based on a mixture of $TiCl_3$ and Al-alkyl.

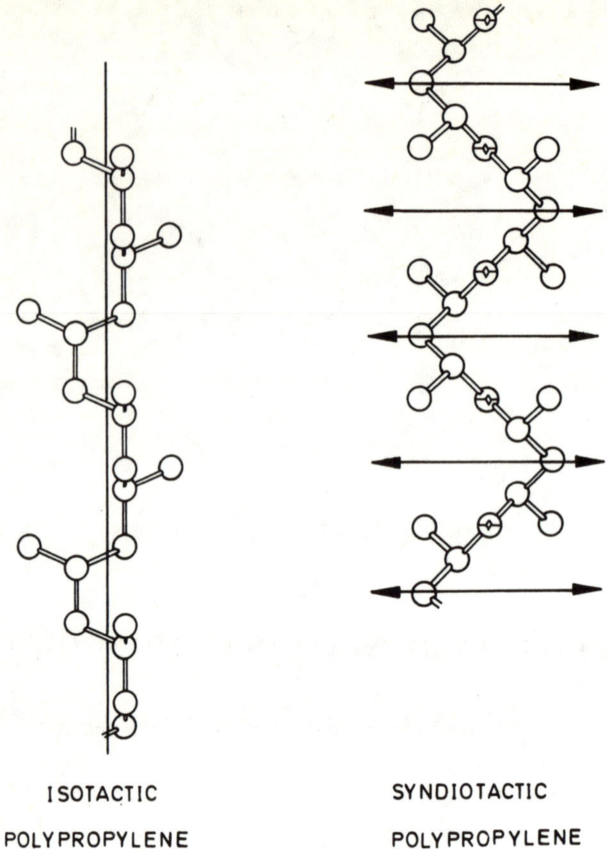

ISOTACTIC POLYPROPYLENE

SYNDIOTACTIC POLYPROPYLENE

Fig. 3.

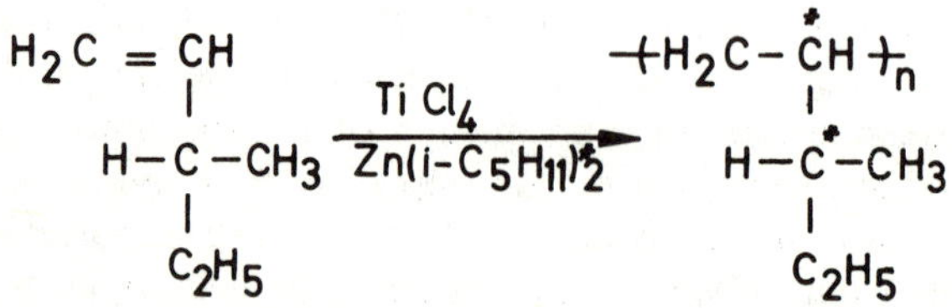

$$H_2C=CH-\overset{*}{C}H(CH_3)-C_2H_5 \xrightarrow[Zn(i\text{-}C_5H_{11}^{*})_2]{TiCl_4} -\!\!\left[H_2C-\overset{*}{C}H\right]_n- \text{ with side group } H-\overset{*}{C}(CH_3)-C_2H_5$$

Fig. 4.

$$-TiCl_4 + AlR_3 \xrightarrow{\text{low. T}} RTiCl_3\ (+AlR_2Cl) \longrightarrow \beta\text{-}TiCl_3\ (+R^0),$$

a brown variety formed of stacks of fiber-like chains of $TiCl_2$, has only limited long-range crystal order.

$$-\mathrm{TiCl_4} + \mathrm{H_2} \xrightarrow{500^\circ} \alpha\text{-}\mathrm{TiCl_3}$$

$$\mathrm{Al} \xrightarrow{>250^\circ} \alpha\text{-}\mathrm{TiCl_3} \quad \text{(containing } \mathrm{AlCl_3}\text{)}$$

$$-\mathrm{TiCl_4} + \mathrm{AlR_3}\ (\text{or } \beta\text{-form}) \xrightarrow{>200^\circ} \gamma\text{-}\mathrm{TiCl_3}$$

These α and γ forms are violet, and have a structure composed of layers of Ti-atoms sandwiched between 2 chloride ion layers. The Ti atoms, in an octahedral coordination geometry, occupy positions in the interstices formed by close packed Cl ions. That packing is hexagonal for the α form and compact cubic for the γ form (One can also obtain a chaotic δ variety by grinding the α or γ forms, possible with $AlCl_3$).

The influence of that crystalline structure on the course of the reaction is decisive: while remarkable stereoselectivity can be achieved (up to 98—99%) with α and γ forms, only limited control (*ca.* 50% isotacticity) is exerted by the β variety.

The productivity of these catalysts has been progressively increased, by controlling the conditions, and also the state of division of $TiCl_3$; modern versions (supported, superactive, high mileage, etc. catalysts) include extremely fine dispersions of $TiCl_3$ on inorganic supports (usually isostructural Mg salts) reacted with different donor ligands (i.e. esters). Using such systems, one can approach productions of 100 Kg and even 1 ton of PE per gram of Ti (depending on the process conditions); that has a very important industrial consequence, i.e. to be allowed to leave these very small amounts of M_T (harmless in terms of oxidation) in the polymer, i.e. to avoid a difficult and costly purification step.

TABLE I
Polyolefins production from 1966 to 1983

10^3 Tons per year	1966	1970	1975	1983
Polyethylene H.D.	800	1600	2500	*ca.* 3500
Polypropylene isotactic	400	1200	2500	*ca.* 4000
E.P. copolymers (E.P. rubbers)	100	250	500	*ca.* 900

1.3. SOLUBLE COMPLEXES

These can also be very active, if not able to promote isotactic stereocontrol. A typical example is the reaction of Cp_2TiCl_2 with AlR_2Cl, giving a ν-bridged binuclear complex $Cp_2TiClR.AlRCl_2$, excellent for C_2H_4 polymerization.

A new interesting development is the discovery and tailoring by Kaminsky of Z.N. type systems composed of titanium cyclopentadienyl complexes, and aluminoxanes obtained by controlled hydrolysis of aluminum alkyls (see [8] and papers of the corresponding symposium held at Akron in June 1986). These soluble and extremely active systems (for ethylene) are even claimed to exhibit a good stereoselectivity in propylene polymerization under homogeneous conditions (probably due to the hindered environment provided by the alumoxane structure).

1.4. ROLE OF THE TWO METALS

From a purely experimental point of view, and although the necessary Al derivative certainly plays an important role, the influence of the M_T is obviously decisive, as illustrated by the fact that it controls:

- the type of reaction (i.e. Ni → butene, Ti → PE);
- the selectivity (i.e. V → *syndio*, Ti → *iso*; Cr → 1, 2, Co → 1,4);
- the relative reactivity in copolymerization reactions;
- the kinetic law: $V_p = k\,[M_T]\,[\text{Ol}]$;

all features rather insensitive to the Al derivative (although it can significantly modify the overall rate, its concentration does not come into the rate expression).

As a final confirmation of that, catalysts containing only M_T have been prepared, i.e. the Union Carbide Silica—O—Cr—Cp system for C_2H_4 polymerization.

2. The Mechanism and Molecular Characteristics of Polymerization Catalysis with $TiCl_3$—AlR_3

On the basis of the above observations, an insertion mechanism has been proposed by Natta's group, and more formally discussed by P. Cossee [2]. Its main features, and further sophistication, are summarized here after.

2.1. THE INITIATION STEP

This will obviously take place on the necessarily existing coordination vacancies at the surface of the crystal, by μ-coordination of the Al derivative and a metathetic exchange implanting R groups which foreshadow the future growing chain (Fig. 5). That has been semiquantitatively confirmed by analytical determination with labeled groups.

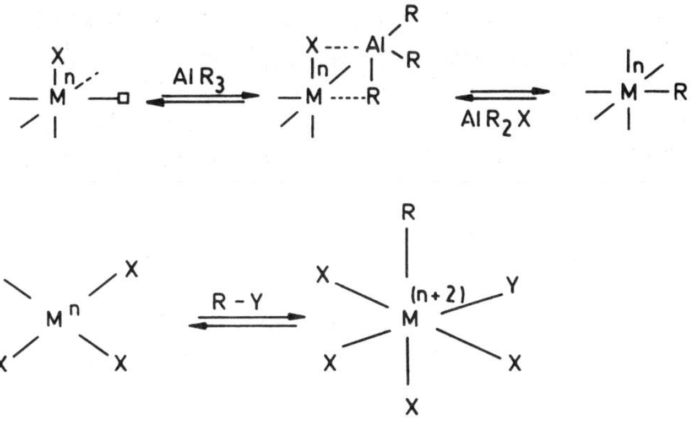

Fig. 5.

2.2. THE PROPAGATION STEPS

This may take place in 2 steps: first a dynamic coordination of the monomer through a σ—π bond (implying back(bonding) on a coordination vacancy: that interaction may be more or less strong. The previous existence of a vacancy is not an absolute prerequisite however.) As illustrated in Fig. 6, that bonding approach is possible, but implies a steric interaction with other groups, important in terms of stereocontrol;

Then an insertion of the monomer in the growing chain, called *cis*-insertion (or *cis*-rearrangement or *cis*-migration), since it involves vicinal

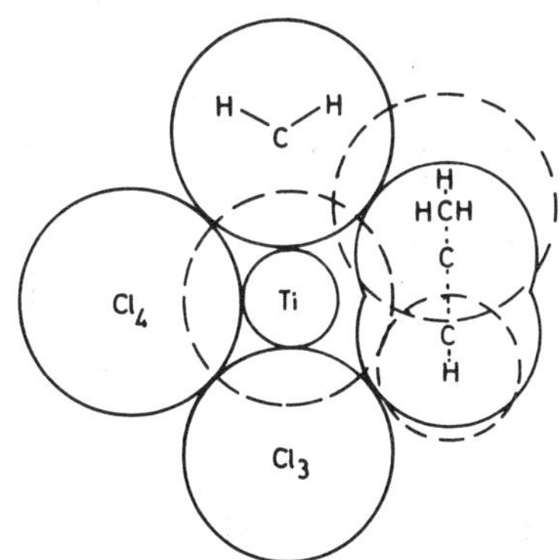

DIAGRAM OF THE COMPLEX OF C_3H_6 WITH THE ACTIVE CENTER. RADII HAVE BEEN DRAWN TO SCALE.

$r_{CH_2} = 2.0$ Å ; $r_{Cl^-} = 1.8$ Å ; $r_{Ti^{3+}} = 0.8$ Å ; $r_C(\text{in } C_2H_4) = 1.7$Å

Fig. 6.

positions on the complex. Promoted by lateral vibrations of the σ-bonded alkyl group leading to its overlap with the π^* olefin orbital, it gives rise to a new alkyl group $(n + 1)$ coordinated now to the vicinal position. The process is obviously iterative (Fig. 7, 8), ensuring the chain growth as that of a hair (polymer) on the head (catalyst), fed at its base (by monomer).

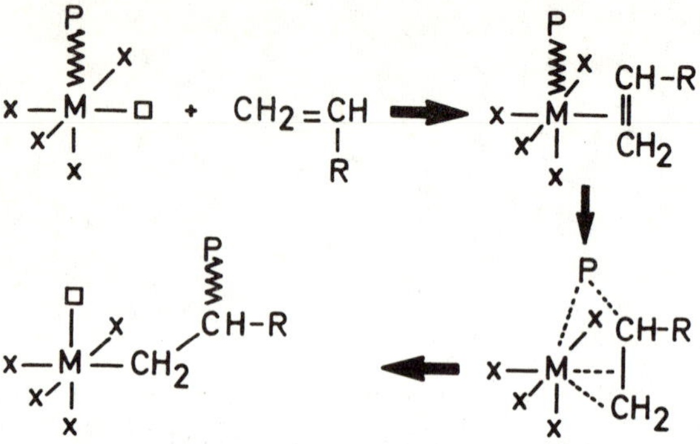

Fig. 7.

Such insertion reactions have been exemplified in organometallic synthesis; their main feature is the permanent control exerted during the whole process by a whole set of hybridized σ, π, p and d orbitals (Figs. 9, 10) explaining the amazing simultaneous achievement of a high degree of polymerization and of stereocontrol. (Displacements smaller than at most 1.9 Å.)

That reaction raised an interesting challenge, since it represents formally a $\pi_s^2 + \sigma_s^2$ concerted pericyclic rearrangement, which is thermally disfavoured in the Woodward—Hoffmann formalism: possible answers (injection of additional electrons from the M_T, or phase fracturation through the metal) may be considered.

2.3. THE ENERGETICS OF THE CHAIN GROWTH

Based on an orbital image of the relevant potential wells, Fig. 10 clearly illustrates how the d-orbital system of the M_T (and the corresponding back-bonding) provide for a broader overlap of $R(\sigma)$ and olefin (π^*) orbitals, thereby appreciably decreasing the activation energy of that determinant step. A simple orbital diagram essentially points to the same conclusion, and simplified quantum chemical calculations (Cossee and Schachtschneider, 4th Int. Congress on Catalysis, Moscow 1969) strongly

H5
H4
C5
154
C4
2.6
H3
C3
11
Cl2
C2
Cl4
26
26
135
Ti
C1
Cl1
Z
26
H2
H1
X
Y
Cl3
(a)
H4
C5
H5
C4
C3
H3
C2
Cl2
Cl4
Ti
C1
Cl1
H2
H1
Cl3
(b)

H4
C5
H5
C4
C3
H3
C2
Cl2
Cl4
Ti
C1
Cl1
H2
H1
Cl3
(c)
C'5
C5
04
154
C'4
H4
19
C4
154
H3
H5
C2
C3
Cl2
154
Cl4
26
Ti
26
C1
Cl1
H2
H1
26
Cl3
(d)

Fig. 8.

HCH
+
+
+
−
−
−
−
Cl4
Ti
Cl1
+
Cl3

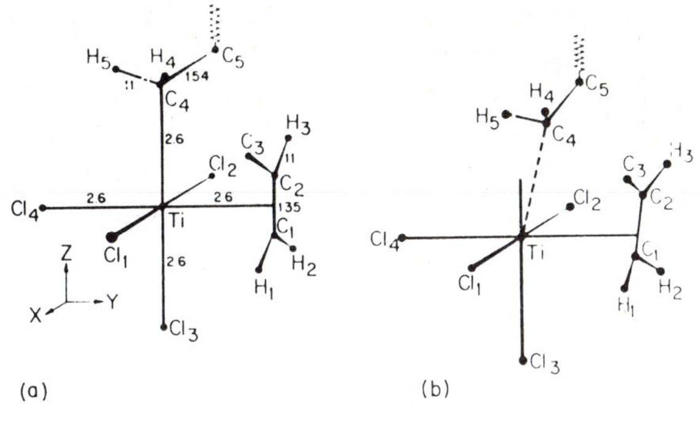

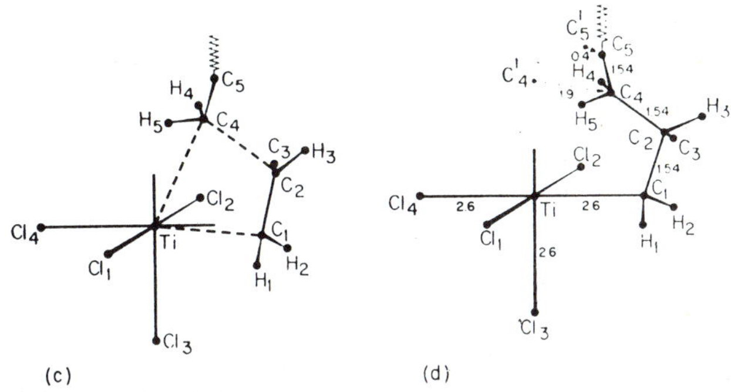

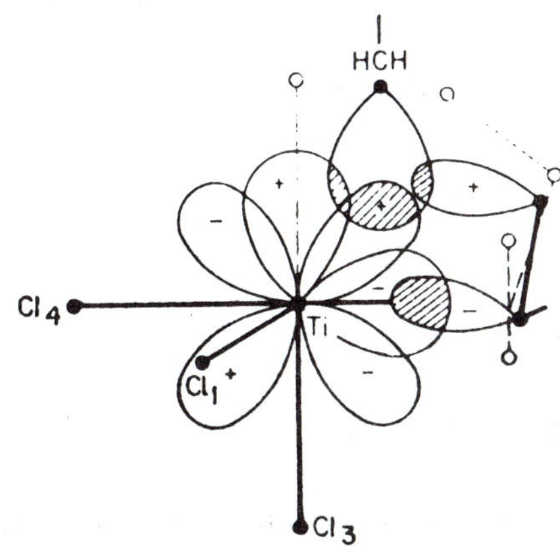

Fig. 9.

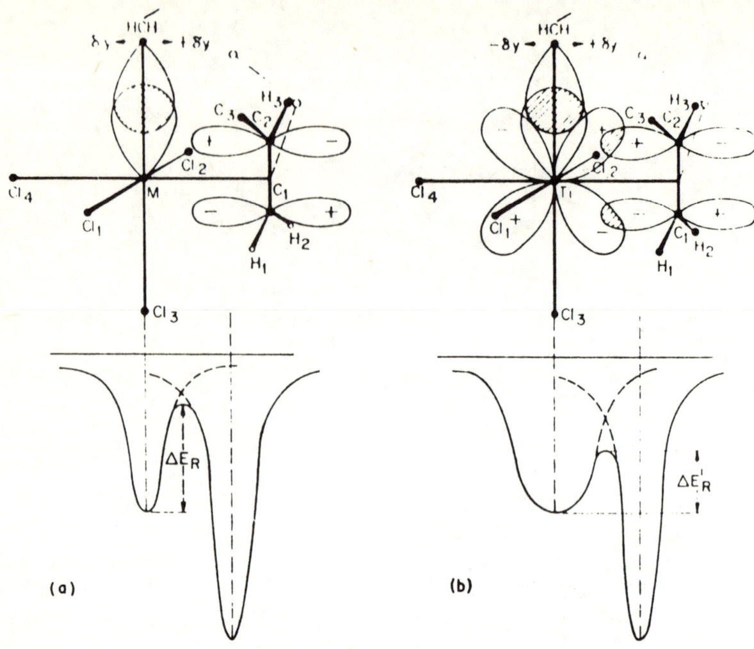

Fig. 10.

indicate, in addition to the increased R group lability, that no significant ion-pair generation takes place.

Moveover, these schemes visualize the fact that it will be possible to 'tailor' a given catalyst by modulating the orbital energy levels through (slight) modifications in the ligand field (see, e.g., the Phillips-type oxide-based catalyst for C_2H_4 polymerization).

2.4. KINETIC FEATURES

2.4.1. *General laws*

The 2 steps of the so-called Cossee scheme can be represented as [2]

$$\mathrm{A} + \mathrm{O} \underset{k_2}{\overset{k_1}{\rightleftharpoons}} \mathrm{AO}, \quad \text{and} \quad \mathrm{AO} \xrightarrow[k_3]{} \mathrm{A}'$$

where A and A′ are similar vacant active sites with different situations for R, and O is the olefin. The experimentally verified stationary state assumption implies $D[AO]/dt = 0$, with $C_{[Ti]} = [A] + [AO]$.

Consequently:

$$[AO] = \frac{k_1[C][O]}{k_1[O] + k_2 + k_3},$$

and since

$$\frac{d[AO]}{dt} = -k_3[AO] - \frac{d[O]}{dt} = 0$$

one finds

$$\frac{-d[O]}{dt} = V_P = \frac{k_1 k_3[C][O]}{k_1[O] + k_2 + k_3},$$

a general equation for coordinatively activated processes, reminiscent of a Michaelis—Menten kinetic equation.

The experimental law for C_2H_7 polymerization on Ti catalysts, though, is simply $V_P = k'$ [C] [O]. This possibly means that k_1 is definitely smaller than k_2, a probable situation in view of the well known instability (low K_f, due to low back bonding) of olefin—Ti complexes (Ballard). It would be highly unlikely indeed that $k_1 \ll k_3$ thus becoming the rate-determining step.

It is interesting to note that a large K_f (high-stability Group VIII metal complexes, for instance should lead to a 'saturation' equation $V_P = k''$ [C], where the actual olefin concentration does not play a role anymore, since in practice there is always a monomer coordinated to the complex.

2.4.2. *Number of active sites*

"Classical" former Z.N. catalysts contained very few actual active centers (*ca.* 1% being a typical value) as inferred from calculations based on product molecular weight and on radiolabelling of the chains.

More detailed investigations have recently been performed (i.e. by P. Tait and N. Haward) using labelled CO or allene: they clearly show that the productivity increase found in the new generation of catalysts was essentially due to an increase in the number of actual active sites and not of k_p. It also seems that stereoselective centers definitely display a higher k_p that nonselective ones (which may be due to more difficult (i.e. hindered) competition from the Al-derivative).

2.5. THE STEREOREGULATION

2.5.1. *Cossee's proposal*

Considering a crystallographic model of the α-$TiCl_3$ surface, Cossee noticed that the layers basic plane was not parallel to it but formed an angle of about 55°; that situation creates a non-equivalence between the 2 vicinal coordination positions located in a "coordination cavity" and very strong steric compression around one of them (Fig. 11). As a consequence:

(a) the monomer will always orient its alkyl substituant "outward" ensuring regioselectivity (99% head-to-tail in isotactic polymers), and
(b) after migration in the *cis*-rearrangements, the alkyl group will have a definite tendency to migrate back (through the same set of orbitals) to the more "outward" less hindered coordination position, in a reaction step schematized as $A' \underset{k_4}{\rightarrow} A$ (see Section 2.4.):

the system would accordingly go back to exactly the same geometry as before insertion, favoring isotactic placement (all substituted carbons with the same relative configuration). Increasing temperature would favor that return (increase in k_4), while decreasing the occupation of the vacancy by the monomer (decrease in $k_f = k_1/k_2$), thus favoring isotacticity. By contrast, a temperature decrease would slow down the shift back and

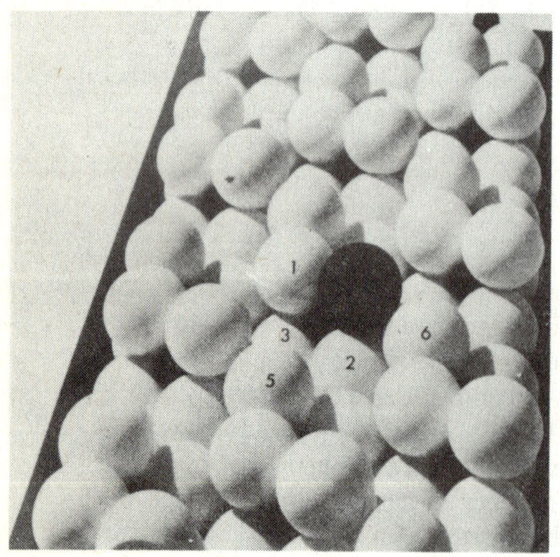

Fig. 11.

favor the coordination of the monomer, a situation favoring alternation of placements, thus syndiotacticity (if at least steric hindrance around the site is high enough to avoid atactic growth). This analysis fits the experimental data.

Since the stereocontrol is essentially due to steric interactions with the site environments, it seems logical that β-$TiCl_3$, a much less regular type of crystal, will give lower performance; in fact, a typical value of 50% isoactivity might seem still quite high (see below).

2.5.2. *The Rodriguez—Van Looy model*

Cossee's model implies a ($TiCl_3R$) type of active center, many of them being *de facto* present at the surface of the $TiCl_3$ crystal (reacted with AlR_3). In a series of striking electron micrographs taken during a polymerization experiment of C_3H_6 over $TiCl_3$, Rodriguez and Van Looy have shown [1] that:

- the appearance of the nice hexagonal α. $TiCl_3$ crystals is not modified by reaction with (gaseous) $Al(CH_3)_3$;
- when C_3H_6 is admitted and polymerized, chains are formed only on separated spots, specifically located on the edges of the crystal (growth hexagonal spiral, Fig. 12);
- in areas of mechanical dislocation (Fig. 13) or on β-$TiCl_3$, the polymer produced covers the surface.

The observations have led Rogriguez to conclude that the more active sites were in fact crystalline defects, i.e. Ti atoms with 2 vacancies ($TiCl_3R\square\square$), one of those being occupied most of the time by the μ-bridged Al-derivative (Fig. 14); these deficient situations will obviously be formed preferentially on the crystal edges (lower Cl extraction energy). The insertion process itself remains the same as formerly, of course. Obviously, Cossee type centers may coexist, but will be much less active, since the Al derivative will strongly compete with the monomer for the vacancy.

This proposal has the merit of explaining a number of experimental facts:

- it allows an easy interpretation of the micrographs;
- it explains the influence of the "ligand" aluminum derivative on the overall rate of reaction by electronic effect through the μ-bridges, and even on the stereoselectivity by its bulkiness (particularly if it bears big

Fig. 12.

counterions, such as iodide), and the driving force it provides for the restoration of the Ti—R bond always in the same position;

- it shows why β-$TiCl_3$ can still have an appreciable number of stereoselective centers (highly hindered with only 1 vacancy available for the monomer) (Figs. 15 to 17);
- it better rationalizes the facts that ZnR_2 (a good alkylating agent but a stronger ligand) gives less efficient catalysts, that addition of Lewis bases increases stereoselectivity (by blocking "loose" excess coordina-

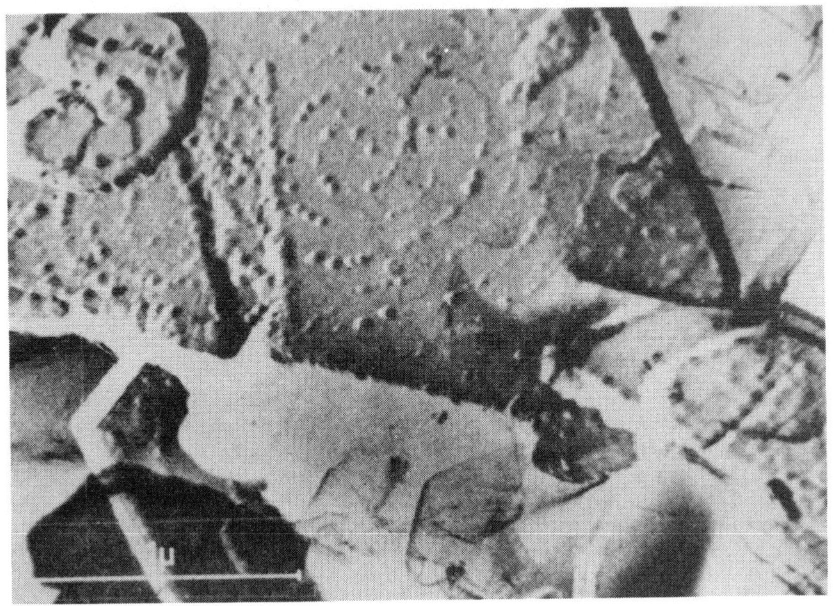

Fig. 13.

$+ AlR_3$

$+ AlR_2Cl$

$+ CH_3-CH=CH_2$

Fig. 14.

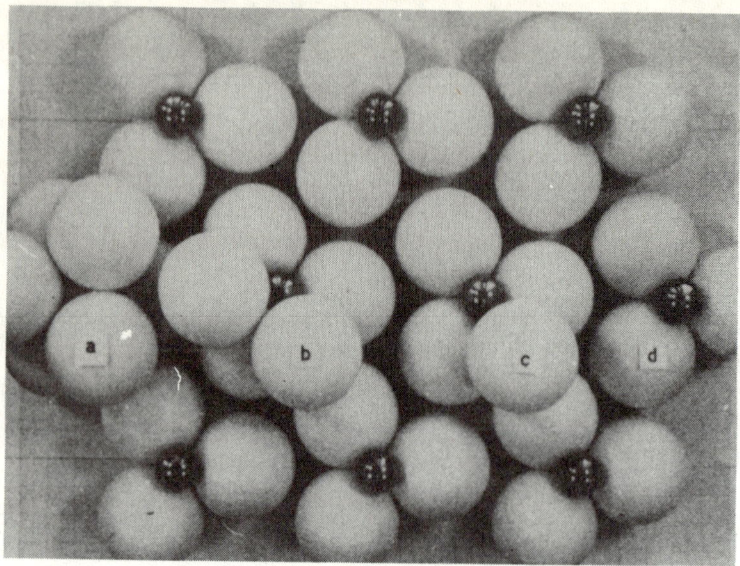

Fig. 15.

tion positions), and that "stereoblock" isotactic chains can be formed by successive exchanges of growing chains from site to site, through the Al-alkyl. It is worthwhile to note that a similar picture has been obtained for catalysts supported on $MgCl_2$.

2.5.3. *More recent approaches*

In summary, the Rodriguez model fits most experimental facts available and is probably a good simple approximation of the "real-life" catalytic situations. It has, of course, to be refined further by taking into account the possibility of having different environments for active sites located on the same catalytic particle; that implies, essentially, differences in steric hindrance (different selectivities), and also differences in "acidity" i.e. electron density (different rates). Important parameters are the values of the rotation ($K_r \rightarrow$ rotamers) and dissociation ($K_d \rightarrow$ diastereomers) equilibrium constants, and insertion rate constants for the different coordination situation; and still more important, the group electronegativity and relative volume of the X, S, L and B ligands (one of them possibly being the Al derivative) around M_T (see Fig. 18), versus that of the first element of the growing chain (with its substituents). Along these lines, a more

(a)

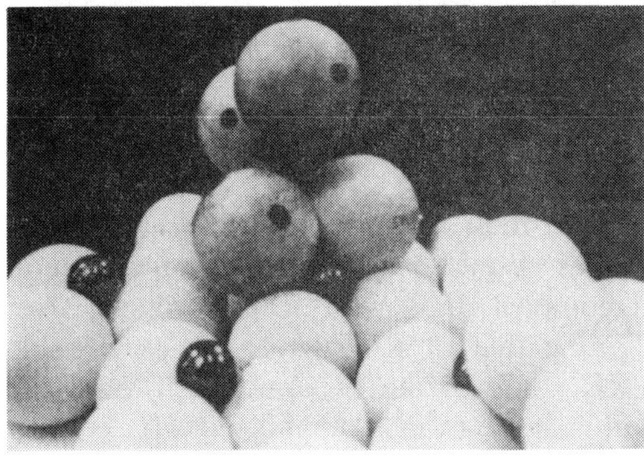

(b)

Fig. 16.

detailed interpretation can be given for the control of regioselectivity (1,2 or 2,1 addition), of stereoselectivity and of enantioselectivity (see the work of Pino and coworkers, particularly in [7] and [9]). Interesting examples include: (1) the polymerization of an olefin monomer racemate (*R, S*) yielding a 50 : 50 mixture of poly-*R* and poly-*S* isomers on $TiCl_3/AlR_3$, but a random "copolymer" of *R* and *S* when using soluble $TiBr_4/AlBr_3$; (2) the polymerization of 4-methylhexene into an optically active polymer using a $TiCl_4/MgCl_2/AlR_3$ catalyst coordinated with 1-menthyl (*p*-methylphenylbenzoate); and (3) the intriguing observation that syndiotactic stereocontrol may imply 2,1 instead of 1,2 addition.

A further refinement of the picture emerges from the recent work of J. J. Eisch on model systems [*J. Am. Chem. Soc.,* **107**, 7219 (1985)]. This

Fig. 17.

suggests that the transition metal derivative might be in the form of a cationic-like complex counterbalanced by a bonded anionic form of the alkylating metal complex. It is worthwhile stressing that this proposal, nicely complementing that of Rodriguez, fits exactly some conclusions based on an NMR study of the dynamics of η^3-allyl-type catalysts for diene polymerization (see Part II, Section 2 and [6]).

2.6. CHAIN TERMINATION

Except for the presence of impurities that kill active centers, spontaneous limitation of the kinetic chain length (in the steady-state) occurs essentially via the already well known β-elimination reaction (Figs. 19, 20). That process, formally a typical transfer to monomer, is favored by a temperature increase, but this parameter does not allow much flexibility in controlling the final product molecular weight. The same holds for transfer to aluminum alkyl (other possible transfer reactions, e.g. to monomer etc., are less important and practically useless in that regard).

Another type of control has accordingly been developed, i.e. oxidative addition of molecular hydrogen, followed by reductive elimination of the chain with one of the resulting hydrido-ligands. Again a typical transfer process, it will yield an alkyl-terminated chain (instead of a vinyl one) as confirmed experimentally. It represents a very convenient way to precisely

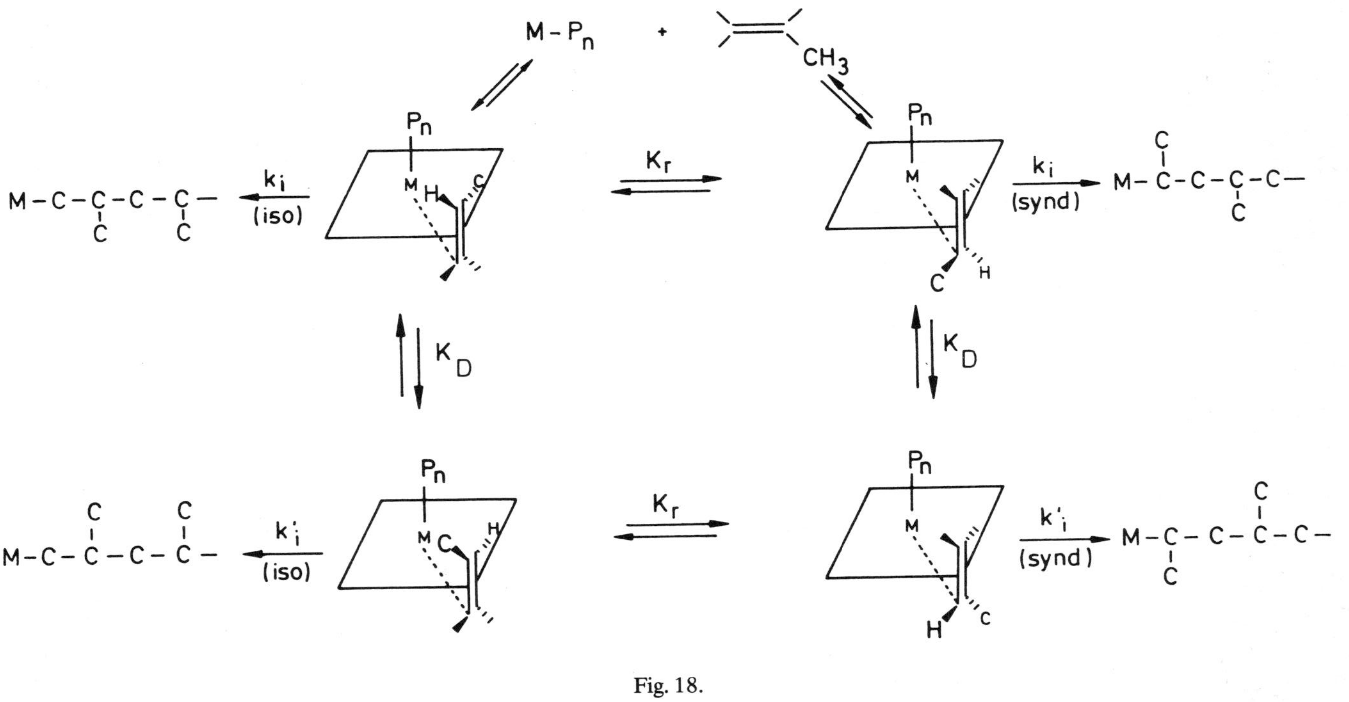

Fig. 18.

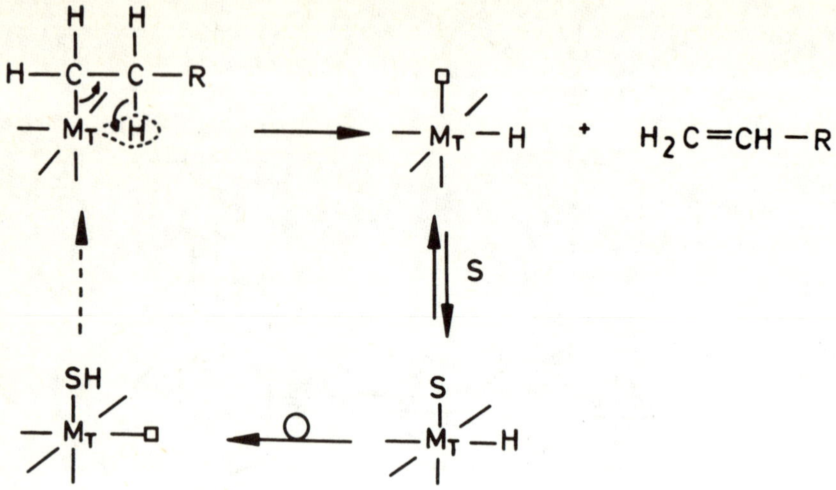

Fig. 19.

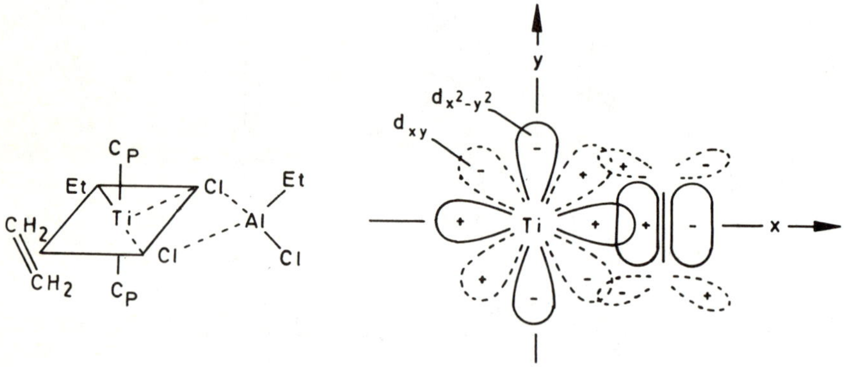

Fig. 20.

control the final molecular weight, since that will depend inversely on the H_2 pressure applied to the reactor, all other conditions remaining unchanged.

3. Comparison with Soluble Catalytic Systems

3.1. ETHYLENE POLYMERIZATION

The complex resulting from the reaction of Cp_2TiCl_2 with AlR_2Cl is a powerful catalyst for homogeneous polymerization of C_2H_4. Its behavior has been first studied by the Olives [*Adv. Polymer Sci.*, **6**, 421 (1969)] who drew a number of interesting conclusions:

— the active entity is a binuclear complex involving both Ti and Al derivative having exchanged one alkyl group (Fig. 21);

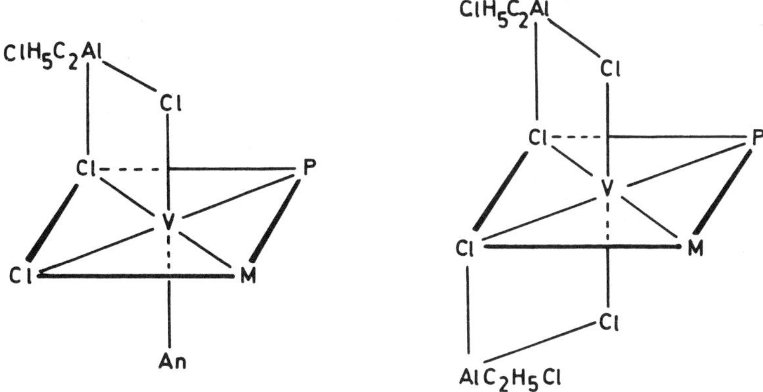

Fig. 21.

— the Ti atom is in an octahedral coordination sphere, containing one vacancy occupied by the monomer. It has to be stressed though that a permanent existence of that vacancy is not essential. Either it can be occupied by another ligand (i.e. solvent, μ-bridged Al-derivative, or even another catalyst molecule), or the Ti complex may be pentacoordinate (i.e. trigonal bipyramid) and switch to an octahedral structure upon monomer binding [D. R. Armstrong, *J. Chem. Soc., Dalton 1*, 1972 (1972)];
— Ti is in a +4 oxidation state as indicated by the parallelism between appearance of an EPR signal due to Ti^{3+} and decrease of catalyst activity (probably due to a bimolecular reduction process);
— a simple kinetic law is obeyed: $V_P = k[Ti][C_2H_4]$, although here again the overall rate changes upon modifying the structure of the Al-alkyl ligand. The polymerization rate increases in the series $Me_2 < MeEt < Et_2$, i.e. in the same relative order as the rate of spontaneous reduction of the complex in absence of the complex in absence of monomer. It is interesting to note that this reduction rate also increases, in the

presence of non polymerizable olefins, in the order trans 2-olefin < 2-*cis*- < 1-olefin, the same as the one of increasing complex formation constants K_f. All these observations indicate of course increasing destabilization of the Ti—R bond. Although the actual situation in that system might be more complicated than described here, these general trends appear to be significant (a recent study by Fink [*Z. Naturforsch.* **40b**, 158 (1985)] has indeed shown that complex stepwise equilibria between active and "dormant" forms might be controlling the chain propagation).

It is obviously rewarding to find here all of the typical features of the Cossee—Rodriguez model, with the additional observation that the actual oxidation state of the M_T is not a determinant parameter (+4 instead of +3 or even +2 for the heterogeneous system) and, on the other hand, that it provided a suitable geometry and electron distribution, which is a well-documented fact in coordination catalysis.

3.2. PROPYLENE POLYMERIZATION

Since the previous titanium-based catalyst was not able to perform α-olefin polymerization, it was obviously interesting to investigate other soluble active systems to further study the problem of stereoregulation. That was done by Zambelli [*Makromol. Chemie,* **112**, 160 (1968)] with the complex obtained from VCl_4 and AlR_2Cl (Fig. 21) in the presence of either an excess of Al derivative or a Lewis base ligand, e.g. anisole.

Again, in agreement with the previously presented models, that coordination structure, which does not exert any important steric compression around the bonded monomer and the growing chain, will yield an atactic polymer under normal conditions (*RT*). However, it yields a stereoregular syndiotactic one at low temperature (*ca.* −70 °C), where the steric barriers due to mutual interaction of the bulky methyl groups carried by the growing chain and the monomers [10] are no longer wiped off by thermally activated rotations; that type of control, however, is only able to produce the lower energy isomer, i.e. the syndiotactic one. Again, it would be interesting to check if that phenomenon is or is not related to a particular type of regioselectivity, i.e. 2,1 versus 1,2 placements.

Finally, it is worthwhile stressing that these catalytic systems have the ability, under well-controlled conditions, to ensure "living" polymerizations of propene, as described by Keii and Doi [9].

3.3. CONCLUSIONS

The bulk of these results have put on the record the general meaning of the mechanistic pictures proposed by different researchers in the field, whatever the system used, soluble or heterogeneous. Of course, the origin (surface or individual ligands) and nature (more or less localized) of the ligand field will differ, offering an additional tool to regulate the energetics of the reaction, as well as its selectivity. They also stress, however, that for simple olefins only solid surfaces have the capability (thanks to a very rigidly imposed steric compression) to produce the higher energy isotactic polymer with a high stereoregularity (98—99%). It is a consistent observation that olefins carrying two potentially coordinating groups, e.g. *o*-methoxystyrene or 2-vinylpyridine, can yield isotactic polymers in solution, and that styrene itself gives a higher than expected proportion of isotactic placements, thanks to metal-aromatic ring interactions; in other words, additional rigidity is ensured here by chelation rather than steric hindrance.

4. Other Related Mechanisms

4.1. ISOMERIZATION POLYMERIZATION

Although Z.N. catalysts do not readily homopolymerize the more abundant 1,2-disubstituted olefins, it has been proposed (G. Lefebvre and Y. Chauvin) to use a dual-function catalyst, $NiX_2/TiCl_3/AlR_3$, able to partly isomerize these monomers into a 1-olefin (on the Ni—Al system) which is then polymerized into a 1-polyolefin in the same reaction medium. This principle works well for 2-butene, but the rate is of course low since it is proportional to 1-olefin concentration, i.e. a few % at thermodynamic equilibrium.

4.2. GREEN'S PROPOSAL

Recently, Malcolm Green has proposed the application to olefin polymerization of the mechanism developed for olefin metathesis reactions (Fig. 22); it is an attractive suggestion, and it might represent the actual mechanism in some particular chain addition processes studied recently. It seems, however, very difficult to apply to classical Ti-based Z.N. heterogeneous catalysts, for a number of reasons (odd Ti intermediate oxidation states, unexplained stereocontrol features, non-relevance for diolefins, etc.).

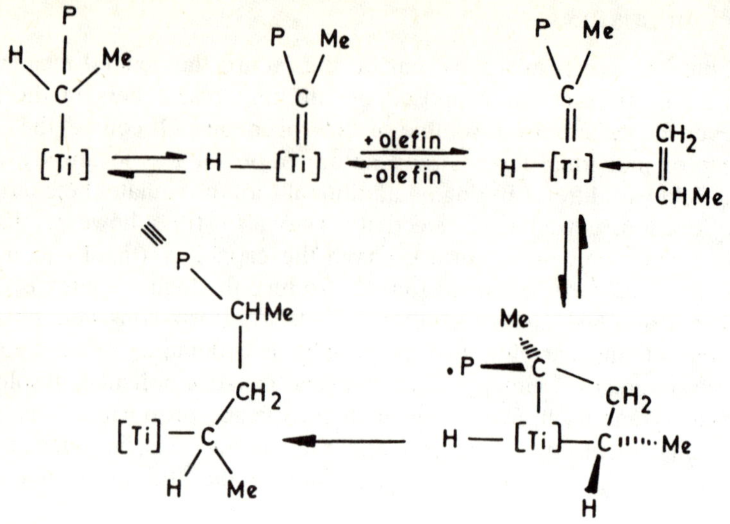

Fig. 22.

PART II: POLYMERIZATION OF DIOLEFINS

1. Polymerization with Z.N. Catalysts

1.1. IMPORTANCE

First patented by S. Horn (B. F. Goodrich), coordination polymerization of diolefins has yielded several industrially important products, as indicated by the approximate data below (Table II).

TABLE II
Polydiene production from 1966 to 1985

Elastomer (10^3 tons/year)	1966	1970	1975	1985
Cis-1,4-polybutadiene (CoX_2/AlR_2Cl)	270	500	725	900
Cis-1,4-polyisoprene ($TiCl_4/AlR_3$) (roughly equivalent to natural rubber from *Hevea* tree)	100	240	500	700

It is interesting to note that besides these two high-performance rubbers, *trans*-1,4-polyisoprene (similar to gutta-percha yielded by the *Balata* tree) has been produced, using a vanadium-based catalyst, as the raw basic material for golf balls. The *cis*-1,4-polymers compete, of course, for politico-economical reasons, with natural rubber, getting a roughly equal share of the market.

1.2. MECHANISM: STRUCTURAL ASPECTS

The four possible stereoregular isomeric polybutadienes have been prepared in a satisfactory state of purity. Their formation can be explained in the frame of a typical Cossee—Rodriguez scheme (see Scheme):

In other words, sites with only 1 coordination vacancy will bind the monomer by one double bond, and it will accordingly put itself in the more stable *S-trans*-conformation, which in turn will be frozen into a non-reversible *trans*-configuration when insertion occurs to C-4 through a 6-membered transition state. If the distance between R and C_2 is shorter (or under the influence of other electronic parameters, i.e. substituents), insertion will proceed through a 4-member transition state to yield a vinyl unit, and that may be either an isotactic or a syndiotactic placement, just as with an α-monoolefin, depending on the amount of steric hindrance (i.e. heterogeneous state, etc.) (see Fig. 23). However, that regioselective choice also depends on electronic factors, such as ligand *trans*-effects and diolefin substitution. Indeed, since the monomer insertion may directly generate an η^3-allyl species, these factors will be determinant in orienting the next M—C σ-bond formation either on C_1 or C_3 (upon returning to the η^1 form under the polymer as illustrated in Fig. 24 [M. Julémont and Ph.Teyssié in *Aspects of Homogeneous Catalysis* (R. Ugo Ed.), D. Reidel Publ. Co, Dordrecht, p. 126 (1981)].

On the other hand, cisoid bidentate coordination of the monomer is well-documented in coordination chemistry, the additional energy required to obtain that conformation being largely compensated by the gain resulting from the formation of two new coordination bonds. A remarkable confirmation of that picture seems to evolve from experiments performed on α-$TiCl_3/AlR_3$ catalysts. The α-$TiCl_3$ surface indeed offers three types of centers: $Cl_4RTi\square$ (1), $Cl_3RTi\square\square$ (2), and $Cl_2RTi\square\square\square$ (3); when occupied further by AlR_2Cl as a ligand, center (1) will be of low activity, while centers (2) and (3) will offer an active situation with 1 and 2 vacant orbitals, respectively. It is rewarding to discover that polymerization on such a catalyst yields a mixture of pure *trans*-1,4 (from site 2) and *cis*-1,4 (from site 3) polybutadienes, with little mixed structure in the same chain,

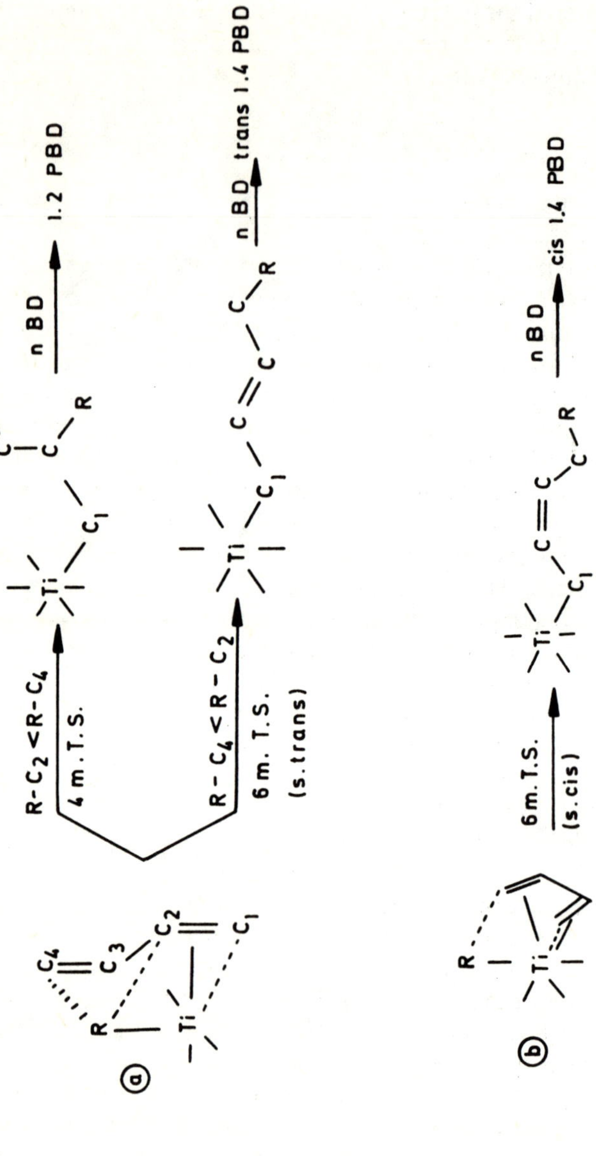

Scheme B

SOLID LINE SINGLE TRANS FORM
DASHED LINE SINGLE CIS FORM
H
C_4 trans
H
H
H trans
H cis
Cl(E)
C_3
C_0
M
H
C_2
H
H
T_i
C_1
H
Cl
Cl

R, Ti AND Cl IN
CHLORINE VACANCIES ABOVE
CHLORINE IONS BELOW
THE PLANE OF DRAWING

C_a
H
H
H
Ti
H
C_4
C_1
C
C
H
H
Cl

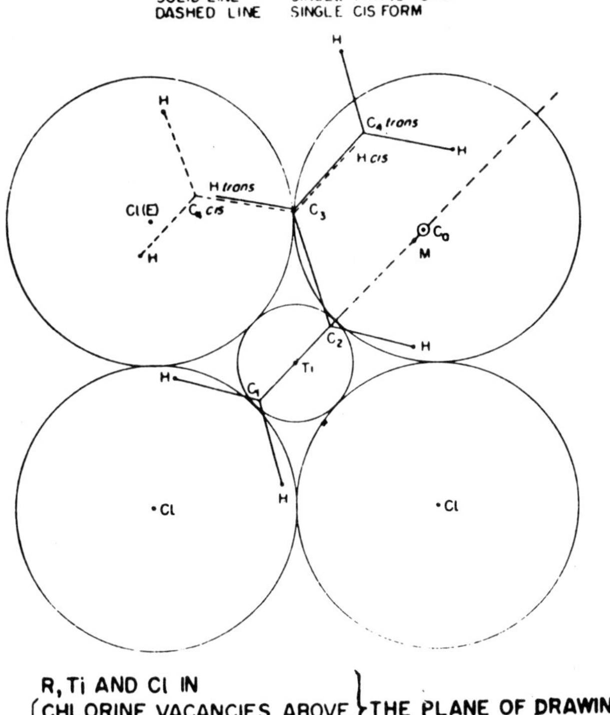

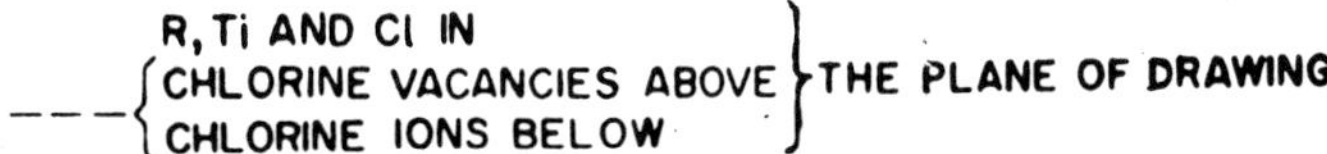

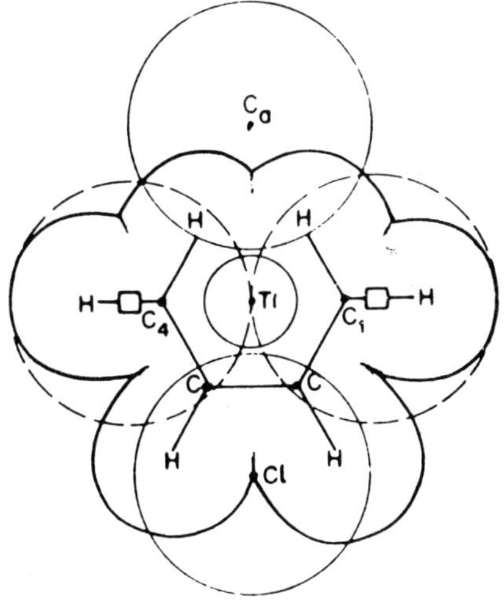

Fig. 23.

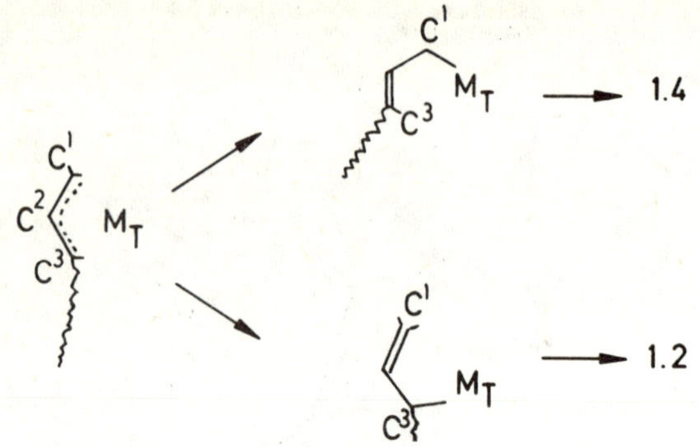

Fig. 24.

while essentially a pure trans-1,4-product is obtained when using the one-vacancy α-$TiCl_3$.

The same approach can obviously be applied to substituted dienes, with however a rapidly increasing complexity: 6 isomers for isoprene, still more isomers, with some having asymmetric carbons in the chain for 1,3 pentadiene, and so on.

1.3. KINETIC ASPECTS

There is, however, a particular aspect of these polymerizations which has to be discussed in more detail. The fact that the first double bond in the growing chain is in a suitable position to coordinate to the required *cis*-vacancy implies that this growing chain will be bound most of the time to the M_T by a so-called π-allyl (or η^3-allyl) structure involving 3 electrons in a bonding and a non-bonding orbital (i.e. 2 coordination positions). That is a stable situation for energetic and entropic reasons, and most of the time the chain will remain in that "dormant" situation; it is only when, statistically, that structure dissociates partially into a π- (or η^1) allyl structure that coordination-insertion of the monomer will take place (Fig. 25). Depending on the mono- or bidentate approach of the monomer an intermediate *syn*- or *anti*-Ti-allyl-isomer will result via the transition state mentioned before [1, 2] giving rise to the corresponding *trans*- or *cis*-isomeric unit after decoordination into the growing chain, unless random isomerization occurs in the mean time. The resulting "dormant" kinetic is typical of these relatively slow polymerizations.

2. η^3-Allyl Model Catalysts and the Concept of 'Chronoselectivity'

In agreement with these basic considerations, isolated η^3-allyl complexes

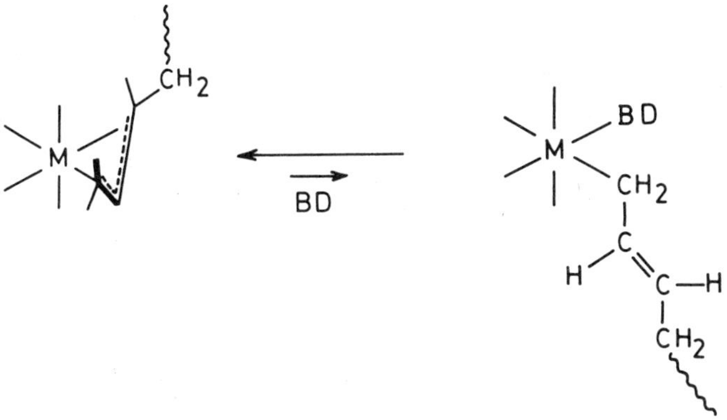

Fig. 25.

were indeed proved to be good initiators for butadiene polymerization, without a need for additional Al-alkyl since the allyl groups already foreshadow the growing chain. Their main structural and kinetic characteristics are summarized here (see also [6] (Fig. 26):

- usually under the form of binuclear complexes (Allyl-M_T^{II}—X)$_2$ they offer no permanent vacancy but can liberate one by going to the corresponding η^1 structure, and still another vicinal one by temporary dissociation of a Ni— bond (as demonstrated by NMR dynamic measurements); the chain growth proceeds by insertion into the Ni—allyl bond, as shown by the use of deuterated monomer;
- the rate of polymerization is essentially proportional (for a given M_T) to the electron withdrawing ability of the counteranion X: the F_3—COO^- anion is particularly efficient;
- the regioselectivity of the polymerization is governed by the nature of the metal: "central" metals like Cr and Mo yield pure 1,2-polybutadiene, while Group VIII metals like Co or Ni yield the 1,4-isomer;
- in a non-interacting medium (paraffinic solvent) one obtains, at a high rate, very pure *cis*-1,4-PBD with the Ni complex (98/99% *cis*, m.p. >0 °C, comparable to the uranium based systems);
- in the presence of electron-donating ligands blocking some coordination positions, i.e. $P(OR)_3$ in stoichiometric ratio to the nickel, one changes the stereochemistry all the way to pure *trans* (again 98—99%; cryst., m.p. *ca.* 140 °C, but produced at a slower rate);
- without any additional ligand but in an interacting solvent, one produces an intriguing, non-thermodynamic, 50 *cis*/50 *trans* "equibinary" microstructure. Still more strikingly, benzene promotes a completely random distribution of the *cis* and *trans* units (Bernoullian statistics),

= "Dormant" Species
pure trans
L = P(OR)$_3$
pure cis
S
C_nH_{2n+2}
k_r
k_i^t
k_i^c
+ BD
k_i^c
k_i^t
Chain propagation
If k_i^x = rates of butadiene insertion,
statistics of placement = f (k_i / k_r)

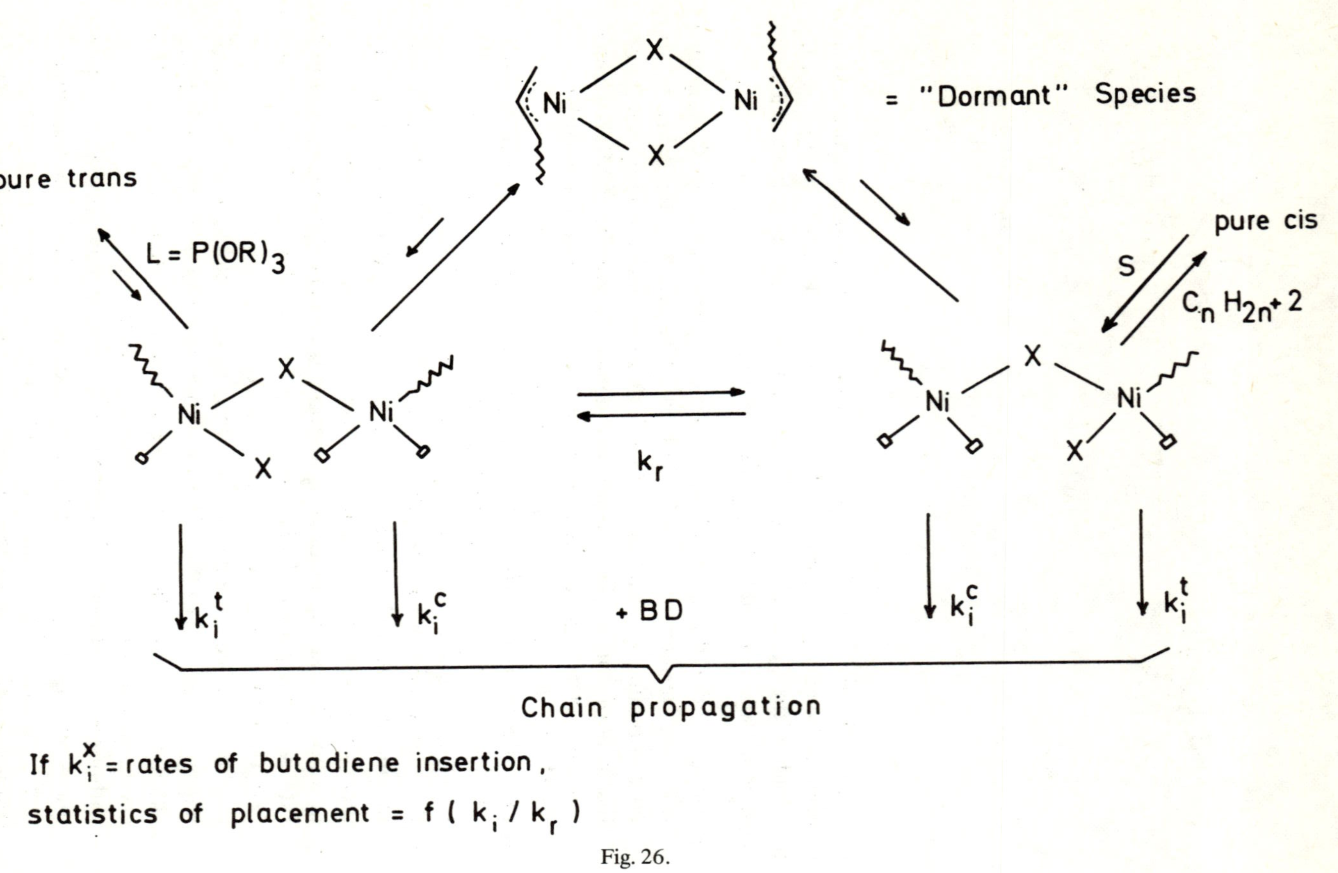

Fig. 26.

i.e. an amorphous polymer, while methylene chloride gives rise to a semicrystalline product which is actually a multistereo-block polymer poly(*cis*-b-*trans*)$_n$-butadiene (1st order Markov). In other words, one sees here the one-step production of a thermoplastic elastomer from one single monomer.

The name "chronoselectivity" has been coined to describe this phenomenon, which represents the first example of a synthetic catalytic "code": indeed, the complex simultaneously controls the regioselectivity, the stereoselectivity, and different statistics of placement (as a function of time) of the chain formation, in a non-thermodynamic manner. A preliminary mechanistic explanation has been suggested (Fig. 26), based on relative insertion and dynamic rearrangement rates in the complex (see also [6]).

— interestingly enough, under well-controlled experimental conditions, these polymerization processes are perfectly "living" and allow the synthesis of interesting block copolymers [Ph. Teyssié *et al.*, *Macromolecules*, **17**, 11 (1984)].

3. Conclusions

All of these experimental results fit very well with the previously proposed mechanisms, and strongly support them; the detailed kinetic and structural data obtained give a pretty clear view of the molecular course of the reaction, although the fundamental reasons underlying the regioselectivity control are still not very well understood.

Other mechanistic pathways have been proposed, in particular a so-called outer-sphere mechanism (similar to a pericyclic allylic transposition) by Hugues and Powell. Although they are probably relevant to some stoichiometric reactions of η^3-allyl complexes with dienes, they do not seem to fit the characteristic experimental features of these dienes coordination polymerizations.

PART III: HOMO- AND COPOLYMERIZATION OR OTHER TYPES OF MONOMERS

Although Z.N. catalysts are very sensitive to polar substituents which tend to clock active sites, coordination polymerization by modified complexes is of course not limited to unsaturated hydrocarbons. A few examples are discussed hereafter which have put in evidence interesting new concepts.

1. Polar Vinyl Monomers

1.1. THE CATALYTIC PROCESS

Although $TiCl_3$- based Z.N. catalysts are in this case inactive, A. Yamamoto has shown [*J. Am. Chem. Soc.*, 5989 (1967)] that it is possible to tailor the ligand field around the transition metal so that it could tolerate strongly coordinating substrates. More precisely, the reaction of $Fe(AcAc)_3$ with AlR_2OR in the presence of bipyridyl has produced a soluble iron alkyl bis(bipyridyl) complex, able to polymerize (meth-)acrylates, vinyl acetate, vinyl ethers and even (meth-)acrylonitrile. From kinetic data (rates and competitions) and structural determinations, it can be concluded that a typical coordinative *cis*-insertion mechanism is operative, wherein chelation of the chain (and maybe of the monomer) ensures stereoselection (i.e. production of isotactic PMMA).

1.2. THE CONTROL OF APPARENT REACTIVITY IN COPOLYMERIZATION

The most interesting finding, however, was the fact that in copolymerization experiments, the apparent reactivity ratios (r_1 and r_2) are directly related (Fig. 27) to the respective equilibrium formation constants of the corresponding complexes (k_f of the monomer—iron complex, as determined independently). In simple words, the monomer which is more strongly bonded to the M_T is incorporated preferentially, even if its rate of insertion is not the higher one. That situation can be summarized as shown in the Scheme.
This is particularly interesting, since it gives the opportunity to purposely modify the relative reactivity ratios of the two comonomers (r_i being a combination of a thermodynamic constant Kf_i, and a kinetic one ki_i) just by changing the ligand field around the metal: indeed, changing ligands, or changing their interaction by competing solvents, will modify Ki and k_i differently. That is a new and unique possibility of controlling apparent relative reactivities, which has also been applied in organic synthesis.

2. Oxiranes

E. Vandenberg [*J. Polymer Sci.*, **A1**(7), 525 (1969)] and ourselves [*Int. Rev. of Sciences, Phys. Chem. Sci.*, **2**(8), 191 (1975)] have successively described new catalytic systems for the ring-opening polymerization of oxiranes into high molecular weight (10^6D) polyethers. Based on Al

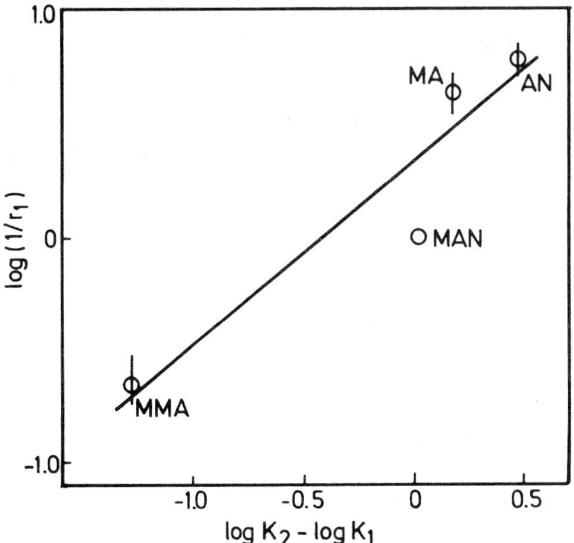

Copolymerization of methacrylonitrile with olefins. M_1 = methacrylonitrile as a reference. K represents the stability of the nickel-olefin π complex.

Fig. 27.

$$[C] \underset{K_{f_1}}{\overset{M_1}{\rightleftharpoons}} [CM_1] \xrightarrow[k_{i_1}]{\bigcirc} [CP_1] \cdots$$

$$[C] \underset{K_{f_2}}{\overset{M_2}{\rightleftharpoons}} [CM_2] \xrightarrow[k_{i_2}]{\bigcirc} [CP_2] \cdots$$

Scheme C

and/or Zn μ-oxo-polynuclear structures, these catalysts obey a typical coordinative mechanism which can be summarized as shown in the Scheme (for the simple Al-derivative).

Although it involves very different monomers and bonding types, it offers a complete mechanistic similitude with the mechanism described in

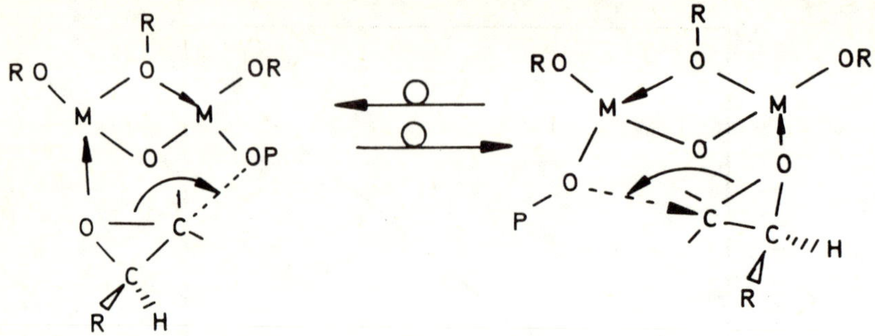

Scheme D

Part I for monoolefins: coordination of the monomer, electronic rearrangement (6-membered transition state) with the growing chain in a vicinal position, flip-flop alternating mechanism, stereoselectivity in the ring-opening (isotactic polymethyloxirane can indeed be produced), and enantioselective choice if optically active R″ groups are used (with the Zn catalysts). It is a general process, which can also be extended to thiiranes and lactones. One must also note that partially crystaline high molecular weight polymethyloxirane might be a very attractive general purpose elastomer, provided a cheap direct synthesis of the monomer can be found.

Here again, relative reactivities can be tailored by changing the coordinative conditions, and a particularly interesting example involves the copolymerization of epichlorohydrin, ECH ($H_2C\overset{O}{\frown}CH{-}CH_2Cl$) with methyloxirane (MO). Although MO homopolymerizes much faster than ECH (300×), the ECH is incorporated preferentially when copolymerization takes place in a non-polar (paraffinic) solvent. Obviously, in practice, the overall process is as slow as ECH homopolymerization. In polar medium, however, (e.g. *o*-dichlorobenzene), that difference is erased, and the more reactive MO is now prefered, in an overall high-rate process. Again, a fine modulation of these relative reactivities is possible, i.e., the use of chlorobenzene promotes a practically azeotropic copolymerization (Fig. 28).

GENERAL CONCLUSION

It is fair to say that coordination polymerization represents one of the

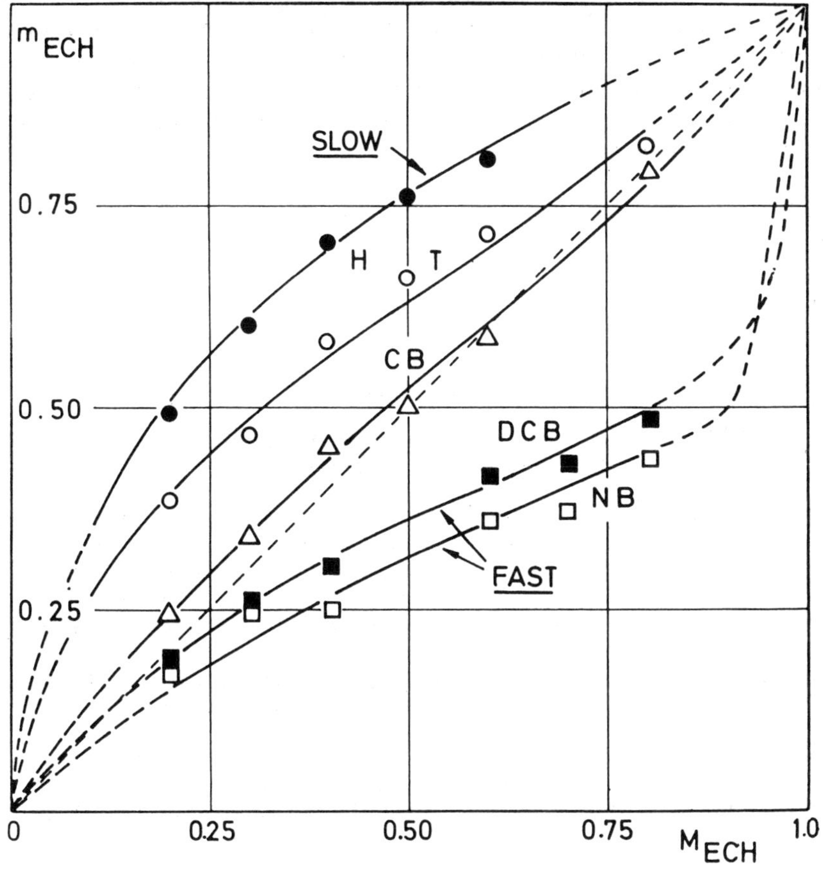

Fig. 28.

most exciting and fruitful chemical ventures of the last three decades, from both basic research and industrial development points of view. In particular, Ziegler—Natta catalysis is one of the most studied reactions in fundamental research and one of the most productive processes in industry (*ca.* 10^7 tons a year of excellent materials produced in 2—3 steps from petroleum).

Despite all that tremendous research and development activity, much remains to be done. Typical areas where substantial progress is still expected include: more detailed visualization of the mechanism (especially through study of the catalyst dynamic behavior *in situ*); better control of the termination and transfer reactions; achievement of higher rates for highly stereoselective systems (also including longer stabilization); mastering of the copolymerization processes in terms of apparent reactivity ratios and monomer distributions; devising catalysts for polar monomers homo- and co-polymerization; etc. Obviously, the wealth of the field is far from being exhausted!

Laboratory of Macromolecular Chemistry and Organic Catalysis
University of Liège
Sart Tilman
B-4000 Liège, Belgium

References

1. L. Rodriguez and H. Van Looy, *J. Polymer Sci.*, **A1**(4) 1905—1971 (1966).
2. P. Cossee, in *Stereochemistry of Macromolecules* (Ed. A. D. Ketley), Vol. 1, M. Dekker (1967).
3. T. Keii, *Kinetic of Ziegler—Natta Polymerizations*, Chapman & Hall (1972).
4. F. Dawans and Ph. Teyssié, in *Stereo Rubbers* (Ed. W. Saltman), J. Wiley (1978).
5. J. Boor, *Ziegler—Natta Catalysts and Polymerizations*, Academic Press (1979).
6. Ph. Teyssié, A. Devaux, P. Hadjiandreou, M. Julémont, J. M. Thomassin, E. Walkiers and R. Warin, in *Preparation and Properties of Stereoregular Polymers* (Ed. R. Lenz and F. Ciardelli), D. Reidel (1980).
8. P. Galli, High yield catalysts in olefins polymerization: general outlook of theortical aspects and industrial uses', 28th *IUPAC Symposium on Macromolecules*, Amherst (Mass.) 1982.
9. Lectures presented at the international symposium *Transition Metal Catalyzed Polymerizations*, R. Quirk Ed., MMI Press Symp. Ser., **4** (1983).
10. Z. Zambelli, I. Pasquon, R. Signorini and G. Natta, *Makromol, Chem.*, **112**, 160 (1968).

See also: papers in the scientific litterature by D. Ballard, J. W. Chien, L. Giannini, G. and S. Olivé, P. Pino, L. Porri, H. Sinn and Kaminsky, P. Tait, etc; industrial laboratories, i.e. Mitsui (in Japan), Montepolimeri (in Italy), Solvay (in Belgium), Union Carbide, Shell, Exxon, Du Pont (in the U.S.A.), etc., also publish interesting patents and papers.

ANDRÉ MORTREUX AND FRANCIS PETIT

OLEFIN METATHESIS AND RELATED REACTIONS

1. Introduction

One of the most interesting reactions studied within the last two decades is the metathesis of unsaturated compounds which, in the case of alkenes, can be written as follows:

$$2\ R_1CH{=}CHR_2 \rightleftarrows R_1CH{=}CHR_1 + R_2CH{=}CHR_2$$

This catalytic reaction alone has been the subject of six international Symposia since 1975; the last was held in Hull in August 1987. Such a great deal of work done by many groups all over the world has finally resolved the more intimate details of this reaction, at least for some of the catalytic combinations among the plethora of systems able to undergo this transformation.

This reaction was discovered in 1931 by Schneider and Frölich [1] who established that, at 725 °C, propylene could be thermally converted into ethylene and butenes:

$$2\ CH_3CH{=}CH_2 \rightleftarrows CH_3CH{=}CHCH_3 + CH_2{=}CH_2$$

In 1964, Bank and Bailey [2] reported the first catalytic reaction on the same substrate, using heterogeneous catalysts derived from molybdenum or tungsten hexacarbonyles supported on alumina.

The first homogeneous catalytic system was discovered in 1967 by Calderon [3] using a 'WCl_6—EtOH—$AlEtCl_2$' mixture, working at ambient temperature, with 2-pentene and β-olefins as substrates.

$$2\ CH_3CH{=}CHC_2H_5 \rightleftarrows CH_3CH{=}CHCH_3 + C_2H_5CH{=}CHC_2H_5$$

The catalytic dismutation of olefins is thermoneutral and results from the rupture and reformation of carbon—carbon double bonds with a statistical redistribution of the alkylidene moities. It has also been extended to substituted (functionnalized) alkenes, diolefins, polyenes and alkynes. The reaction with alkenes, to produce macrocyclic polyenes and polymers, is recognized as a special case of disproportionation.

A. Mortreux and F. Petit (Eds.), Industrial Applications of Homogeneous Catalysis, 229—256.

At the same time, widespread interest from the organometallic and mechanistic viewpoints has been reflected by an increasing number of publications relevant to the synthesis of organometallic complexes in connection with the alkylidene exchange observed during catalysis.

The object of this paper is to describe the scope of this reaction, and to enumerate the theories which have been advanced to explain this catalysis, with particular emphasis on the nature of the catalysts and their formation in homogeneous systems. Finaly a brief description of some industrial applications of this reaction will be given.

2. Scope of the Reaction

2.1. ACYCLIC MONOOLEFINS

Although β-olefins are generally inert towards homopolymerization involving addition across the double bond (Ziegler—Natta Catalysis), they are metathesized almost as readily as α-olefins. This reaction thus provides a useful route for the synthesis of new alkenes according to the general equilibrium.

$$2\,R_1CH{=}CHR_2 \rightleftarrows R_1CH{=}CHR_1 + R_2CH{=}CHR_2$$

A terminal or an internal acyclic olefin will, on reaction with itself, give rise to symmetrical internal alkenes. Removal of ethylene from the final mixture of a terminal olefin affords high yields of the other alkene product.

$$2\,CH_2{=}CHR \rightleftarrows CH_2{=}CH_2\uparrow + RCH{=}CHR$$

The rate of reaction of olefins in the metathesis reaction is sterically controlled, as demonstrated by the decrease along the series [4]:

$$H_2C{=} > RCH_2CH{=} > R_2CH{-}CH{=} > R_2C{=}.$$

Also the "cross" metathesis between acylic olefins and bulky cycloalkenyl olefins has been reported.

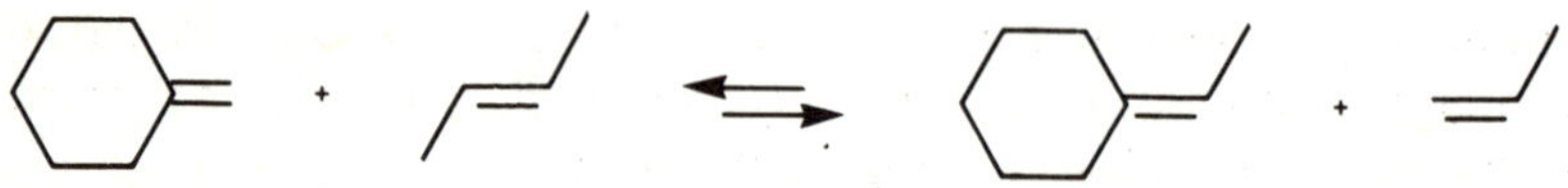

2.2. ACYCLIC POLYOLEFINS

Acyclic dienes allow the possibilities of inter- and intramolecular reactions, the extent of each being governed by the relative thermodynamic stabilities of ring and chain products.

For intramolecular reactions[5]:

$$CH_2{=}CH-(CH_2)_4-CH{=}CH_2 \longrightarrow C_2H_4 + \text{cyclohexene}$$

$$(CH_2)_4(CH{=}CH_2)_2 \longrightarrow \left[(CH_2)_4 \begin{matrix} CH\cdots CH_2 \\ \vdots \quad\quad \vdots \\ CH\cdots CH_2 \end{matrix}\right] \longrightarrow C_2H_4 + \text{cyclohexene}$$

For intermolecular reactions:

$$2\ CH_2{=}CH(CH_2)_4CH{=}CH_2 \longrightarrow CH_2{=}(CH_2)_4CH{=}CH(CH_2) + CH{=}CH_2 + C_2H_4.$$

Conjugated dienes have also been claimed to be metathesized in the heterogenous phase over 'WO_3/SiO_2' catalysts at high temperature [6].

$$2\ \text{butadiene} \xrightarrow{WO_3/SiO_2} \text{ethylene} + (\text{hexatriene}) \downarrow C_6H_6 + \text{cyclohexadiene}$$

2.3. CYCLIC OLEFINS

Cycloalkenes should theoretically yield cyclic dienes initially and then produce further large macrocyclic polyenes. However, the presence of trace amounts of acyclic olefins explains why the high molecular weight fractions observed are acyclic hydrocarbons, obtained by cross metathesis, rather than cyclic polymers [7].

Cyclic alkenes, except cyclohexene, from C_4 to C_{12}, yield polyalkenes which vary from amorphous elastomers to crystalline materials. The ring-opening of cyclopentene can be controlled to yield polypentene, in either a high *cis* or a high *trans* form. Moreover, perfectly alternating copolymers can be produced from substituted cycloolefins [7].

A detailed study of the polymerization of cyclooctene using 'WCl_6—$AlEtCl_2$' as catalyst showed that the mixtures contained low as well as high molecular weight species, recognized as macrocycles. Further catalytic treatment of the high molecular weight species results in the production of low molecular weight macrocycles. Thus an equilibrium involving macrocyclization of the parent compound must be envisaged.

$$2\ (CH_2)_n\begin{matrix} \diagup CH_2 \diagdown CH \\ \parallel \\ \diagdown CH_2 \diagup CH \end{matrix} \rightleftharpoons \begin{matrix} CH_2-CH=CH-CH_2 \\ (CH_2)_n \qquad\qquad (CH_2)_n \\ CH_2-CH=CH-CH_2 \end{matrix}$$

The polyalkene species obtained by ring opening polymerization of cyclic olefins range from amorphous elastomers to crystalline materials, depending on the structure of the repeat units and the configuration about the C=C bond.

For example, ring opening of 5-methylcyclooctene by 'WCl_6—$AlEtCl_2$' gives a polymer with the following repeating unit:

$$\mathord{+}(CH_2-CH=CH-CH_2)-(CH_2-CH_2)-(\underset{\displaystyle CH_3}{\underset{|}{CH}}-CH_2)\mathord{+}_n$$

This is equivalent to the alternating copolymer of butadiene, ethylene, and propylene [7].

Under suitable conditions (high dilution), low molecular weight macrocycles can be obtained by ring opening polymerization. This provides a useful route to intermediates in the preparation of perfume base [8].

2.4. CYCLIC DIOLEFINS

Provided the double bonds are not conjugated, cyclic dienes undergo ring opening polymerization [9]. The metathesis of cyclopentadiene, 1,3-cyclooctadiene and 1,3,5-cycloheptatriene has not been achieved.

$$\longrightarrow \; -\!\!\left[CH=CH-(CH_2)_3\right]_n\!\!-$$

2.5. CROSS METATHESIS BETWEEN CYCLIC AND ACYCLIC ALKENES

This reaction is a convenient method for the synthesis of polyenes. Generally the acyclic olefin used is ethylene, which by interaction with cyclopentene, cyclooctene or cyclohexene gives 1,6-heptadiene, 1,9-decadiene and 1,7-octadiene, respectively. Most studies in this field have involved heterogeneous systems like '$Mo(CO)_6$—Al_2O_3' or 'CoO—MoO_3—Al_2O_3' [10].

With homogeneous systems 1,5,9-decatriene can be obtained by treatment of 1,5-cyclooctadiene with ethylene, and telomers are produced by interaction of cyclooctene with propylene [5].

$$CH_2\!=\!\left[CH(CH_2)_6CH\right]_n\!=\!CH_2$$

$$CH_3-CH\!=\!\left[CH(CH_2)_6CH\right]_m\!=\!CH-CH_3$$

$$CH_2\!=\!\left[CH(CH_2)CH\right]_p\!=\!CH-CH_3$$

2.6. ALKYNES

Disproportionation of alkynes has also been observed. The first catalysts used were heterogeneous, essentially WO_3 and MoO_3 deposited on silica [11].

$$2\,R_1C{\equiv}CR_2 \rightleftharpoons R_1C{\equiv}CR_1 + R_2C{\equiv}CR_2$$

Under these conditions, terminal acetylenic hydrocarbons cyclotrimerize, the selectivity of metathesis products being very low. However, a cobalt containing heterogeneous catalyst has been claimed to metathesize terminal as well as internal alkynes [12].

$$2\,C_6H_5C{\equiv}CH \rightleftarrows C_6H_5C{\equiv}C{-}C_6H_5 + HC{\equiv}CH$$

In addition, metathesis of 2-methyl-1-buten-4-yne occurs both at the double and the triple bond with the same catalyst.

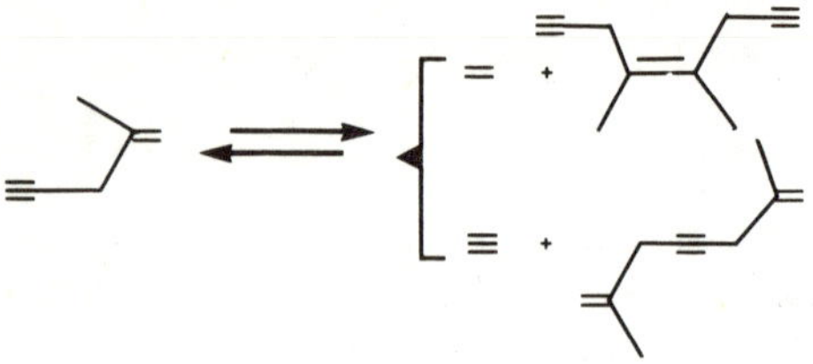

Only internal alkynes are presently metathesized with homogeneous systems, formed from Mo, W and Re complexes or salts. These reactions give an equilibrium mixture of the disubstituted alkynes arising from a statistical distribution of the alkylidyne moieties [13].

2.7. FUNCTIONAL OLEFINS

Some examples of the metathesis of olefins bearing functional groups have been reported. A patent has claimed that the reaction of acrylonitrile with propene gives crotonitrile and ethylene [14].

$$CH2{=}CHCN + CH_2{=}CHCH_3 \rightleftarrows CH_2{=}CH_2 + NCCH{=}CHCH_3$$

Apart from halogen substituted alkenes, heteroatoms normally deactivate catalytic systems. However the synthetic utility of such reactions has encouraged further research in this field. Chlorine substitution at vinylic positions deactivates the double bond but halogen substituted alkenes in which the double bond is in an ω position undergo cross metathesis with internal alkenes. For example, 5-bromo-1-pentene undergoes cross metathesis with 2-pentene. Unsaturated compounds containing ester groups also react, e.g. methyl-9-octadecenoate is converted to 9-octadiene and dimethyl-9-octadecenediodate by the '$WCl_6{-}Sn(CH_3)_4$' catalytic combination [15].

$$2\ CH_3(CH_2)_7CH{=}CH(CH_2)_7COOCH_3 \rightleftarrows CH_3(CH_2)_7CH{=}CH(CH_2)_7CH_3 + CH_3OOC(CH_2)_7CH{=}CH(CH_2)_7COOCH_3$$

Thus two factors influence the reactivity of such compounds, the nature of the functional group and the position of the heteroatom. At present, the double bond must be separated from the functional group by at least one methylene unit for sucessful results. Even in these cases the turnover numbers and the reaction rates are generally low (TN < 200).

Of great potential synthetic value is the cometathesis of various unsaturated esters with ethylene to produce ω-unsaturated esters. Thus, methyl-9-decenoate was obtained by co-metathesis of ethylene and methyloleate [16]:

$$CH_3(CH_2)_7CH{=}CH(CH_2)_7CO_2CH_3 + C_2H_4 \rightleftarrows CH_3(CH_2)_7CH{=}CH_2 + CH_2{=}CH(CH_2)_7CO_2CH_3$$

Other examples of reactions between ethylenic hydrocarbons and olefinic esters include the ring opening of cyclooctene [17].

$$\text{cyclooctene} + CH_3CH{=}CHCH_2CO_2Et \rightleftharpoons CH_3CH{=}CH(CH_2)_6CH{=}CHCH_2CO_2Et$$

Unsaturated ethers, nitriles, amines and silylated olefins can also be metathesized but the activity of these reactions is low compared to that observed with normal, α- and β-olefins.

3. Catalysts

One of the reason why metathesis reactions have received so much attention during the last 20 years is that they can be performed with either heterogeneous or homogeneous catalysts, thus providing a useful tool for research into the relationships between these two types of catalysis. Most of these catalytic systems contain tungsten, molybdenum or rhenium compounds in different oxidation states.

3.1. HETEROGENEOUS CATALYSTS

Banks and Bailey [2] discovered the first heterogeneous catalyst in 1964: molybdenum hexacarbonyl deposited on alumina. Since this time a plethora of other systems have been discovered.

A basic heterogeneous catalyst consists of a transition metal (Mo, W or Re) carbonyl or oxide deposited on a high specific area support such as alumina or silica. Pretreatment of the catalyst by heating it to a high temperature (120 °C for carbonyls and 550°C for oxides) in a prepurified nitrogen atmosphere is essential. Similarly, the olefin must be very pure in order to insure a good catalyst lifetime. The reaction temperatures vary from 180—200 °C ('MoO_3—Al_2O_3' and '$Mo(CO)_6$—Al_2O_3' catalysts) to 400 °C ('WO_3—SiO_2' catalysts). However, 'Re_2O_7—Al_2O_3' combinations, prepared as described above, are active at temperature as low as 25 °C.

In order to avoid side reactions, particularly isomerization, alkaline ions can be added during the impregnation stage of catalyst preparation. By this method, 1-butene production can be greatly reduced during metathesis of propylene and thus avoid the synthesis of other olefins by cross disproportionation between the "normal" metathesis products and the isomerized products.

3.2. HOMOGENEOUS CATALYSTS

The first homogeneous catalyst described in the literature is 'WCl_6—EtOH—$AlEtCl_2$' discovered by Calderon in 1967 [3]. Since that time, numerous other combinations have been found. Each system comprises a transition metal salt or compound ([WCl_6], [$W(CO)_6$], [$MoCl_5$], [$ReCl_5$] etc.) associated with a cocatalyst, generally a reducing agent (RLi, RMgX, AlR_3, AlR_xCl_{3-x}, $LiAlH_4$, $Sn(CH_3)_4$ etc.).

With the 'WCl_6—EtOH—$AlEtCl_2$' system as catalyst, equilibrium is quickly obtained using a W : Al ratio of 1 : 4 and W : olefin ratio of 1 : 10 000. Neither constituent of this mixture catalyses metathesis alone. The same observation has also been made with the other systems, which are generally less reactive but more stable.

It must also be noticed that generally tungsten hexachloride derived catalysts are reactive towards β-olefins, whereas molybdenum salts are most widely used with α-olefins. In each case however, the main observation to be made (as with heterogeneous catalysts) is that it is necessary to have a cocatalyst besides the transition metal salt. This fact must be taken in account where mechanistic studies are undertaken on the generation of the active species for these reactions.

4. Mechanism of the Reaction

4.1. TRANSALKYLATION OR TRANSALKYLIDENATION?

The first question which arises about the mechanism of metathesis reactions is whether or not the double bond is broken, and thus whether the reaction is a transalkylidenation or a transalkylation process.

Early work using ^{14}C and deuterium labelled propene and butene quickly established that two olefinic bonds are switched between four carbon atoms, the integrity of all other bonds within the alkenes remain intact [18].

$$2\,C{-}\overset{*}{C}{=}C \rightleftarrows C{=}C + C{-}\overset{*}{C}{=}\overset{*}{C}{-}C$$

A similar mechanism has been postulated for reactions involving cycloalkenes. It has been shown that a copolymer of cyclooctene and cyclopentene (labelled on its double bond with ^{14}C), is predominantly labelled in the C_5 fragment, this result being obtained by reductive ozonolysis of the polymer [9]. The radioactivity found in the C_5 diol proves that ring opening polymerization also proceeds *via* scission on the double bonds rather than cleavage of carbon—carbon single bonds.

$$\text{cyclooctene} + \text{cyclopentene}^{*} \longrightarrow \left[CH(CH_2)_6CH{=}\overset{*}{C}H(CH_2)_3\overset{*}{C}H(CH_2)_6CH \right]_n$$

A transalkylidynation process must again be envisaged for the disproportionation of alkynes, as shown by ^{13}C NMR experiments, the same mechanism being indicated for both heterogeneous ('MoO_3—SiO_2') and homogeneous ('$Mo(CO)_6$—C_6H_5OH') catalysts [20, 13].

$$2\,Ar{-}\overset{*}{C}{\equiv}C{-}Bu \rightleftarrows Ar{-}\overset{*}{C}{\equiv}\overset{*}{C}{-}Ar + Bu{-}C{\equiv}C{-}Bu$$

4.2. PAIRWISE MECHANISMS

To account for this transalkylidenation process a concerted pairwise exchange of alkylidene moieties was initially proposed, a "quasi cyclobutane" transition state being envisaged. According to Woodward—Hoffman rules, the synchronous formation and breaking of double bonds in this manner is thermally forbidden. However, the interaction of the metal *d* orbitals with the alkene orbitals in the transition state, was once thought to reduce the symmetry requirements of such a transformation.

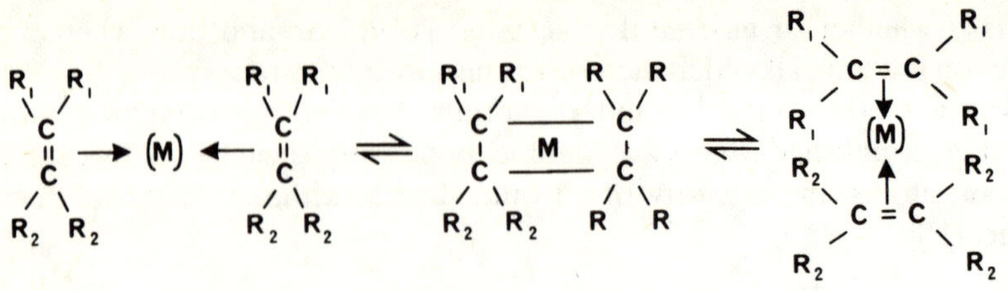

A reaction once though to proceed by such a mechanism is the $[Rh(CO)Cl]_2$ catalysed isomerization of quadricyclene into norbornadiene [22]. However, this reaction now appears to proceed by a stepwise mechanism involving oxidative addition of the substrate to a Rh(I) species.

A mechanism of this type would suggest the possibility of producing cyclobutanes in metathesis reactions since thermodynamic calculations do not preclude their formation. However cyclobutanes have not been observed in such reactions. As a consequence of these arguments a second concerted pairwise mechanism has been proposed where the transition state can best be described as a tetraalkylidene complex [23].

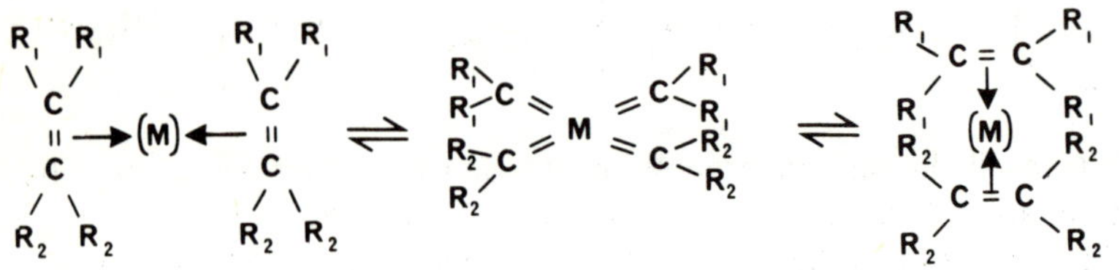

4.3. METALLACYCLOPENTANES AS INTERMEDIATES

A third mechanism has been suggested by Grubbs *et al.* [24] on the basis of reactions conducted between $[WCl_6]$ and deuterated dilithiobutane.

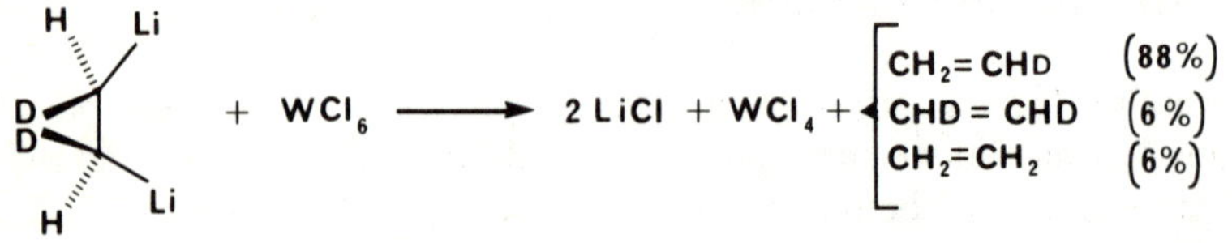

$$+ WCl_6 \longrightarrow 2\,LiCl + WCl_4 + \begin{cases} CH_2{=}CHD & (88\%) \\ CHD{=}CHD & (6\%) \\ CH_2{=}CH_2 & (6\%) \end{cases}$$

The major organic product observed is the expected monodeuterated ethylene, formed from the decomposition of a metallacyclopentane thought to be the initial product of the reaction.

$$2\ CH_2{=}CHD \rightleftharpoons \begin{array}{cc} CHD & - & CHD \\ | & & | \\ CH_2 & & CH_2 \\ & (W) & \end{array} \rightleftharpoons \begin{array}{c} CH_2=CH_2 \\ + \\ CHD=CHD \end{array}$$

$$\begin{array}{ccc} CH_2 & & CH_2 \\ \| \rightarrow & (M) & \leftarrow \| \\ CHR & & CHR \end{array} \rightleftharpoons \begin{array}{ccc} CH_2 & - & CH_2 \\ | & & | \\ RCH & & CHR \\ & (M) & \end{array} \rightleftharpoons \begin{array}{ccc} CH_2 & & CHR \\ \| \rightarrow & (M) & \leftarrow \| \\ CH_2 & & CHR \end{array}$$

The small but significant quantities of undeuterated and dideuterated ethylene also produced can be explained by the formation and subsequent decomposition of a second metallacyclopentane complex.

4.4. NON-PAIRWISE MECHANISMS

These mechanisms involve an odd number of alkylidene units in the intermediates. Here, metallacarbenes react with alkenes to give intermediate metallacyclobutanes. These interconversion require the minimum number of elementary steps for transalkylidenation.

$$\begin{array}{cc} CHR_1 & \\ \| & CHR_2 \\ (M) \leftarrow & \| \\ & CHR_3 \end{array} \rightleftharpoons \begin{array}{ccc} CHR_1 & - & CHR_2 \\ | & & | \\ (M) & - & CHR_3 \end{array} \rightleftharpoons \begin{array}{c} CHR_1=CHR_2 \\ \downarrow \\ (M)=CHR_3 \end{array} \rightleftharpoons \begin{array}{c} CHR_1=CHR_2 \\ + \\ (M)=CHR_3 \end{array}$$

$$\begin{array}{c} (M)=CHR_3 \\ \nwarrow \\ CHR_2=CHR_3 \end{array} \rightleftharpoons \begin{array}{ccc} (M) & - & CHR_3 \\ | & & | \\ CHR_2 & - & CHR_3 \end{array} \rightleftharpoons \begin{array}{c} (M)=CHR_2 \\ \nwarrow \\ CHR_3=CHR_3 \end{array} \rightleftharpoons \begin{array}{c} CHR_3=CHR_3 \\ + \\ (M)=CHR_2 \end{array}$$

$$\begin{array}{c} (M)=CHR_2 \\ \uparrow \\ CHR_3=CHR_2 \end{array} \rightleftharpoons \begin{array}{ccc} (M) & - & CHR_2 \\ | & & | \\ CHR_3 & - & CHR_2 \end{array} \rightleftharpoons \begin{array}{c} (M)=CHR_3 \\ \nwarrow \\ CHR_2=CHR_2 \end{array} \rightleftharpoons \begin{array}{c} (M)=CHR_3 \\ + \\ CHR_2=CHR_2 \end{array}$$

This mechanism was first proposed by Y. Chauvin [25]. Lappert *et al.* [26] subsequently showed that the electron rich olefin **1** could be disproportionated with rhodium catalysts and the alkylidene complexes **2** isolated.

$$\left[\begin{array}{c}R_1\\ | \\ N \\ \end{array}\right]C=C\left[\begin{array}{c}R_2\\ | \\ N\end{array}\right] \quad (N-R_1,\ N-R_2) \qquad \mathbf{1}$$

$$RhCl(PPh_3)=C\left[\begin{array}{c}N-R\\ N-R'\end{array}\right] \qquad \mathbf{2} \qquad R=R'=R_1 \text{ or } R_2$$

This mechanism possesses many attractive unifying features.

4.4.1. *Generation of stable carbenes*

It is possible to synthesise stable alkylidene complexes from the reaction between transition metal salts and metal alkyl reducing agents. These conditions are similar to those in some metathesis reactions. For example a Ta alkylidene can be produced in the following manner [27, 28]:

$$\left[Ta(CH_2{}^tBu)_2(O{}^tBu)X_2\right] \xrightarrow[CH_2Cl_2]{2PMe_3} \left[Ta(CH_2{}^tBu)(O{}^tBu)(PMe_3)_2X_2\right] \quad \mathbf{3}$$

$$\mathbf{3} + 2\,LiO{}^tBu \longrightarrow \left[Ta(=CH{}^tBu)(PMe_3)(X)(O{}^tBu)_2\right] \quad \mathbf{4}$$

Complex **4** has been shown to metathesize 2-pentene to 2-butene and 3-hexene.

4.4.2. *Model reactions*

Model reactions have been performed on tungsten carbene complexes, proving that a reaction between a carbene complex and an olefin can produce metathesis products [29]:

$$(CO)_5W=C(Ph)(Ph) + CH_2=C(Ph)(OCH_3) \longrightarrow (CO)_5W=C(Ph)(OCH_3) + CH_2=C(Ph)(Ph)$$

with α-olefins, cyclopropanes are also produced.

$$(CO)_5W{=}C(Ph)_2 + CH_2{=}CR_1R_2 \longrightarrow \text{cyclopropane } (CH_2, CR_1R_2, CPh_2) + CH_2{=}CPh_2 + W(CO)_5$$

In each case, the reaction involves the scission of the alkene with transfer of an alkylidene group *via* the formation of a metallacyclobutane.

$$(CO)_5W \text{ metallacyclobutane } (CR_1R_2, CPh_2)$$

4.4.3. *Cross metathesis*

The products obtained by reaction between a cycloalkene and a symmetric olefin cannot be explained by a pair wise mechanism [30], since the symmetric products **6**, **7** are observed immediately the reaction commences and not as a result of further metathesis of **5** after a period of time. Thus theses reactions are good evidence in support of the carbene cyclobutane mechanism.

$$(CH_2)_n\left<\begin{matrix}CH\\ \|\\ CH\end{matrix}\right. + \begin{matrix}R_1CH\\ \|\\ R_2CH\end{matrix} \rightleftharpoons (CH_2)_n\left<\begin{matrix}CH{=}CH{-}R\\ \\ CH{=}CH{-}R'\end{matrix}\right.$$

5 $R = R_1$, $R' = R_2$

6 $R = R' = R_1$

7 $R = R' = R_2$

4.4.4. *Ring opening polymerization*

Ring opening polymerization is a chain reaction with the carbene as chain carrier. Polymers are not formed by consecutive macrocyclization, but instead cyclic oligomers are obtained when "the carbene end" reacts with a double bond within the polymer chain, by intramolecular metathesis [31].

$$RCH{=}\!\left[CH(CH_2)_mCH{=}\right]\!CH(CH_2)_mCH{=}(M) \rightleftharpoons \begin{matrix} RCH{=}(M)\\ + \\ CH{=}\!\left[CH(CH_2)_mCH{=}\right]\!CH \\ \diagdown (CH_2)_m \diagup \end{matrix}$$

The theory predicts a bimodal molecular weight distribution of polyoctenamers and polydodecenamers prepared by metathesis of cyclooctene. The low molecular weight products were assumed to be linear polymers [32]. Further support for this carbene chain reaction is that the ratio of macrocycles obtained by metathesis of 1,5-cyclooctadiene, C_{12}/C_{16}, C_{20}/C_{16}, C_{24}/C_{16}, C_{28}/C_{16} and C_{32}/C_{16} are essentially constant throughout the ring opening polymerization. The constancy of the C_{12}/C_{16} ratio is especially significant in this respect, since the pairwise mechanism would have required cross metathesis of the cyclooctadiene and cyclobutene, which is unlikely [33].

4.4.5. *Reactions of carbenes and carbynes*

The alcoxycarbene $[W({=}CMeOEt)(CO)_5]$ is used as a catalyst precursor for polymerization of cyclopentene, in the presence of the cocatalyst $EtAlCl_2$ [34] but other carbene complexes e.g. $[W({=}CPh_2)(CO)_5]$, can be used without a cocatalyst for the same reaction [35].

In accordance with this “alkylidene—metallacyclobutane” mechanism, metathesis of alkynes has been performed using tungsten metallacarbyne complexes [36]. The analogy between the reactions of olefins and acetylenic hydrocarbons has led to the assumption that metallacyclobutadienes could be intermediates for this reaction.

$$M{\equiv}R + R_1{-}C{\equiv}C{-}R_2 \rightleftharpoons \text{[metallacyclobutadiene: M, R, } R_1, R_2] \rightleftharpoons \text{[metallacyclobutadiene: M, R, } R_1, R_2] \rightleftharpoons M{\equiv}C{-}R_2 + R{-}C{\equiv}C{-}R_1$$

Further support for this mechanism has also been provided by Schrock *et al.* [37] who have synthesized stable tungstacyclobutadiene complexes.

5. Stereoselectivites

5.1. TERMINAL AND INTERNAL ACYCLIC OLEFINS

Molybdenum catalyst are usually efficient for the metathesis of terminal olefins, whereas tungsten complexes are not. However tungsten complexes are active for the metathesis of internal olefins. One possible explanation of this observation is that metallacarbenes combine with terminal alkenes selectively according to:

$$\text{R(H)C}{=}\text{(M)} + \text{R}_1\text{(H)C}{=}\text{CH}_2 \longrightarrow \text{R(H)C}{=}\text{CH}_2 + \text{R}_1\text{(H)C}{=}\text{(M)}$$

rather than

$$\begin{matrix}R\\ \end{matrix}\!\!\searrow C{=}(M) + R_1\!\searrow C{=}CH_2 \longrightarrow R\!\searrow C{=}C\!\swarrow R_1 + H\!\searrow C{=}(M)$$

$$(\text{R}(\text{H})C{=}(M) + R_1(\text{H})C{=}CH_2 \longrightarrow R(\text{H})C{=}C(\text{H})R_1 + H(\text{H})C{=}(M))$$

depending on the metal and the ligand used.

This possibility has been investigated by using a mixture of [1,1-D_2]-1-octene and 1-hexene with different catalyst and considering the ratio of the rate constants for formation of 1-octene and [1,1-D_2]-1-hexene (→ k_n, non-productive rate constant) with 7-tetradecene and 5-decene (→ k_p, productive rate constant).

Values of k_n/k_d were 25 ± 6 for a molybdenum catalyst and much higher, 75—155, for tungsten catalysts [38]. The only reaction observed using [$W(=CPh_2)(CO)_5$] and a mixture of CH_2=CHMePh and CH_2=CMePr was exchange of the "methylene groups" [39].

Another study of CH_2 and CD_2 exchange between 1-hexene and [1,1-D_2]-1-pentene showed that this reaction was approximatively 10^3 times faster than productive metathesis in 'WCl_6—BuLi' or 'WCl_6—$AlEtCl_2$' systems [35]. The structural selectivity of metathesis reactions can be summed up as follows: Degenerate exchange of =CH_2 groups between terminal olefin > cross metathesis between terminal and internal alkenes > metathesis of internal olefins > non-degenerate metathesis of terminal alkenes.

5.2. *CIS*- AND *TRANS*-ACYCLIC OLEFINS

There is a marked effect on the resulting stereochemistry of the products of metathesis reactions depending on the transition metal. By careful analysis of the *cis/trans* ratios at very low conversion of 2-butenes, obtained by metathesis of *cis*-2-pentene, different behaviour is observed for tungsten and molybdenum complexes [40, 41].

Type of catalyst:	$M(CO)_5PPh_3$ + $EtAlCl_2$ + O_2		MCl_x (x = 5 or 6) + $(CH_3)_3Al_2Cl_3$	
metal	W	Mo	W	Mo
trans/cis ratio	0,73	0,60	0,80	0,22

The retention of stereoselectivity can be explained by considering the energy levels of the two possible metallacycles leading to *cis* or *trans* isomers. The ligands of the precursor complexes do not govern the stereoselectivity simply by their own steric requirement, but also by their electronic effects on the stabilities of the metallacyclobutane intermediates produced.

In contrast to linear alkenes, cycloalkenes can be metathesize stereospecifically. By proper choice of catalyst combinations, cyclopentene can be polymerized to high *cis*- or *trans*-polypentene. For example a predominantly *cis* (85%) or *trans* (95%) polypentene product can be obtained from a 'WF_6—$(C_2H_5)_3Al_2Cl_3$', system by variation of the molar ratio of Al/W from 1 to 7 [42]. The temperature is also critical to the selectivity. For example, with a '$WCl_6/Sn(CH_3)_4/(C_2H_5)_2O$' catalytic system, *cis*-polypentene (92%) is obtained at −30 °C, whereas at 0 °C, *trans*-polypentene (80%) is observed [43]. A 97% *cis*-polypentene is obtained with $[W(=CPh_2)(CO)_5]$ as catalyst while with conventional catalysts a high *trans*-polymer is formed (the *trans*-form being thermodynamically favoured) [44].

6. Initial Production of Carbene

According to the catalyst type, several mechanisms have been invoked to account for the synthesis of the initial carbene. Two different types of catalysts can be envisaged: those which require activation with an organometallic cocatalyst, and those which are active without cocatalysts.

6.1. ALKYLIDENE GENERATION *VIA* REACTION WITH A METAL ALKYL COCATALYST

The formation of the initial alkylidene sometimes occurs *via* an equilibrium of the type:

$$(M)-CH_3 \rightleftharpoons (M)\genfrac{}{}{0pt}{}{/H}{\backslash\!\backslash CH_2}$$

6.1.1. *Alkylidene generation by chemical routes*

In these reactions, an alkylation of the metal is first proposed, followed by hydrogen abstraction in order to produce the alkylidene moiety. This is the proposed method of alkylidene formation in catalytic combinations such as 'WCl_6—$M(CH_3)_x$' (M = Li, $x = 1$; M = Zn, $x = 2$; M = Sn $x = 4$) were methane or HCl are evolved. For example with $[Zn(CH_3)_2]$ [45].

$$WCl_6 + Zn(CH_3)_2 \longrightarrow ZnCl_2 + \underset{Cl}{\overset{CH_3}{>W<}} \longrightarrow HCl + \;>W=CH_2$$

When $[WCl_6]$ and $Sn(CH_3)_4$ or $LiCH_3$ are used for disproportionation of 2-alkenes at low olefin/W ratios, significant amounts of 1-alkenes corresponding to the combination of 'CH_2' with the appropriate alkylidene derivative of the olefin are obtained, together with large amounts of methane and ethane.

A system containing $[Mo(NO)_2Cl_2(PPh_3)_2]$ and $Al_2(CH_3)_3Cl_3$ as catalyst gives in 2,8-decadiene metathesis an initial burst of propene, the formation of which is explained by the following sequence [46]:

$$L_nM \xrightarrow{\overset{*}{C}H_3-M} L_nM-\overset{*}{C}H_3 \xrightarrow{M\overset{*}{C}H_3} CH_4\uparrow + L_nM=\overset{*}{C}H_2$$

$$L_n=\overset{*}{C}H_2 + CH_3-CH=CH-(CH_2)_4-CH=CH-CH_3 \longrightarrow L_nM\diamond(-CH_3)(C_7H_{13})$$

$$\rightleftharpoons L_nM=CH-C_7H_{13} + \overset{*}{C}H_2=CH-CH_3$$

$$\downarrow$$

normal metathesis products.

Good evidence for this mechanism has been obtained by the reversible interconversion of methyl and methylene ligands in tungsten complexes as well tantalum complexes such as $[Ta(=CH^tBu)(CH_2^tBu)_3]$ and $[Ta(=CH_2)(CH_3)(\eta\text{-}C_5H_5)_2]$. These tantalum complexes have been prepared by a route which probably involves hydrogen abstraction.

6.1.2. *Generation by electrochemical route* [47]

It was found that under electrochemical conditions with dichloromethane as solvent and with a sacrificial aluminium anode, a metallacarbene could be directly synthesized from $[WCl_6]$ or $[MoCl_5]$.

Cathodic reaction:

$$[WCl_6] + e^- \longrightarrow [WCl_6]^-$$

Anodic oxidation:

$$[WCl_6]^- \xrightarrow{Al,\ CH_2Cl_2} [WCl_4(=CH_2)]$$

Strong evidence has been obtained for the participation of the solvent in this reaction, since when using CD_2Cl_2, an initial burst of $[D_2]$ propene and $[D_2]$ 1-butene is observed during metathesis of 2-pentene. Subsequently only butenes and hexenes are formed as metathesis products. The initial propene and 1-butene arise from the combination of 2-pentene derived alkylidene moieties with a methylidene group. Carbon-13 NMR spectroscopy also indicate the possible presence of a methylidene group.

6.2. CARBENE FORMATION WITHOUT ALKYL CONTAINING COCATALYST

The mechanism of alkylidene formation and thus chain initiation, in systems that contain no alkyl ligands either on the metal or cocatalyst still remains unknown.

This situation is particularly relevant to most heterogeneous catalysts. In these cases, the initial carbene can only be formed from the olefin and the inorganic catalyst. M. L. H. Green *et al.* [48] have found that nucleophilic attack by H^- ($NaBH_4$) on $[W(\eta\text{-}C_3H_5)(\eta\text{-}C_5H_5)_2][PF_6]$ at the central carbon atom of the allyl group affords the metallacyclobutane:

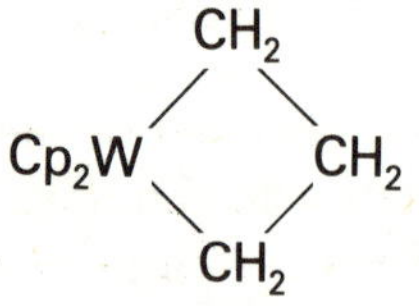

$$Cp_2W{<}(CH_2)_2{>}CH_2 \qquad Cp = \eta\text{-}C_5H_5$$

This complex gives cyclopropane, and propene on heating, but upon irradiation ethylene and some methane is evolved:

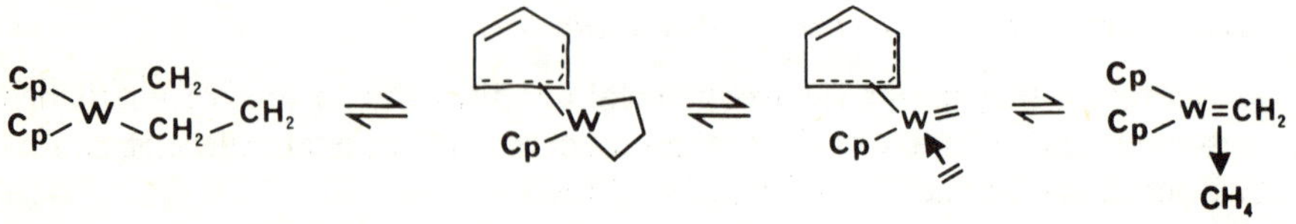

$$Cp_2W{<}(CH_2)_2{>}CH_2 \rightleftharpoons \cdots \rightleftharpoons \cdots \rightleftharpoons Cp_2W{=}CH_2 \rightarrow CH_4$$

Schematically the following path is proposed:

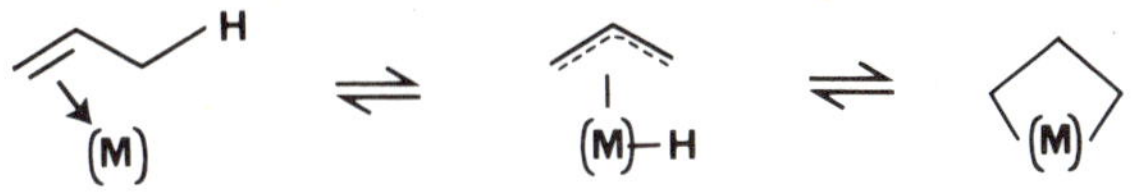

However, this mechanism could not be extended to, for example, norbornene where η^3-allyl formation is not possible. Another possible mechanism is the direct, reversible conversion of a coordinated olefin to a carbene in one step *via* a 1,2 hydride shift [49].

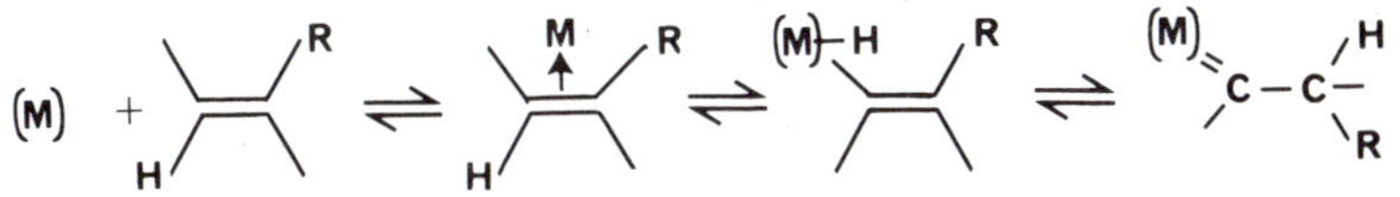

Recent evidence for this type of mechanism has been provided by Miller *et al.* [50] who have directly synthesised an ethylidene moiety from a π-olefin complex.

$$[Cp_2HW(CH_2{=}CH_2)]^+ + I_2 \longrightarrow [Cp_2W{=}CHMe]^+ + HI$$

7. Metallacarbenes as Catalysts

More recent works have been focused towards the synthesis of stable metallacarbenes which could be used directly as metathesis catalysts. A general feature of these reactions is that, in order to get good catalytic results, the metal must be in a high oxidation state and surrounded by bulky ligands, typically alkoxides. Examples of this type of complex are $[M({=}CH^tBu)(O^tBu)_2(PPh_3)Cl]$ (M=Nb, Ta) or tungsten oxo compounds such as $[W(O)({=}CHR)(PEt_3)_2Cl_2]$ (R = H, Pr, Ph) [28].

The complex $[W({=}CHR)(OCH_2{}^tBu)_2Br_2]$ is not an active catalyst on its own, but turnover rates of 1000 min^{-1} are observed in the presence of one equivalent of $AlBr_3$ [51]. The nature of the initiating and propagating species present in this reaction have been established by NMR spectrocopy (^1H, ^{13}C ...) at −30 °C and could explain the role of the acidic cocatalyst in conventional systems. $[W({=}CH^tBu)(OCH_2{}^tBu)_2Br_2]$ **8** rapidly forms a strong 1 : 1-adduct **9** with $AlBr_3$ at −30 °C. On warming to −10 °C, it converts to **10** which also forms the adduct **11** in the presence of $AlBr_3$.

$(^tBuCH_2O)_2Br_2W{=}C(H)R$ (**8**) $\xrightarrow[-30^\circ C]{AlBr_3}$ $(^tBuCH_2O)_2Br_2W{=}C(H)R \cdot AlBr_3$ (**9**) $\xrightarrow{-10^\circ C}$

$(^tBuCH_2O)Br_3W{=}C(H)R$ (**10**) $\longrightarrow$ $(^tBuCH_2O)Br_3W{=}C(H)R \cdot AlBr_3$ (**11**)

Addition of *cis*-2-pentene to a mixture of the carbene complexes **9** and **11** yielded a mixture of compounds which when analyzed by ^1H NMR spectroscopy, show signals corresponding to the expected methyl and ethyl substituted carbenes.

$[W]{=}C(H)Me$ $[W]{=}C(H)Et$

Hence, the chain carrying carbene complexes, in their standing concentrations, are observed in this highly active system and the binding of the Lewis acid seems necessary for both initiation and propagation of this reaction.

8. Role of Oxygen

8.1. ROLE OF OXYGEN

The importance of the contamination of metathesis mixtures by H_2O or O_2 has been shown by Muetterties *et al.* [52], who observed that $[WCl_6]$, in combination with different alkyl metal compounds, was inactive at room temperature for the metathesis of acyclic internal alkenes, when the reaction was performed in an atmosphere free of water and oxygen.

The importance of oxo ligands has been previously suspected by the numerous examples of oxygen containing activators. These reactions clearly demonstrated the crucial role of an oxo ligand. This activity could well be based on the double bond character of the M=O linkage. This may well facilitate carbene formation *via* a sequence of metal hydride reactions involving internal oxidative addition and reductive elimination.

$$\mathrm{RCH{=}CHR'}\rightarrow\mathrm{(M){-}OH} \rightleftharpoons \mathrm{RCH{=}CHR'}\rightarrow\mathrm{(M)(H){=}O} \rightleftharpoons \mathrm{RCH{-}CH_2R'}\,/\,\mathrm{(M){=}O} \rightleftharpoons \mathrm{R{-}C(CH_2R'){=}(M)(H){=}O} \rightleftharpoons \mathrm{R(CH_2R')C{=}(M){-}OH}$$

8.2. APPLICATION

The only alternative explanation for carbene formation in 'M=O' systems is that 'M=O' is directly involved in carbene formation.

$$\mathrm{(M){=}O} + \mathrm{CHR{=}CHR'} \longrightarrow \mathrm{(M){=}O\cdots CHR{=}CHR'} \longrightarrow \mathrm{(M){-}O{-}CHR{-}CHR'{-}(M)}\ \text{(four-membered ring)}$$

Such a mechanism would relate metathesis to the Fleming and MacMurry [53, 54] reaction for the reductive dimerization of aldehydes or ketones to alkenes,

$$2\,\mathrm{RR'C{=}O} + \mathrm{(M)} \longrightarrow \mathrm{(M)O_2} + \mathrm{RR'C{=}CRR'}$$

particularly since 'W(CO)$_6$—AlCl$_3$' or 'WCl$_6$—BuLi' mixtures are reagents for this reaction as well as 'TiCl$_3$—LiAlH$_4$'. However, the strength of the [M]=CR$_2$ bond is probably too large for the forward reaction to be significant except perhaps, with strained olefins.

9. Industrial Applications

The potential industrial applications of this reaction can be classified into three categories: polymer synthesis, olefin synthesis and fine chemicals synthesis.

9.1. POLYMERIZATION AND RING OPENING POLYMERIZATION

The polymerization of cyclic olefins with Calderon's catalyst 'WCl$_6$—EtOH—EtAlCl$_2$' was used as a probe of the ability of many unsaturated compounds to participate in the olefin metathesis reaction. A significant decrease in molecular weight of the polymers obtained was considered to be evidence of metathesis activity.

The addition of small quantities of functionally active alkenes to the cyclic olefin monomer changes the course of the reaction. These coreactants can be classified in to four groups [55]:

(i) inhibitors (phenylacetylene, acrylonitrile);
(ii) inert solvents (perchloroethylene);
(iii) activating participants (e.g. vinyl bromide or esters);
(iv) active participants (e.g. unsaturated esters, ethers, haloalkenes, alkenyl silanes or stannates).

The use of this last group of compounds in the synthesis of functional polymers as well as the use of group (iii) coreactants in the polymerization or copolymerization of cyclic olefins and in the cross metathesis of polydodecene and *cis*-1,4-polybutadiene has been pioneered commercially by the Huls company. A 80% *trans*-polyoctenamer is now commercially available under the tradename Vestenamer 8012.

However, the most important cycloolefin is probably polypentene, owing to the availability of cyclopentene from naphta cracking. A 93% *cis*-isomer can be prepared which melts at −41 °C [56] or a 99% *trans*-isomer, which melts at 20 °C [57]. The *cis*-polymer is competitive with low temperature elastomers like silicone rubber and the *trans*-polymer is a good substitute for natural rubber in the tyre industry. Norbornene type ring compounds can also be readily polymerized by ring opening, thus numerous functionnalized polymers have been prepared by the following sequence [58]:

n X Y ⟶ [X Y]n

Norbornene itself can be polymerized on a [$RuCl_3$] catalyst in alcohol to give a compound commercially called Norsorex [59] (CDF Chimie, 20 000 t/year in 1979).

9.2. SYNTHESIS OF MONO AND POLYOLEFINS [60]

Most of these come from heterogeneous catalytic processes.

(i) The first industrial application of olefin metathesis chemistry was for the production of high purity ethylene and normal butenes from propylene (the Triolefin Process) [61].

$$2\ C_3H_6 \longrightarrow C_2H_4 + C_4H_8$$

A demand for very high purity 1-butene as a comonomer in polyethylene production has created renewed interest in this application of alkene metathesis. Propylene is converted to ethylene and 2-butenes which are then isomerized to high purity 1-butene.

(ii) Neohexene, an intermediate in the synthesis of musk perfume, is produced commerically in a plant completed in 1980, the reaction involves cometathesis of diisobutylene with ethylene [62].

$$C{-}C(C)(C){-}C{=}C(C)C \xrightarrow[C=C]{\text{metathesis}} C{-}C(C)(C){-}C{=}C + C{=}C(C)C$$

(neohexene)

$$C{=}C{-}C(C)(C){-}C + \text{(p-cymene)} \longrightarrow \text{(tetralin)} \xrightarrow{CH_3COCl} \text{(acetyl tetralin)}$$

(iii) A ring-substitued symmetrical alkene, 1,2-bis(3-cyclohexenyl)ethylene, which was evaluated as an intermediate in the production of flame retardants, has been produced in developmental quantities by self metathesis of 4-vinylcyclohexene, which can be synthesized from butadiene cyclodimerization.

$$2 \text{ (butadiene)} \xrightarrow{Fe(NO)_2} \text{(4-vinylcyclohexene)} \xrightarrow{\times 2} \text{(1,2-bis(3-cyclohexenyl)ethylene)} + {=}$$

(iv) Developmental quantities of α,ω-diolefins, intermediates in the preparation of agricultural chemicals are synthesized by the reaction of ethylene with selected cyclic olefins.

$$(CH_2)_n\,\| + \| \longrightarrow (CH_2)_n{=}$$

Multi-ton quantities of 1,5-hexadiene have been manufactured in alkene metathesis "semi-works" units and the process has been studied in a small specialty olefin pilot plant by the Phillips Company. Feedstocks for this process are the commerically available butadiene derivatives (1,5-cyclooctadiene and 1,5,9-cyclododecatriene). The use of excess ethylene and a tungsten oxide—silica catalyst, has resulted in high per pass conversions and yields greater than 65% of 1,5-hexadiene [69].

2 (butadiene) $\xrightarrow[L]{Ni^0}$ (1,5-cyclooctadiene) $\xrightarrow{C_2H_4}$ (1,5-hexadiene) + (1,5,9-decatriene)

9.3. SYNTHESIS OF FUNCTIONALLY SUBSTITUTED OLEFINS

The metathesis of olefins bearing functional groups provides potential routes to many valuable compounds. Metathesis catalysts, most of which contain a Lewis acid, are unfortunately poisoned by polar and basic compounds. As yet, only a few catalytic systems have proved to be active in metathesizing functionalized alkenes. In the homogeneous disproportionation of these substrates, the catalytic combination 'WCl_6—$Sn(CH_3)_4$' is still the best known [63]. In addition to its well known application in the conversion of unsaturated oxygen containing olefins, such as fatty acid esters and unsaturated ethers, its effectiveness for the homogeneous metathesis of unsaturated amines has also been described [64].

In fact, heterogeneous catalysts, such as 'ReO_7—Al_2O_3' promoted by a small amount of an alkyltin compound, are better than homogeneous ones. This heterogeneous system is very active and selective in the metathesis of unsaturated esters [65].

$$2\ CH_2{=}CH(CH_2)_2COOCH_3 \longrightarrow CH_3OOC(CH_2)_2CH{=}CH(CH_2)_2COOCH_3 + C_2H_4$$

At 50 °C a conversion of 51% was reached within 1 h with very high selectivity (99%) using a molar ratio of ester/Re/Sn of 219/25/1. It was also reported that the catalyst 'Re_2O_7—Al_2O_3' alone showed no activity for the metathesis of polar compounds [65].

Of great potential value is the cometathesis of various unsaturated esters with ethylene to produce α,ω-unsaturated esters. Thus, methyl-9-decenoate, a key intermediate in the synthesis of oxo *trans*-9-decenoic acid, was obtained by co-metathesis of ethylene and methyl oleate with two different catalyst systems, *viz.* 'WCl_6—$(CH_3)_4Sn$' and 'Re_2O_7—Al_2O_3—$(CH_3)_4Sn$'. Both catalyst systems show good activity at 70 °C, the latter even at room temperature [66]. This substance is a honey bee phenomone.

Allylethylether could be metathesized with the same system at 20 °C. After 6 h, a conversion of 40% with a selectivity of 98% was obtained [67].

$$2\,CH_2{=}CH(CH_2)_2C(=O)CH_3 \longrightarrow CH_3C(=O)(CH_2)_2CH{=}CH(CH_2)_2C(=O)CH_3 + C_2H_4$$

Again, with this supported catalyst, unsaturated ketones (conversion 35%, selectivity 97%, after 5 h at 20 °C) and allylbromide (conversion 85%, selectivity 95%, after 40 min. at 20 °C) could be metathesized [67].

$$2\,CH_2{=}CHCH_2OEt \longrightarrow EtOCH_2CH{=}CHCH_2OEt + C_2H_4$$

Thus, the system 'Re_2O_7—Al_2O_3—$(CH_3)_4Sn$' is active and very selective in the dismutation of unsaturated esters and other alkenes containing hetero atoms even in situations where homogeneous systems fail. New interesting systems have been described for metathesis of unsaturated nitriles. These consist of the previously known [WCl_6] precursor, associated with diphenylsilane [68]. However, to be of interest for the chemical industry, it is necessary to develop catalysts which are more active and more stable than those presently known.

10. Conclusion

This review paper clearly demonstrates the exciting possibilities of alkene metathesis reactions for both new alkene and polymer synthesis as indeed does the vast quantity of work published by so many groups throughout the world.

However, for metathesis of functionally substituted acyclic alkenes, there are at present no catalysts which are both sufficiently active and stable to make these routes of interest to the chemical industry. Therefore, for practical applications, it is important to develop catalysts which are several orders of magnitude more active towards these substrates.

Laboratoire de Catalyse Hétérogène et Homogène
(UA CNRS 402) Université de Lille-Flandres Artois
(Laboratoire de Chimie Organique Appliquée)
ENSC Lille BP 108.
59652 Villeneuve d'Ascq France

References

1. V. Schneider and P. K. Frölich, *Ind. and Eng. Chem.*, **23**, 1405 (1931).

2. R. L. Banks and G. C. Bailey, *Ind. and Eng. Chem., Prod. Res. and Develop.*, **3**, 170 (1964).
3. N. Calderon, H. Y. Chen and K. W. Scott, *Tet. Lett.*, 3327, (1967).
4. N. Calderon, *Accounts Chem. Res.*, **5**, 127 (1972).
5. E. A. Zuech, W. B. Hughes, D. H. Kubicek and E. I. Kittelman, *J. Amer. Chem. Soc.*, **92**, 528 (1970).
6. L. F. Heckelsberg, R. L. Banks and G. C. Bailey, *J. Catalysis*, **13**, 99 (1969).
7. N. Calderon, E. A. Ofstead and W. A. Judy, *J. Polym. Sci.*, **5A**, 2209 (1967).
8. E. Wasserman, D. A. Ben Efraim and R. J. Wolovsky, *J. Amer. Chem. Soc.*, **90**, 3286 (1968) *cf. CA* **71**, (1969) 38438 c and 80807 x.
9. N. Calderon, *J. Macromol. Sci.*, **C7**, 105 (1972).
10. G. C. Ray and D. L. Crain, *Fr. Pat.*, 1511 381 (1968).
11. F. Pennella, R. L. Banks and G. C. Bailey, *J. Chem. Soc., Chem. Commun.*, 1548 (1968).
12. A. V. Mushegyan, V. H. Ksipteridis and G. A. Chukhadzhyan, *CA* **84**, 16693 (1976).
13. A. Mortreux, F. Petit and M. Blanchard, *Tet. Lett.*, **49**, 4967, (1978).
14. *Ger. Offen.* 2063150, 1971, *CA* **75**, 631726 (1971).
15. P. B. Van Dam, M. C. Mittelmeijer and C. Boelhouwer, *J. Chem. Soc., Chem. Commun.*, 1221 (1972).
16. J. C. Mol, *J. Mol. Catal.*, **15**, 35 (1982).
17. S. Hosaka and J. Tsuji, *Tetrahedron* **27**, 3821 (1971).
18. J. C. Mol, J. A. Moulijn and C. Boelhouwer, *Coll. Czech. Chem. Comm.*, 633 (1963).
19. G. Dall'Asta and G. Motroni, *Eur. Polym. Journal,* **7**, 707 (1971).
20. A. Mortreux, F. Petit and M. Blanchard, *J. Mol. Catal.* **8**, 97 (1980).
21. F. Mango and J. Schachtschneider, *J. Amer. Chem. Soc.*, **93**, 1123 (1971).
22. H. Hogeveen and H. C. Volger, *J. Amer. Chem. Soc.*, **89**, 2486 (1967).
23. G. S. Lewandos and R. Pettit, *J. Amer. Chem. Soc.*, **93**, 7087 (1971).
24. R. H. Grubbs and T. K. Brunck, *J. Amer. Chem. Soc.* **94**, 2538 (1972).
25. J. L. Herisson and Y. Chauvin, *Makromol. Chem.*, **141**, 161 (1970).
26. D. J. Cardin, M. J. Doyle and M. F. Lappert, *J. Chem. Soc., Chem. Commun.* 927 (1972).
27. R. Schrock, S. Rocklage, J. Wengrovius, G. Rupprecht, J. Fellmann, *J. Mol. Catal.* **8**, (1980).
28. S. Rocklage, J. D. Fellmann, G. A. Rupprecht, L. W. Messerle and R. R. Schrock, *J. Amer. Chem. Soc.* **103**, 1440 (1981).
29. C. P. Casey and T. J. Burkhardt, *J. Amer. Chem. Soc.* **96**, 7808, (1974).
30. T. J. Katz and J. M. Ginnis, *J. Amer. Chem. Soc.* **97**, 1592 (1975).
31. H. Höcker, W. Reimann, K. Riebel and Z. Szentivanyi, *Makromol. Chem.* **177**, 1707 (1976).
32. W. Scott, N. Calderon, E. A. Ofstead, W. A. Judy and J. P. Ward, *Rubber Chem. Techn.* **44**, 134 (1971).
33. W. J. Kelly and N. Calderon, *J. Makromol. Sci.* **A9**, 911 (1975).
34. *Ger. Pat.* 2151662.
35. E. L. Mutterties and M. A. Busch, *J. Amer. Chem. Soc.* **98**, 1283 (1976).
36. J. H. Wengrovius, J. Sancho and R. R. Schrock, *J. Amer. Chem. Soc.* **103**, 3932 (1981); J. Sancho and R. R. Schrock, *J. Mol. Catal.*, **15**, 75 (1982).
37. L. G. McCullough, R. R. Schrock, *J. Amer. Chem. Soc.* **106**, 4067 (1984).

38. J. McGinnis, T. J. Katz and S. Hurwitz, *J. Amer. Chem. Soc.* **98**, 605 (1976).
39. C. P. Casey, H. C. Tuinstra and M. C. Saeman, *J. Amer. Chem. Soc.* **98**, 608 (1976).
40. M. Lecomte and J. M. Basset, *J. Amer. Chem. Soc.* **101**, 7236 (1979).
41. M. Lecomte, Y. Ben Taarit, J. L. Bilhou and J. M. Basset, *J. Mol. Catal.* **8**, 263 (1980).
42. P. Günther, F. Haas, G. Marwede, K. Nutzel, W. Oberkirch, G. Pampus, N. Schon and J. Witte, *Angew. Makromol. Chem.* **14**, 87 (1970).
43. G. Pampus and G. Lehnert, *Makromol. Chem.* **175**, 2605 (1974).
44. T. J. Katz and W. H. Hersch, *Tet. Letters* **6**, 585 (1977).
45. E. L. Muetterties, *Inorg. Chem.* **14**, 951, (1975).
46. R. H. Grubbs and C. R. Hoppin, *J. Chem. Soc., Chem. Commun* 634 (1977).
47. M. Gilet, A. Mortreux, J. C. Folest and F. Petit, *J. Amer. Chem. Soc.* **105**, 3876 (1983).
48. M. Ephritikine, M. L. H. Green and R. E. McKenzie, *J. Chem. Soc., Chem. Commun.* 619 (1976).
49. E. O. Fischer and W. Held, *J. Organomet, Chem.* **C59**, 112 (1976).
50. G. A. Miller, N. Y. Cooper, *J. Amer. Chem. Soc.,* **107**, 709 (1985).
51. J. Kress, M. Wesolek and J. A. Osborn, *J. Chem. Soc., Chem. Commun.* 514 (1982).
52. M. T. Molella, R. Rovner and E. L. Muetterties, *J. Amer. Chem. Soc.* **98**, 4689 (1976).
53. J. E. McMurry and M. P. Fleming, *J. Amer. Chem. Soc.* **96**, 4708 (1974).
54. M. Petit, A. Mortreux and F. Petit, *J. Chem. Soc., Chem. Commun.* 342 (1984).
55. R. Streck, *J. Mol. Catal.* **15**, 3 (1982).
56. F. Haas, K. Nutzel, G. Pampus and D. Theisen, *Rubber Chem. Techn.* **43**, 1116 (1970).
57. G. Dall'Asta and P. Scaglione, *Rubber Chem. Techn.* **42**, 1235 (1969).
58. K. F. Castner and N. Calderon, *J. Mol. Catal.* **15**, 47 (1982).
59. *Informations Chimie.* **83**, 179 (1978).
60. R. L. Banks, D. S. Banasiak, P. S. Huson and J. R. Norell, *J. Mol. Catal.* **15**, 21 (1982).
61. *Phillips Petroleum Company, Hydrocarbon Processing.* **46**(11), 232 (1967).
62. E. Reusser and D. L. Crain, *U.S. Pat.* 3 729 524 (1973).
63. J. C. Mol. *Chemtech,* **13**, 250 (1983).
64. J. P. Laval, A. Lattes, R. Mutin and J. M. Basset, *J. Chem. Soc., Chem. Commun.* 502 (1977).
65. E. Verkuijlen, F. Kapteijn, J. C. Mol and C. Boelhower, *J. Chem. Soc., Chem. Commun.* 198 (1977).
66. R. H. A. Bosma, G. C. N. Van Den Aardweg and J. C. Mol., *J. Chem. Soc., Chem. Commun.* 1132 (1981).
67. J. C. Mol and E. F. G. Woerlee, *J. Chem. Soc., Chem. Commun.* 330 (1979).
68. J. Levisalles, H. Rudler, D. Cuzin and T. Rull, *J. Mol. Catal.* **26**, 231 (1984).
69. The Shell company has also developed a similar plant at Berre (France) which also uses metathesis in the Shop process (See article by Y. Chauvin in this book).

For Reviews see:

G. C. Bailey, *Catalysis Reviews* **3**(1), 37 (1969).

N. Calderon, *Accounts of Chemical Research* **5**, 127 (1972).
W. B. Hughes, *Organomet. Chem. Synth.* **1** 341 (1972).
R. J. Haines, *Chem. Reviews* **4**, 155 (1975).
R. Streck, *Chemiker Zeitung* **10**, 397 (1975).
J. C. Mol and J. A. Moulijn, *Adv. Catal.*, **24**, 131 (1975).
J. J. Rooney and A. Stewart, *Catalysis* **1**, 277 (1977).
R. H. Grubbs, *Prog. Inorg. Chem.*, **24**, 1 (1978); *Adv. Organomet. Chem.*, **17**, 449 (1979).

Seven Metathesis Symposia have been held in:

— Mainz (West Germany) 1976
— Amsterdam (The Netherlands) 1977
— Lyon (France) 1979
— Belfast (N. Ireland) 1981
— Graz (Austria) 1983
— Hamburg (West Germany) 1985
— Hull (U.K.) 1987

The proceedings of the Amsterdam Symposium have been published in the *Journal of the Royal Netherlands Chemical Society* **96**, 279—292 (1977).

4 special issues of *Journal of Molecular Catalysis* have currently been devoted to metathesis. From the IIIrd to the VIth Symposium: **8** (1980), **15** (1982), **28** (1985) and **36** (1986).

MICHEL EPHRITIKHINE

ACTIVATION OF ALKANE CH BONDS BY ORGANOMETALLICS

1. Introduction

During the '70s, considerable attention was paid to the chemical transformation of alkanes, one of the most important feedstocks for the synthesis of organic molecules. Most functionalisation of hydrocarbons on an industrial scale, in particular the synthesis of olefins, uses heterogeneous catalysis [1].

Homogeneous reactions of alkanes frequently involve the radical abstraction of hydrogen atoms (chlorination and autoxidation are classical examples) and in many cases, the formation and the reactions of these radicals are mediated by transition metal complexes [2, 3]. Other reagents, such as superacids and carbenes, can react with alkanes, apparently by non-radical mechanisms [4].

The major problem with all these reactions is not the lack of reactivity of the alkane molecule itself (long reputed to be inert) but the difficulty in achieving selective reactions; nevertheless, drastic experimental conditions are often required. These reactions mostly give mixtures. It is however possible to bring about preferential rupture of tertiary CH bonds. Secondary and primary hydrogens are more difficult to activate; reactions of methane are rare. Extensive research on these classical reactions of alkanes has given us a new insight into their mechanism (for example a better knowledge of surface chemistry) [1] and the development of new systems capable of reacting with saturated hydrocarbons under milder conditions (see the radical reactions of porphyrin transition metal complexes which serve as models for cytochrome P450) [3].

At the same time, the interest in CH bond activation also prompted many organometallic chemists to look for reactions of alkanes involving new mechanisms. As a result some unprecedented reactions between soluble transition metal systems and alkanes were observed; thus encouraged, an ever growing number of workers are investigating the possibilities of this still relatively young field.

This paper deals with these discoveries which represent important

A. Mortreux and F. Petit (Eds.), Industrial Applications of Homogeneous Catalysis, 257–275.

advances in the fundamental understanding of reactions between alkanes and transition metal complexs [5]. We will successively examine the insertion reactions of transition metal systems into CH bonds (oxidative addition) and the electrophilic reactions of organoactinides with saturated hydrocarbons [6].

2. Oxidative Addition of Alkane CH Bonds to Organometallics

2.1. BACKGROUND. OXIDATIVE ADDITION OF ACTIVATED sp^3 CH BONDS [7]

Oxidative addition is a fundamental organometallic primary reaction in which an X—Y bond adds to a metal complex which is formally oxidized by two units: [M] + X—Y → X—[M]—Y. The metal coordination number is also increased after this insertion into the X—Y bond. Such additions occur essentially to d^8 and d^{10} metal centres, and several studies showed that they take place both by a concerted mechanism of by successive elementary processes [8].

Many examples of intramolecular oxidative addition of CH bonds have been found during the past 10 years [9]. In these cyclometallation reactions, which are apparently favoured by proximity effects, the insertion takes place into a CH bond of a ligand. Specific examples [10, 11] are illustrated by reactions 1 and 2.

$$IrCl_3 + (tBu)_2P-(CH_2)_5-P(tBu)_2 \longrightarrow$$ P(tBu)$_2$, Ir, Cl, H, P(tBu)$_2$ (1)

L_2Pt → L_2Pt (2)

L = trialkylphosphine

Whitesides found a large positive value for the entropy of activation for the cyclometallation of neopentyl groups on Pt(II) (reaction 2) [11]; these data would indicate a rate determining reductive elimination of neopentane from the Pt(IV) hydrido alkyl complex formed by oxidative addition of the γ-CH bond.

Intermolecular oxidative addition of sp^3 CH bonds are rare [7] and, until recently, were limited to organic molecules with CH bonds "activated" by electronic effects (reactions 3—5) [12, 13, 14]. A particular example is the hydrogen transfer reaction (reaction 4) [13] catalyzed by transition metal complexes.

$$(dmpe)_2M(Ar)(H) \xrightarrow{CH_3COCH_3} (dmpe)_2M(CH_2COCH_3)(H) \qquad (3)$$

dmpe = $Me_2PCH_2CH_2PMe_2$ M = Fe, Ru

$$\text{dioxane} + \text{cyclopentene} \xrightarrow{RhCl(Ph_3P)_3} \text{dioxene} + \text{cyclopentane} \qquad (4)$$

$$Cp_2WH_2 \xrightarrow{Me_4Si,\ h\nu} \text{HW(Cp)}(\mu\text{-}C_5H_4)_2\text{W(Cp)}(CH_2SiMe_3) \qquad (5)$$

These reactions are effected by relatively electron rich metal centres; for example, tungstenocene, the probable intermediate in reaction 5 [14], inserts into the weakly activated CH bond of Me_4Si but not into a CH bond of neopentane.

This difference between internal and external CH oxidative addition has been discussed in terms of kinetic and thermodynamic factors. Oxidative addition of RH is the reverse process of reductive elimination, which is well known; Halpern suggested that the difficulty of oxidative addition is due to thermodynamic rather than kinetic factors [15]. In view of the free energies of formation of the CH bond (*ca.* 100 Kcal/mole) and the M—H bond (*ca.* 60 Kcal/mole), the free energy of the M—C bond would have to be *ca.* 40 Kcal/mole to make the reaction thermodynamically feasible. The limited data on M—C bond energies (20—30 Kcal/mole) is not encouraging and suggests that, if any such oxidative addition of an alkane CH bond could be observed, it would be with highly unstable and reactive species. Since, in going down a transition metal group, the M—H and M—C (alkyl) bond strengths increase, third row transition metals would be expected to insert more easily into alkane CH bonds.

While oxidative addition reactions were well known for metallic ions in the gas phase [16] or for metal atoms in alkane matrices [17] or in the vapour phase [18] or in a surface [1], such reaction was not observed, until recently, with soluble transition metal complexes.

2.2. DIRECT OBSERVATION OF THE OXIDATIVE ADDITION REACTION

$$[M] + RH \longrightarrow [M]\langle^{R}_{H}$$

Three years after Green discovered the tetramethyl silane CH bond activation (reaction 5) [14], Bergman *et al.* [19] and Graham *et al.* [20] found independently, in 1982, the first examples of oxidative addition reaction of alkane CH bonds with formation of hydrido alkyl complexes (reactions 6 and 7).

$$\underset{\underset{\sim}{1}}{Cp'Ir(CO)_2} \xrightarrow{h\nu} \underset{\underset{\sim}{2}}{[Cp'(CO)Ir]} \xrightarrow{RH} \underset{\underset{\sim}{3}}{Cp'(CO)Ir\langle^{R}_{H}} \quad (6)$$

$R = C_6H_{11}$, CH_2tBu

$$\underset{\underset{\sim}{4}}{Cp'IrH_2(PMe_3)} \xrightarrow{h\nu} \underset{\underset{\sim}{5}}{[Cp'(PMe_3)Ir]} \xrightarrow{RH} \underset{\underset{\sim}{6}}{Cp'(PMe_3)Ir\langle^{R}_{H}} \quad (7)$$

$R = C_6H_{11}$, CH_2tBu

In the Graham system, irradiation of the carbonyl iridium complex **1** extrudes a CO molecule and leads presumably to the coordinatively unsaturated intermediate **2** which then inserts into the CH bonds of cyclohexane or neopentane giving the compounds **3**. Similarly, Bergman found that the iridium dihydride **4** loses H_2 upon irradiation; when the reaction takes place in cyclohexane or neopentane, the hydrido alkyl compounds **6** are formed, resulting from the oxidative addition of a CH bond to the intermediate **5**.

Free or solvent-caged radicals are not involved in the formation of compounds **6**; this was demonstrated by crossover experiments and by relative reactivity studies [21]. Thus, the products of irradiation of **4** in a mixture of neopentane and cyclohexane-d_{12} were the hydrido neopentyl and the deuterio (perdeuteriocyclohexyl) iridium complexes, with very small quantities of the cross products. Moreover, it was found that irradiation of **4** in liquid cyclopropane leads only to the CH insertion product (reaction 8).

$$Cp'(PMe_3)IrH_2 \xrightarrow{h\nu,\ \triangle} Cp'(PMe_3)Ir(\triangle)(H) \quad (8)$$

Addition of the strong CH bond of cyclopropane (106 Kcal/mole) is favoured over insertion into the relatively weak C—C bond; this result rules out hydrogen abstraction reaction of radicals which favour CH bond with low bond energies. In fact, cyclopropane is the cyclic hydrocarbon with the highest reactivity towards CH insertion (Figure 1).

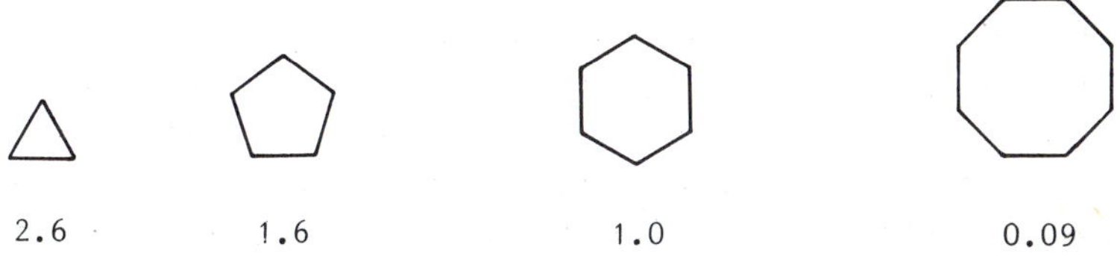

Fig. 1. Relative rates of reactions with **5** of one CH bond in each of the cycloalkane.

These unprecedented results were tentatively explained by steric factors: the CH bonds in the smaller rings should be more accessible, because the two carbon atoms adjacent to the reacting carbon are more tightly "tied back".

Bergman also studied competition reactions between the different CH bonds located in the same alkane molecule, and found, here again, an anti radical selectivity, insertion taking place into primary CH bonds about twice as fast as into secondary; tertiary CH bonds were not reactive (reaction 9).

$$Cp'(PMe_3)IrH_2 \xrightarrow{h\nu,\ \text{butane}} Cp'(PMe_3)Ir(n\text{-butyl})H + Cp'(PMe_3)Ir(\text{isobutyl})H \quad (9)$$

4

primary / secondary rate ratio : 1.51

A greater selectivity was observed with the complex $Cp'(PMe_3)RhH_2$ **7**, the rhodium analogue of **4**, which was independently studied by Bergman [22] and Jones [23]. Apparently only primary CH bonds of acyclic alkanes are activated by the rhodium complex (reaction 10); the hydrido alkyl products are more unstable towards reductive elimination of RH than their iridium analogues and must be prepared and characterized below −20 °C.

$$Cp'(PMe_3)RhH_2 \xrightarrow{h\nu,\ \text{propane}} Cp'(PMe_3)Rh(n\text{-propyl})H \quad (10)$$

7

These peculiarities of the rhodium system (greater selectivity and easier reductive elimination of its hydrido alkyl complexes) are in agreement with a lower exothermicity of its CH insertion reaction. Irradiation of the rhenium complex CpL_3Re also gave hydrido alkyl compounds $CpL_2Re(R)H$ (L = PMe_3; R = methyl, *n*-hexyl, cyclopropyl) with the corresponding alkane; cyclohexane can be used as an inert solvent [24].

This trend of selectivity suggested that methane itself could be activated by these systems. Everybody realized that methane activation is interesting because of its very high bond dissociation energy (104 Kcal/mole) and its abundance in natural gas. Graham found that irradiation of **1** in the inert solvent perfluorohexane under methane pressure leads to the formation of the hydrido methyl compound **8** (reaction 11) [25].

$$Cp'(CO)_2Ir \xrightarrow{h\nu,\ CH_4} Cp'(CO)Ir(Me)H \quad (11)$$

1 8

The unsubstituted cyclopentadienyl analogue of **1** $Cp(CO)_2Ir$ similarly gave $Cp(CO)Ir(CH_3)H$. The hydridomethyl compound **9** was not prepared directly from the dihydride **4** due to experimental difficulties. However, Bergman observed that primary alkyl complexes are thermodynamically more stable than secondary complexes [26]. For example, the reaction of **4** with *n*-propane (reaction 9) or *n*-pentane gave mixtures of primary and secondary insertion products (with preferential formation of primary compounds). Treating these mixtures at 110 °C converts the secondary complexes into the more stable primary derivatives; this conversion takes place via reversible reductive elimination from the secondary hydrido alkyl isomers. This very interesting result allowed the preparation of selectively pure primary insertion products of linear alkanes and the achievement of methane activation (reaction 12). Thus, heating the secondary hydrido cyclohexyl compound **6** in cyclooctane under CH_4 pressure caused the formation of the methyl product **9** which is the "thermodynamic sink" for the system.

$$\underset{\mathbf{6}}{Cp'(PMe_3)Ir(C_6H_{11})(H)} \xrightarrow{110°,\ CH_4} \underset{\mathbf{9}}{Cp'(PMe_3)Ir(Me)(H)} \quad (12)$$

2.3. REACTIONS OF ALKANES BY OXIDATIVE ADDITION

The important discovery of the iridium and rhodium systems, which clearly shows the feasibility of insertion reaction of a soluble transition metal species into an alkane CH bond, lent credibility to the results of some investigators who had claimed to have found homogeneous alkane activation involving CH bond oxidative addition. A few systems were known, before the Graham and Bergman work, which described the transformation of alkanes by means of soluble transition metal complexes. Many at the time argued that free radicals or carbonium ions were involved in these reactions. As Crabtree put it: "We had great difficulty persuading the people that our work was real" [27].

A long controversy also grew up over the work of Shilov *et al.* [28]. In 1969 these authors published the first example of homogeneous CH bond activation in alkanes: the hydrogen–deuterium exchange reaction catalyzed by platinum(II) complexes (reaction 13).

$$RH \xrightarrow{PtCl_6^{=},\ AcOD,\ D_2O} RD \quad (13)$$

These reactions also lead to the formation of alkyl chlorides and acetates. Cycloalkanes are more reactive than acyclic alkanes; primary CH bonds were found to react faster than secondary CH bonds. Because of the presence of platinum metal, which sometimes precipitates, these reactions were suspected to be heterogeneous. Shilov demonstrated that Pt(IV) and Pt(O) are inactive in these exchanges; it is now accepted that these reactions may involve an oxidative addition step.

Crabtree *et al.* discovered the first system which reacted with saturated hydrocarbons to give stable and isolated organometallic products with a ligand coming from the alkane molecule [29]. The cationic iridium complex **10a** reacts in dichloroethane in the presence of 3,3-dimethylbutene with cyclopentane, cycloheptane and cyclooctane to give the compounds **11a–13a**, respectively (reaction 14). Other alkanes did not react. 3,3-Dimethylbutene, which acts as a hydrogen acceptor, must be present to observe the reactions. Convincing evidence that these reactions are homo-

$[L_2(CH_3COCH_3)_2IrH_2][A]$ + (3,3-dimethylbutene) → (cyclopentane) $[L_2IrH][A]$ (11); → (cycloheptane) $[L_2IrH][A]$ (12); → (cyclooctane) $[(C_8H_{12})IrL_2][A]$ (13) **(14)**

10 11 12 13

a : L = Ph_3P, A = BF_4

b : L = $(pF\text{-}C_6H_4)_3P$, A = SbF_6

geneous was provided; in particular these dehydrogenations proceed in the presence of mercury, a poison for heterogeneous catalysis.

In a subsequent report [30], the related complex **10b** was used instead of **10a**. The reactions with **10b** were performed in pure alkanes; thus, the yields were greatly improved and a variety of previously unreactive hydrocarbons were found to react under these new conditions. Methyl and ethylcyclopentane gave $[L_2Ir(R—C_5H_4)H][SbF_6]$(R = Me, Et; L = $(p\text{-}F—C_6H_4)_3P$). Activation of cyclohexane is exemplified by reaction 15. Methylcyclohexane gave only a cyclohexadienyl complex.

$$\text{C}_6\text{H}_{12} \xrightarrow[\text{(tBuCH=CH}_2\text{)}, 85°\text{C}, 20\text{h}]{\mathbf{10b}} [L_2IrH][A] \;(5\%) + [L_2Ir][A] \;(45\%) + \text{C}_6\text{H}_6 \;(32\%) \quad (15)$$

$L = (pF\text{-}C_6H_4)_3P$, $A = SbF_6$

Another modification to this system was the addition of a base to the reaction mixture. Whereas the cyclooctadiene compound **13b** was obtained from the reaction of **10b** with cyclooctane, the catalytic formation of cyclooctene (8 turnovers) was observed when the reaction was performed in the presence of 1,8-(dimethylamino)naphthalene. It is presumed that the base deprotonates the metal, and the olefin, which would be less strongly attached to the resulting neutral species, would then be released. More recently, the complex $[IrH_2(CF_3CO_2)(PCy_3)_2]$ (Cy = cyclohexyl) was found to catalyse the photochemical dehydrogenation of cyclooctane both in the presence and absence of a hydrogen acceptor [31]; it is assumed that the photochemical energy provides the thermodynamic driving force necessary for dehydrogenation to take place.

Baudry, Ephritikhine and Felkin found the first soluble organometallic system capable of functionalizing alkanes selectively and catalytically under mild conditions. The bis(phosphine)rhenium heptahydrides L_2ReH_7 **14** react with cycloalkanes C_nH_{2n} (n = 6, 7, 8) in the presence of dimethylbutene to give the corresponding cycloolefins (reaction 16) [32, 33]; these are not further dehydrogenated to dienes or benzene. The reaction with cyclopentane leads to the formation of the cyclopentadienyl compound **15** [33, 34].

$$\text{CH}_2\text{—CH}_2\,(\text{CH}_2)_{n-2} \xrightarrow{L_2ReH_7,\ \mathbf{14}} \text{CH=CH}\,(\text{CH}_2)_{n-2}$$

(16)

$$C_5H_{10} \xrightarrow{L_2ReH_7,\ \mathbf{14}} (C_5H_5)L_2ReH_2\ \mathbf{15} + C_5H_8$$

$n = 6, 7, 8$; L = Ph_3P, $(pMe\text{-}C_6H_4)_3P$, $(pF\text{-}C_6H_4)_3P$

These reactions were generalized by Felkin *et al.*, who studied the dehydrogenation of alkanes catalyzed by ruthenium and iridium polyhydrides L_3RuH_4 and L_2IrH_4 (L = tertiary phosphines) [35, 36]. The dehydrogenation of cyclooctane by means of these complexes required higher temperatures and longer reaction times than with the rhenium heptahydrides **14**, but the yields are much increased: with $[(p\text{-F—}C_6H_4)_3P]_2ReH_7$, 80 °C, 10 min, 9 turnovers; with $(i\text{-}Pr_3P)_2IrH_5$, 150 °C, 5 days, 70 turnovers.

The catalytic cycle presented in Figure 2 was proposed to explain these dehydrogenation reactions. The postulated active intermediates are the coordinatively unsaturated $[M]H_{n-2}$ **16** and $[M]H_{n-4}$ **17** which would be formed by stripping hydrogen atoms from the polyhydrides $[M]H_n$. It was assumed that the key step of the reaction would be the insertion of the 14-*e* species **17** into the CH bond of the alkane molecule RH to give the hydrido alkyl intermediate **18**. This latter could undergo a β elimination reaction leading to the olefin R(—H) via the olefin complex **19**.

These dehydrogenation reactions are essentially hydrogen transfer reactions between the alkane and dimethylbutene. In the case of the rhenium system, the species **17**, L_2ReH_3, was invoked to account for the products of the reactions of **14** with various organic substrates [38]. The same species L_2ReH_3 can also be responsible for the activation of

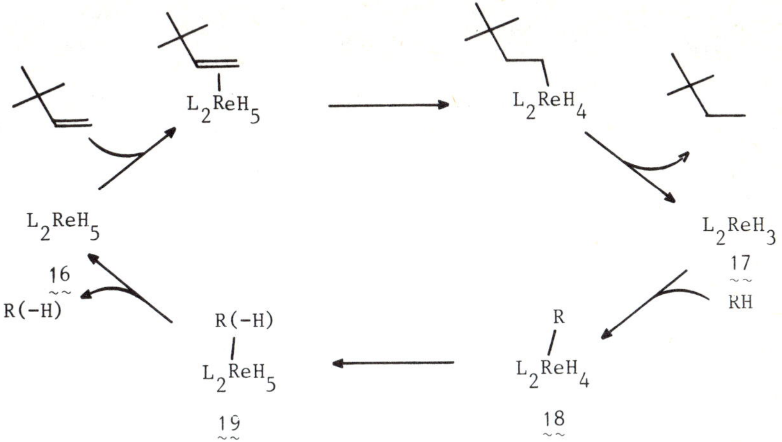

Fig. 2. Postulated catalytic cycle for the dehydrogenation of alkanes by L_2ReH_7 [35, 37].

cyclopentane, found by Caulton, [39] by irradiation of the complex $(Ph_3P)_3ReH_5$ in the presence of dimethylbutene (reaction 17).

$$+ \ L_3ReH_5 \xrightarrow{h\nu,} L_2ReH_2 \qquad (17)$$

$L = Ph_3P$

The catalytic reaction described in Figure 2 will follow a different course when the olefin R(—H) is strongly attached to the metal in the complex **19**. This is certainly what happens with cyclopentene which is a better ligand than the larger cycloalkenes; here, further dehydrogenation of **19** gives the cyclopentadienyl product **15**.

Linear alkanes also react with the rhenium heptahydrides **14** (L = Ph_3P) to give organometallic compounds: *n*-pentane gave the trihydridopentadiene compound **20** (reaction 18) [40]; the larger homologues were converted into the mixtures of the diene complexes **21** (it was demonstrated that these isomers are in equilibrium) [41]. Treatment of **20** or each of the mixtures **21** with trimethyl phosphite led selectively and quantitatively to the formation of the corresponding terminal olefins.

The selectivity of the dehydrogenation of methylcyclohexane by the

L_2ReH_7, 14

L_2ReH_3 20

$P(OMe)_3$

R

(18)

R, R, R

L_2ReH_3 + L_2ReH_3 + L_2ReH_3

21

R

L = Ph_3P ; R = H, Me, Et

rhenium [33] or the iridium [36] polyhydrides showed that the order of the reactivity of the various CH bonds towards these systems is primary > secondary > tertiary. Competitive experiments conducted with (i-$Pr_3P)_2IrH_5$ indicated that CH bonds of methyl CH groups in n-hexane (which is converted into a mixture of hexenes) are 8 times more reactive than methyl CH bonds in methylcyclohexane [36]. As expected, the iridium pentahydride was found to activate methane CH bonds; catalytic deuteriation of CH_4 was observed under mild conditions using C_6D_6 as the deuterium source [42]. Catalytic H—D exchange between deuteriobenzene and methane also occurs upon irradiation in the presence of $Cp(PPh_3)_2ReH_2$ [43].

3. Activation of Alkanes by Organoactinides

One year after Bergman and Graham published their work on the first definite examples of oxidative addition of an alkane CH bond, two new activation reactions of methane were reported (reactions 19 and 20).

$$Cp'_2Lu(CH_3)\cdots CH_3-LuCp'_2 \rightleftarrows Cp'_2LuCH_3 \xrightarrow{^{13}CH_4} Cp'_2Lu-{}^{13}CH_3 \quad \underline{via} \quad Cp'_2Lu\begin{matrix}{}^{13}CH_3\\ \vdots \quad H\\ CH_3\end{matrix} \qquad (19)$$

22 23

$$Cp'_2Th(CH_2CMe_3)_2 \longrightarrow Cp'_2Th(\text{thoracyclobutane}) \xrightarrow{CH_4} Cp'_2Th(CH_2CMe_3)Me \quad (20)$$

24 25 26

Watson observed by NMR the remarkable exchange of methane on the methyl lutetium complex **22** (reaction 19) [44]. Cyclohexane, which is the solvent of the reaction, does not react; the linear alkanes react more slowly, but the alkyl lutetium products decompose by β hydrogen transfer. Marks *et al.* found that methane reacts with the thoracyclobutane **25** (obtained by thermolysis of **24**) [45] to give the methyl neopentyl product **26** (reaction 20) [46]. The series of reactions **24** → **25** → **26** was devised from thermochemical considerations.

These organoactinide reactions are unlikely to proceed by oxidative addition of a CH bond into the Th(IV) or Lu(III) metal centres (in their highest oxidation state) and represent a new type of hydrocarbon activation. These reactions must be connected to the intramolecular cyclometallation reactions occuring at d^0 metal complexes of the early d block elements. The recent and very interesting report by Rothwell [47] reviews the literature on this subject and presents data which rule out a free radical mechanism and directly support a pathway involving a four centre transition state (**23** in reaction 19). We will not develop these arguments further; we will simply note the difference between the reactions 20 and 2. By contrast to the platinum complex [11], a negative value for the entropy of activation was found in the case of the cyclometallation of the bis(neopentyl)thorium compound **24** [45]. This entropy change suggests the presence of a highly ordered transition state and a multicentre pathway could be implicated.

Similarly to the oxidative addition reactions, for a long time the only intramolecular reactions known followed a multicentre pathway (α and β hydride abstractions at d^0 metal centers are classical examples). Reactions 19 and 20 constitute the first examples of such intermolecular reactions with saturated hydrocarbons.

4. Concluding Remarks

In view of these recent and important advances in the field of alkane CH bond activation, the future for hydrocarbon chemistry appears particularly bright. A series of new problems arise from these discoveries.

Most of the systems known at present give stoichiometric reactions: the polyhydride systems which catalytically dehydrogenate alkanes are far from having a practical application. Development of efficient, selective and catalytic functionalization of alkanes remains a challenging goal and faces a number of difficulties, one of which is the undesired side reactions of the active intermediates. Robust catalysts have to be found.

On the fundamental side, several questions are emerging. It is not clear why some metal complexes favour intermolecular reactions, whereas many other systems undergo intramolecular cyclometallation or do not react at all with alkanes. Kinetics and thermodynamics of inter and intramolecular CH bond activation were determined by Jones comparing the reaction of the 16-*e* intermediate **27** with propane solvent or with the phosphine propyl group (reaction 21) [48].

Cp'–Rh(H)(Me_2P⌒) ⟵ [Cp'–Rh–PMe_2Pr] (**27**) ⟶ (propane) Cp'–Rh(PMe_2Pr)(propyl)(H) (21)

Surprisingly, the study of this system indicated that the kinetic selectivity for aliphatic CH bond oxidative addition favours intermolecular reaction over intramolecular when neat hydrocarbon is the solvent; there is, however, a moderate thermodynamic preference for the intramolecular reaction. Therefore, thermodynamic but not kinetic terms would favour the unimolecular cyclometallation reactions and it is possible that some systems may activate alkane CH bonds intermolecularly but that the products may not have been observed because of its thermal instability.

Another problem is to understand the basis for the remarkable selectivities between CH bonds, primary bonds being more reactive than secondary and tertiary bonds. It was proposed that this selectivity could be due to steric factors, the CH bonds in the smaller cycloalkanes [21] and the methyl CH bonds in methyl cyclohexane [33] being more accessible. On the other hand, Parshall and Watson suggested that a polarized transition state like **23** would explain the lower reactivity of higher alkanes.

In connection with the preceding questions, it would be interesting to know what kind of interaction does exist between the metal centre and the alkane molecule. A number of electron deficient transition metal complexes are known which exhibit an interaction between the metal centre and a nearby CH bond [49]. The geometry of the M...H...C fragment varies from a linear to a triangular type situation (Figure 3).

M...H...C M.....H⋮C

Fig. 3. Interactions of CH bonds with metal centres.

Kinetic and thermodynamic study of the reversible CH activation/reductive elimination of alkanes at iridium suggested the possibility of a σ-complex intermediate between **6** (R = C_6H_{11}) and Cp′(PMe_3)Ir plus cyclohexane [50].

It seems likely that in reaction 19, the interaction of **22** with methane would be similar to that found in the unsymmetrical methyllutetium dimer where a methyl group of one monomer donates electron density to the second monomer. On the other hand, orbital analysis of oxidative addition of methane on d^6ML_5, d^8ML_4 and CpML fragments indicated that, because of steric effects, the CH bond would more easily approach the metal with the M...H...C... atoms linear [51]. A trajectory for the oxidative addition reaction was also proposed by Crabtree *et al.* [52], by studying the crystal structures of agostic complexes. It was assumed that a succession of static structures can give information about the dynamics of a reaction, each structure being considered to be a point on the reaction profile. The resulting trajectory (Figure 4) shows the CH bond approaching the metal with an M—H—C angle aroung 130°. The CH bond then elongates and rotates so as to close the angle before finally breaking. The selectivity of alkane CH bond activation is also explained on the basis of this trajectory and it is suggested that sterically uncongested metal species would favour external alkane activation rather than cyclometallation.

Reactions involving a multi-centre transition state occur with electrophilic metals; the rates of reactions of methane with the derivatives

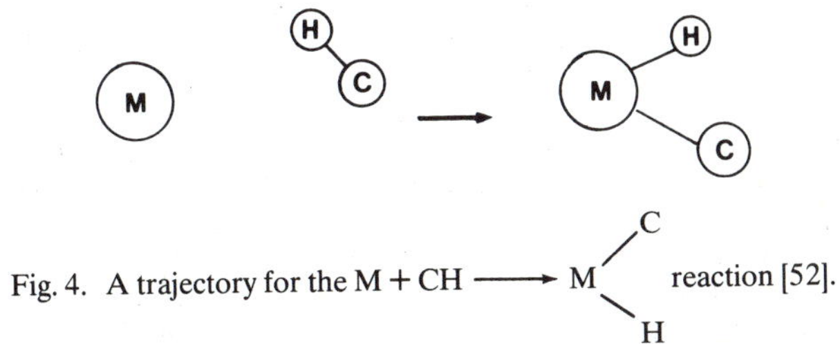

Fig. 4. A trajectory for the M + CH ⟶ M(C)(H) reaction [52].

Cp'_2MCH_3 (M = Sc, Lu, Y) increase with the electrophilicity of the metal [44]. The situation is less clear for oxidative addition. It was long believed that very electron rich metal centres would insert more easily in CH bonds; this idea must be revised as we know, for example, that both species **2** and **5** effect oxidative addition.

The results described in this paper represent very important advances in the story of CH bond activation and they are an incentive to devise new, more efficient systems which will improve our knowledge of these exciting reactions. The diversity of these first experiments suggests that several ways will be envisaged to develop the chemistry of alkanes.

Service de Chimie Moléculaire
IRDI/DESICP/DPC CNRS UA 331
CEA-CEN Saclay, 91191 Gif Sur Yvette, France

References

1. E. L. Muetterties: *Chemical Society Reviews.* **11**, 283 (1982); F. G. Gault: *Gazetta Chimica Italiana* **109**, 255 (1979); E. L. Muetterties: *Angew. Chem. Int. Ed. Engl.* **17**, 545 (1978); R. Ugo: *Catal. Rev.* **11**, 225 (1975).
2. R. A. Sheldon and J. K. Kochi: *Metal Catalyzed Oxidations of Organic Compounds*, Academic Press, New-York, vol. V, p. 136—149, vol. VII, p. 206—215 (1981); J. A. Sofranko, R. Eisenberg and J. A. Kampmeier: *J. Amer. Chem. Soc.* **102**, 1163 (1980); R. Davis, I. F. Croves and C. C. Rowland: *J. Organomet. Chem.,* **239**, C 9 (1982); H. Mimoun, L. Saussine, E. Daire, M. Postel, J. Fisher and R. Weiss: *J. Amer. Chem. Soc.,* **105**, 3101 (1983); A. Onopchenko and J. G. D. Schulz: *J. Org. Chem.,* **38**, 909 and 3729 (1973); J. Hanotier, P. Camerman, M. Hanotier-Bridoux and P. de Radzitzky: *J. Chem. Soc. Perkin II*, 2247 (1972); R. D. Bestre, E. R. Cole and G. Crank: *Tetrahedron Letters* **24**, 3891 (1983).
3. A. M. Khenkin and A. A. Shteinman: *J. Chem. Soc. Chem. Commun.* 1219 (1984); R. E. White and M. B. McCarthy: *J. Amer. Chem. Soc.* **106**, 4922 (1984); J. T. Groves and T. E. Nemo: *ibid.* **105**, 6243 (1983); J. A. Smegal and C. L. Hill: *ibid.* **105**, 3515 (1983); C. L. Hill, J. A. Smegal and T. J. Henly: *J. Org. Chem.* **48**, 3277 (1983); D. Mansuy, M. Fontecave and J. F. Bartoli: *J. Chem. Soc. Chem. Commun.*, 253 (1983); I. Tabushi and A. Yazaki: *J. Amer. Chem. Soc.,* **103**, 7371 (1981); C. L. Hill and J. A. Smegal: *Nouveau Journal de Chimie* **6**, 287 (1982); J. T. Groves, W. J. Kruper and R. C. Haushalter: *J. Amer. Chem. Soc.* **102**, 6375 (1980); C. L. Hill and B. C. Schardt: *ibid.* **102**, 6374 (1980).
4. N. Yoneda, T. Fukuhara, Y. Takahashi and A. Suzuki: *Chem. Letters*, 17 (1983); A. G. Olah, D. G. Parker and N. Yoneda: *Angew. Chem. Int. Ed. Engl.,* **17**, 909 (1978); J. Sommer, M. Muller and K. Laali: *Nouveau Journal de Chimie* **6**, 3 (1982); S. Rozen, C. Gal and Y. Gaust: *J. Amer. Chem. Soc.,* **102**, 6861 (1980); D. H. R. Barton, M. J. Gastiger and W. B. Motherwell: *J. Chem. Soc. Chem. Commun.*, 41 and 731 (1981); D. H. R. Barton, R. S. H. Motherwell and W. B. Motherwell: *J. Chem. Soc. Perkin I.*, 445 (1983); D. H. R. Barton, R. S. H. Motherwell and W. B.

Motherwell: *Tetrahedron Letters* **24**, 1979 (1983); D. H. R. Barton, J. Boivin, N. Ozbalik and K. M. Schwartzentruber: *Tetrahedron Letters* **25**, 4219 (1984).
5. For an extensive review see: R. H. Crabtree: *Chem. Rev.* **85**, 245 (1985).
6. The major part of this paper was published as: M. Ephritikhine: *Nouveau Journal de Chimie* **10**, 9 (1986).
7. G. W. Parshall: *Catalysis: Specialist Periodical Reports* The Chemical Society, London Vol. 1, p. 335 (1977); G. W. Parshall: *Acc. Chem. Research* **8**, 113 (1975); D. E. Webster: *Adv. Organomet. Chem.* **15**, 147 (1976); C. Masters: *Homogeneous Transition Metal Catalysis; A Gentle Art*, Chapman and Hall, London, (1981); A. Shilov and A. A. Shteinman: *Coord. Chem. Rev.*, **24**, 97 (1977).
8. J. A. Labinger and J. A. Osborn: *Inorg. Chem.*, **19**, 3230 (1980); F. R. Jensen and B. Knickel: *J. Amer. Chem. Soc.*, **92**, 6339 (1971).
9. R. Mason and D. W. Meek: *Angew. Chem. Int. Ed. Engl.* **17**, 183 (1978); M. I. Bruce: *ibid.* **16**, 73 (1977); S. Hietkamp, D. F. Stufkens and J. Vrieze: *J. Organomet. Chem.* **122**, 419 (1976); *ibid.* **139**, 189 (1977); *ibid.* **152**, 347 (1978); P. W. Clark: *J. Organomet. Chem.*, **137**, 235 (1977); G. R. Clark, M. A. Mazid, D. R. Russel, P. W. Clark and A. Jones: *ibid.* **166**, 109 (1979); W. Y. Youngs and J. A. Ibers: *Organometallics*, **2**, 979, (1983).
10. H. D. Empsall, E. M. Hyde, R. Markham, W. S. McDonald, M. Norton, B. L. Shaw and B. Weeks, *J. Chem. Soc. Chem. Commun.*, 589 (1977).
11. R. Di Cosimo, S. S. Moore, A. F. Sowinski and G. M. Whitesides, *J. Amer. Chem. Soc.*, **104**, 124 (1982); P. Fowley, R. Di Cosino and G. M. Whitesides: *J. Amer. Chem. Soc.*, **102**, 6713 (1980); J. A. Ibers, R. Di Cosino and G. M. Whitesides: *Organometallics*, **1**, 13 (1982).
12. S. D. Ittel, C. A. Tolman, A. D. English and J. P. Jesson: *J. Amer. Chem. Soc.*, **100**; 7577 (1978); A. D. English and T. Herskovitz: *ibid.* **99**, 1648 (1977); S. D. Ittel, C. A. Tolman, A. D. English and J. P. Jesson: *ibid.* **98**, 6073 (1976).
13. C. Masters, A. A. Kiffen and J. P. Visser: *J. Amer. Chem. Soc.*, **98**, 1357 (1976); H. Imai, T. Nishiguchi and K. Fukuzumi: *J. Org. Chem.*, **41**, 665 (1976).
14. M. Berry, K. Elmitt and M. L. H. Green: *J. Chem. Soc. Dalton Trans.* 1950 (1979).
15. J. Halpern: *Acc. Chem. Res.* **15**, 332 (1982); J. Halpern: *Inorganica. Chim. Acta* **100**, 41 (1985).
16. D. B. Jacobson and B. S. Freiser: *J. Amer. Chem. Soc.* **105**, 736 and 5197 (1983); T. J. Carlin, L. Sallans, C. J. Cassady, D. B. Jacobson and B. S. Freiser: *ibid.* **105**, 6320 (1983); P. B. Armentrout and J. L. Beauchamp: *ibid.* **103**, 784 (1981); R. B. Freas and D. P. Ridge: *ibid.* **102**, 7129 (1980); J. Allison, R. B. Freas and D. P. Ridge: *ibid.* **101**, 1332 (1979).
17. G. A. Ozin, D. F. McIntosh and S. A. Mitchell: *J. Amer. Chem. Soc.* **103**, 1574 (1981); W. E. Billups, M. M. Konarski, R. H. Hauge and J. L. Margrave: *ibid.* **102**, 7394 (1980); K. J. Klabunde and Y. Tanaka: *ibid.* **105**, 3544 (1983).
18. M. L. H. Green and D. O'Hare: *J. Chem. Soc. Chem. Commun.* 355 (1985); M. L. H. Green, D. O'Hare and G. Parkin: *ibid.* 356 (1985); M. L. H. Green and G. Parkin: *ibid.* 1467 (1984); J. A. Bandy, F. G. N. Cloke, M. L. H. Green, D. O'Hare and K. Prout: *ibid.* 240 (1983).
19. A. H. Janowicz and R. G. Bergman: *J. Amer. Chem. Soc.* **104**, 352 (1982).
20. J. K. Hoyano and W. A. G. Graham: *J. Amer. Chem. Soc.* **104**, 3723 (1982).
21. A. H. Janowicz and R. G. Bergman: *J. Amer. Chem. Soc.* **105**, 3929 (1983).
22. R. A. Periana and R. G. Bergman: *Organometallics* **3**, 508 (1984).
23. W. D. Jones and F. J. Feher: *Organometallics* **2**, 562 (1983).

24. R. G. Bergman, P. F. Seidler and T. T. Wenzel: *J. Amer. Chem. Soc.*, **107**, 4358 (1985).
25. J. K. Hoyano, A. D. McMaster and W. A. G. Graham: *J. Amer. Chem. Soc.* **105**, 7190 (1983).
26. M. J. Wax, J. M. Stryker, J. M. Buchanan, C. A. Kovac and R. G. Bergman: *J. Amer. Chem. Soc.* **106**, 1121 (1984).
27. T. H. Maugh: *Science* **220**, 1261 (1983).
28. N. F. Gol'dshleger, M. B. Tyabin, A. E. Shilov and A. A. Shteinman: *Zh. Fiz. Khim* **43**, 2174 (1969); A. E. Shilov: *Soviet Scientific Reviews, Section B*, Chemistry Reviews, **4**, 71 (1982); L. A. Kushch, V. V. Lavrusko, Y. S. Misharin, A. P. Moravsky and A. E. Shilov: *Nouveau Journal de Chimie* **7**, 729 (1983). see also: R. J. Hodges, D. E. Webster and P. B. Wells: *J. Chem. Soc.* (A) 3230 (1971); R. J. Hodges, D. E. Webster and P. B. Wells: *J. Chem. Soc. Dalton. Trans.* 2571 (1972).
29. R. H. Crabtree, J. M. Mihelcic and J. M. Quirk: *J. Amer. Chem. Soc.*, **101**, 7738 (1979); R. H. Crabtree, M. F. Mellea, J. M. Mihelcic and J. M. Quirk: *ibid.* **104**, 107 (1982); R. H. Crabtree, P. C. Demou, D. Eden, J. M. Mihelcic, C. A. Parnell, J. M. Quirk and G. E. Morris: *ibid.* **104**, 6994 (1982).
30. M. J. Burk, R. H. Crabtree, C. P. Parnell and R. J. Uriarte: *Organometallics*, **3**, 816 (1984).
31. M. J. Burk, R. H. Crabtree and D. V. Mc Grath: *J. Chem. Soc. Chem. Commun.* 1829 (1985).
32. D. Baudry, M. Ephritikhine and H. Felkin: *J. Chem. Soc. Chem. Commun.* 606 (1982).
33. D. Baudry, M. Ephritikhine, H. Felkin and R. Holmes-Smith: *J. Chem. Soc. Chem. Commun.* 785 (1983).
34. D. Baudry, M. Ephritikhine and H. Felkin: *J. Chem. Soc. Chem. Commun.* 1243 (1980).
35. H. Felkin, T. Fillebeen-Khan, Y. Gault, R. Holmes-Smith and J. Zakrzewski: *Tetrahedron Letters* **25**, 1279 (1984).
36. H. Felkin, T. Fillebeen-Khan, R. Holmes-Smith and Y. Lin: *Tetrahedron Letters* **26**, 1999 (1985).
37. N. Aktogu, D. Baudry, D. Cox, M. Ephritikhine, H. Felkin, R. Holmes-Smith and J. Zakrzewski: *Bull. Soc. Chim. France* **381** (1985).
38. D. Baudry, P. Boydell and M. Ephritikhine: *J. Chem. Soc. Dalton Trans.* 525 (1986); D. Baudry, M. Ephritikhine and H. Felkin: *J. Organomet. Chem.* **224**, 363 (1982); D. Baudry, J. C. Daran, Y. Dromzee, M. Ephritikhine, H. Felkin, Y. Jeannin and J. Zakrzewski: *J. Chem. Soc. Chem. Commun.* 813 (1983).
39. M. A. Green, J. C. Huffmann, K. G. Caulton, W. K. Rybak and J. Ziolkowski: *J. Organomet. Chem.* **218**, C 39 (1981).
40. D. Baudry, M. Ephritikhine, H. Felkin and J. Zakrzewski: *J. Chem. Soc. Chem. Commun.* 1235 (1982).
41. D. Baudry, M. Ephritikhine, H. Felkin and J. Zakrzewski: *Tetrahedron Letters* **25**, 1283 (1984); D. Baudry, M. Ephritikhine, H. Felkin and J. Zakrzewski: *J. Organomet. Chem.*, **272**, 391 (1984).
42. C. J. Cameron, H. Felkin, T. Fillebeen-Khan, N. J. Forrow and E. Guittet: *J. Chem. Soc. Chem. Commun.* 801 (1986).
43. W. D. Jones and J. A. Maguire: *Organometallics* **5**, 590 (1986).
44. P. L. Watson: *J. Amer. Chem. Soc.* **105**, 6491 (1983); G. W. Parshall and P. L. Watson: *Acc. Chem. Res.* **18**, 51 (1985).

45. J. W. Bruno, T. J. Marks and V. W. Day: *J. Amer. Chem. Soc.,* **104**, 7357 (1982).
46. J. W. Bruno, T. J. Marks and I. R. Moss: *J. Amer. Chem. Soc.* **105**, 6824 (1983).
47. I. P. Rothwell: *Polyhedron* **4**, 177 (1985).
48. W. D. Jones and F. J. Feher: *J. Amer. Chem. Soc.* **107**, 620 (1985).
49. M. Brookhart and M. L. H. Green: *J. Organomet. Chem.* **250**, 395 (1983).
50. J. M. Buchanan, J. M. Stryker and R. G. Bergman: *J. Am. Chem. Soc.,* **108**, 1537 (1986).
51. J. Y. Saillard and R. Hoffmann: *J. Amer. Chem. Soc.* **106**, 2006 (1984).
52. R. H. Crabtree, E. M. Holt, M. Lavin and S. M. Morehouse: *Inorg. Chem.* **24**, 1986 (1985).

JEAN-PIERRE SAUVAGE

COORDINATION PHOTOCHEMISTRY: PHOTOINDUCED ELECTRON TRANSFER AND REDOX PHOTOCATALYSIS

1. Introduction

Although industrial applications of photochemistry are rather limited, the field of inorganic photochemistry has expanded enormously in the past fifteen years. Classical photochemistry was devoted to the study of chemical transformations of given compounds under light irradiation:

$$\text{starting material} \xrightarrow{h\nu} \text{product.}$$

Numerous such reactions can be found in the exhaustive book of Balzani and Carassiti on the photochemistry of coordination compounds, published in 1970 [1]. This monograph was of extreme importance in the development of this research area. It is noteworthy that it contains relatively little photo-redox chemistry: it is only about ten years ago that electron transfer in the excited state was considered to be important. Interestingly, a special issue of the *Journal of Chemical Education* was recently devoted to inorganic photochemistry [2]: electron transfer reactions are clearly among the most studied and applied photochemical reactions of today.

Several reasons might be put forward in order to explain the "boom" of photoinduced redox inorganic chemistry. Two important factors are the following:

(1) in contrast to the vast majority of organic molecules, coordination compounds are usually coloured.
(2) inorganic photocatalysts are chemically more resistant than their organic analogues. Therefore, they seem to be ideally suited for *photocatalysis under solar light irradiation.*

The two factors mentioned above lead to the start of an active research area, whose long range aim is the *photochemical storage of light energy*, related to the possibility of converting solar energy to a storable fuel.

A. Mortreux and F. Petit (Eds.), Industrial Applications of Homogeneous Catalysis, 277–292.

2. Properties of the Excited State

An electronically excited state of a molecule can be considered as a new species whose chemical properties might differ considerably from those of the same molecule taken in its ground state. Although some of the properties of the excited state are now well understood, mainly those involving thermodynamic aspects, it is extremely difficult to predict other parameters such as the lifetime of the excited state and the rate of electron transfer processes [3—5].

2.1. KINETIC ASPECT

Let us consider a coordination compound A, whose excited state *A is susceptible to various reactions as shown in the following scheme:

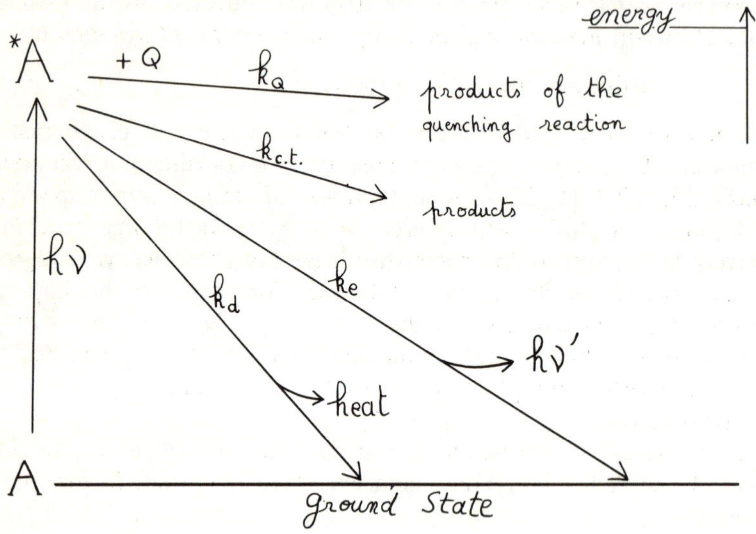

The various rate constants are indicated beside the corresponding arrows. k_d, k_e and $k_{c.t.}$ are the first-order rate constants for radiationless deactivation of the excited state, photonic emission and intramolecular chemical transformation respectively. The intrinsic lifetime τ_0 of the excited state *A is:

$$\tau_0 = \frac{1}{k_d + k_e + k_{c.t.}}$$

In the presence of an added reactant Q, the reaction *A + Q → products has to be taken into account. Clearly, this bimolecular quenching corresponds to the *reaction of interest* if photocatalytic processes are to be induced. Such a reaction might take place if the intrinsic lifetime τ_0 is long enough to allow the two reactants *A and Q to meet each other. If the quenching process is operative, the lifetime τ of *A decreases in the presence of Q:

$$\tau = \frac{1}{k_d + k_e + k_{c.t.} + k_Q[\mathrm{Q}]}$$

where k_Q = bimolecular rate constant of quenching. The Stern-Volmer equation:

$$\frac{I_0}{I} = 1 + k_Q \tau_0 [\mathrm{Q}]$$

is of great importance for evaluating the kinetic aspects of a quenching reaction. Experimentally, if the photoactive component is luminescent in fluid solution, τ_0 and k_Q are relatively easily accessible. Unfortunately, the vast majority of inorganic compounds are non-luminescent and have only extremely short-lived excited states.

2.2. REDOX PROPERTIES OF THE EXCITED STATE

By considering a very simple orbital diagram, we shall demonstrate the following paradoxical statement: *an excited state is, at the same time, a better oxidant and a better reductant than the corresponding ground state.*

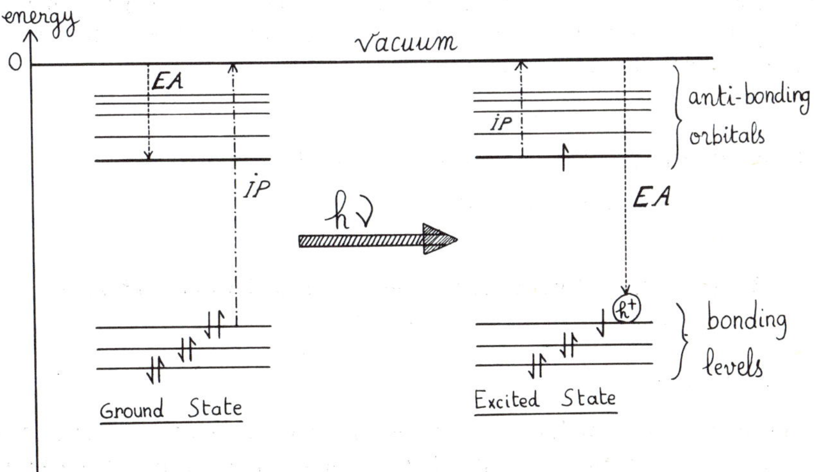

As shown in the diagram, let us assume that an electron is promoted from a low energy bonding orbital of A to an antibonding orbital. Such a process creates a positive charge (hole) at a low energy level and thus increases the electron affinity, EA, of the molecule. EA is directly related to the oxidizing character of the species: molecules or atoms with a large EA are strong oxidants. At the same time, electronic excitation diminishes the ionization potential, IP, and therefore, increases the electron donor character of the excited state *A, providing the latter with strong reducing properties.

We can now roughly estimate the redox potentials of several couples and relate them to ζ_{0-0}, the energy level of the excited state with respect to the ground state:

$$A \xrightarrow{h\nu} {}^{*}A \qquad \zeta_{0-0}$$

where ζ_{0-0} = energy difference between the O vibronic states of the ground state and the excited state.

$$
\begin{array}{lll}
{}^{*}A + e^{-} \longrightarrow A^{-} & & E^{0}_{{}^{*}A/A-} \\
{}^{*}A \longrightarrow A^{+} + e^{-} & & E^{0}_{A+/{}^{*}A} \\
A + e^{-} \longrightarrow A^{-} & & E^{0}_{A/A-} \\
A \longrightarrow A^{+} + e^{-} & & E^{0}_{A+A}
\end{array}
$$

Determination of redox potentials of couples involving the ground state and the determination of ζ_{0-0} by photophysical measurement allows us to construct the following redox diagram.

If one knows the redox potentials of couples involving the quencher Q, it is possible to predict whether a photochemical reaction is thermodynamically possible. For instance, from the above diagram it can be seen that the following electron transfer is allowed:

$${}^{*}A + Q \longrightarrow A^{+} + Q^{-}$$

whereas the same reaction performed in the dark is impossible:

$$A + Q \not\longrightarrow A^{+} + Q^{-}$$

3. Examples of Coordination Compounds with Charge Transfer Transitions

3.1. VARIOUS TRANSITIONS

Transition metal complexes are able to undergo various electronic transitions, leading to the corresponding excited states. If the electron is

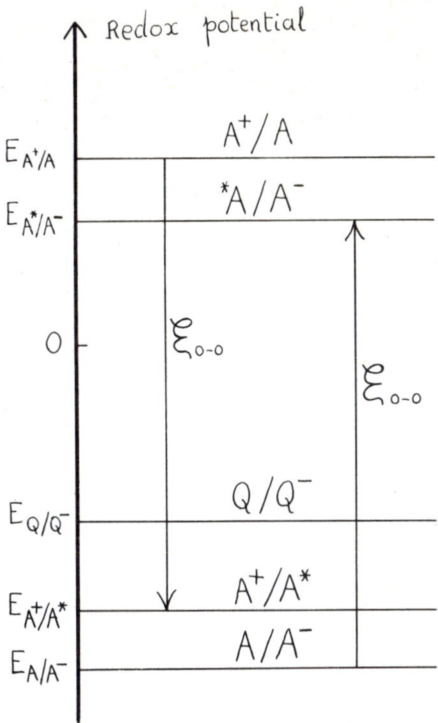

promoted from an orbital located on the metal to another orbital of metal character, the transition is called metal-centered. It originates from the splitting of the five *d* orbitals and corresponds to a *d-d* or ligand-field transition. Another type of transition involves orbitals located on the ligand; this is particularly true for aromatic ligands with low lying π^* orbitals. Other examples of lesser importance can be found: ion pairing charge transfer, charge transfer to solvent, etc.

The most interesting excited states for photoredox chemistry are intramolecular charge transfer (CT) in nature, the direction of the CT being extremely important. Transition metal complexes containing an electron rich ligand and a metal centre in a high oxidation state lead to ligand-to-metal charge-transfer excited states (LMCT). On the other hand, complexes whose metal centre is in a low oxidation state and containing ligands with low-lying unoccupied orbitals, are likely to give rise to metal-to-ligand charge-transfer excited states (MLCT). In particular, this is the case for bipy or phen complexes of d^6 (Cr(0), Mn(I), Re(I), Ru(II), Os(II), . . .) or d^{10} (Cu(I)) metals (bipy = 2,2′-bipyridine; phen = 1,10-phenanthroline).

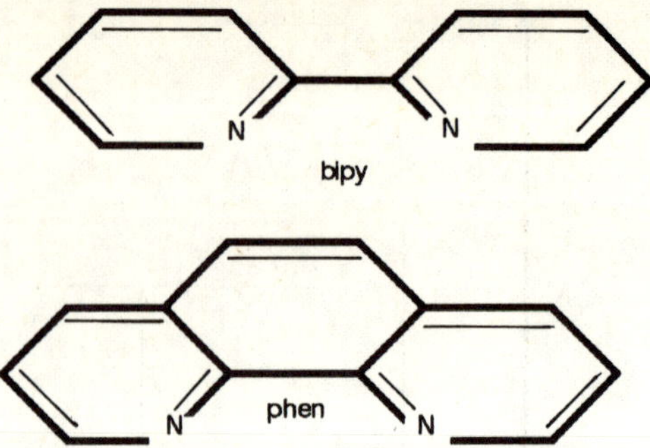

3.2. $Ru(bipy)_3^{2+}$, A COMPLEX WITH A LONG-LIVED MLCT EXCITED STATE

The properties and applications of $Ru(bipy)_3^{2+}$ have recently been discussed in an exhaustive review [6]. The most important spectral and photophysical properties of this compound are the following:

- the lowest excited state is MLCT in nature which makes $Ru(bipy)_3^{2+}$ an attractive candidate as redox photocatalyst. The transition is $d_{Ru(II)}$ (t_{2g}) → π*(bipy), as depicted on the simplified representation of the electronic situation for an octahedral d^6 complex, given below:

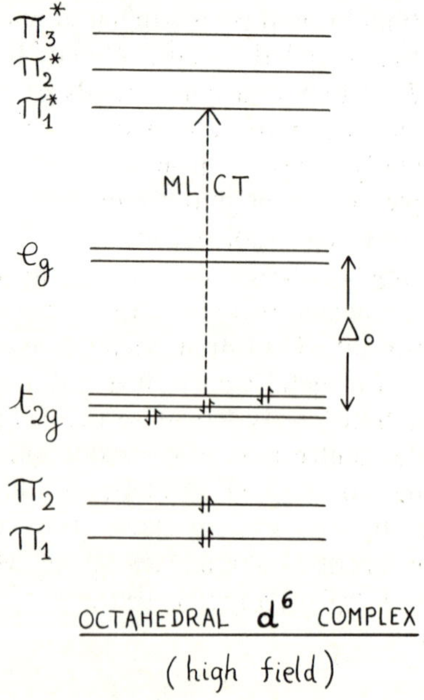

- the energy of the MLCT excited state is 2.1 eV above that of the ground state. In other words, $Ru(bipy)_3^{2+}$ is a highly coloured complex, with a broad absorption bond around 450 nm; it covers a reasonable fraction of the solar spectrum.
- of great importance is the lifetime of the excited state: the MLCT excited state lives for ~0.6 μs at room temperature in solution (H_2O for instance), which is exceptionally long for this type of compound. In addition, the complex is highly luminescent (emission quantum yield ~5%) which facilitates kinetic studies.

$$Ru^{II}(bipy)_3^{2+} \xrightarrow{h\nu} Ru^{III}(bipy)_2(bipy)^{-\cdot})^{2+}$$

The above reaction is an approximation, but it has the advantage of showing the MLCT character of the transition. The intramolecular electron—($bipy^{-\cdot}$)—hole—[Ru(III)] recombination is sufficiently slow to allow bimolecular reactions:

$$Ru^{(III)}(bipy)_2(bipy^{-\cdot})^{2+} \longrightarrow Ru(bipy)_3^{2+}$$

$$k_d = \frac{1}{\tau_0}; \ \tau_0 = 0.6\ \mu s$$

$$Ru(bipy)_3^{2+} + Q \xrightarrow{kQ} \text{quenching products}$$

- in addition, more trivial conditions such as chemical, photo- and electrochemical stability, accessibility, etc. seem to be fulfilled by $Ru(bipy)_3^{2+}$.

In order to understand and to predict the redox behaviour of the MLCT excited state $^*Ru(bipy)_3^{2+}$, it is necessary to keep in mind the redox diagram represented below.

Clearly, $^*Ru(bipy)_3^{2+}$ is both a strong reductant and a moderate oxidant. On thermodynamic grounds only, it is reasonable to suppose that $^*Ru(II)(bipy)_3^{2+}$ might oxidise water to O_2 (with formation of $Ru(bipy)_3^{+}$) and/or reduce water to H_2 (with simultaneous generation of $Ru(bipy)_3^{3+}$). Such considerations have led to a formidable interest in $Ru(bipy)_3^{2+}$ and its derivatives.

3.3. OTHER EXAMPLES: d^6 or d^{10} COMPLEXES

In order to have MLCT excited states separate in energy from metal-centred excited states and, as a consequence, to avoid fast deactivation, the best situation corresponds to d^6 or d^{10} transition metals. In an octahedral d^6 complex, if the ligand field is strong enough, the e_g level will

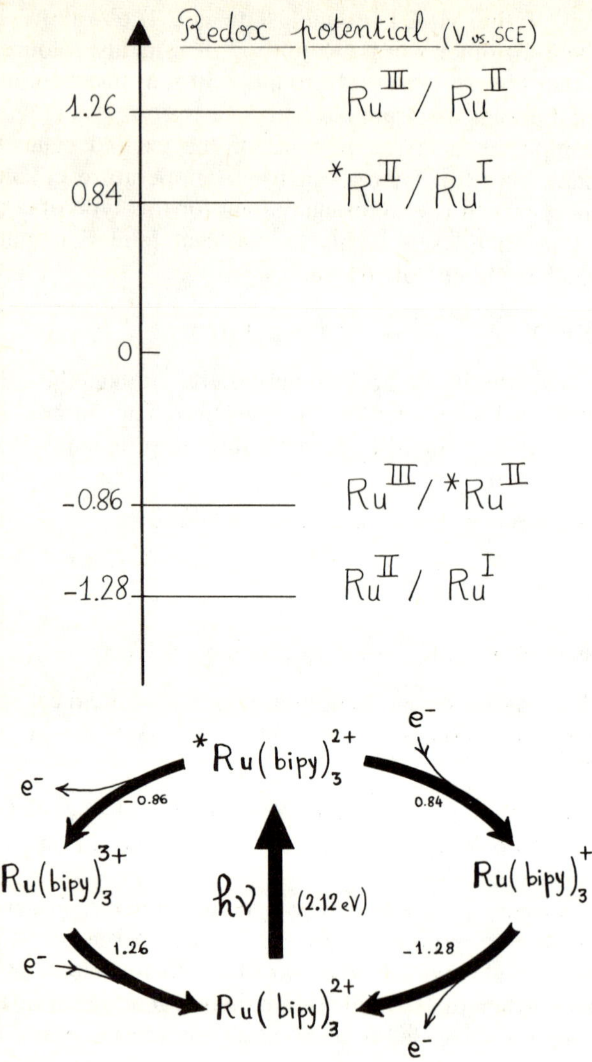

be far above the t_{2g} one, making the $t_{2g} \rightarrow e_g$ electronic transition much higher in energy than MLCT with aromatic polyimines like bipy or phen.

- the $M(CO)_4$bipy or $M(CO)_4$phen complexes (M = Cr, Mo, W) are emissive for instance at room temperature, in benzene [7]. The use of such compounds as photocatalysts might be considered, although their photochemical stability is probably not excellent.

- Re(I)$(CO)_3$(Cl)phen (facial isomer) is luminescent in fluid solution [8]; the MLCT excited state is long lived (~ 0.6 s), its energy being ~ 2.3 eV above that of the ground state. The redox diagram of $Re(CO)_3$(Cl)phen and its MLCT excited state is represented below:

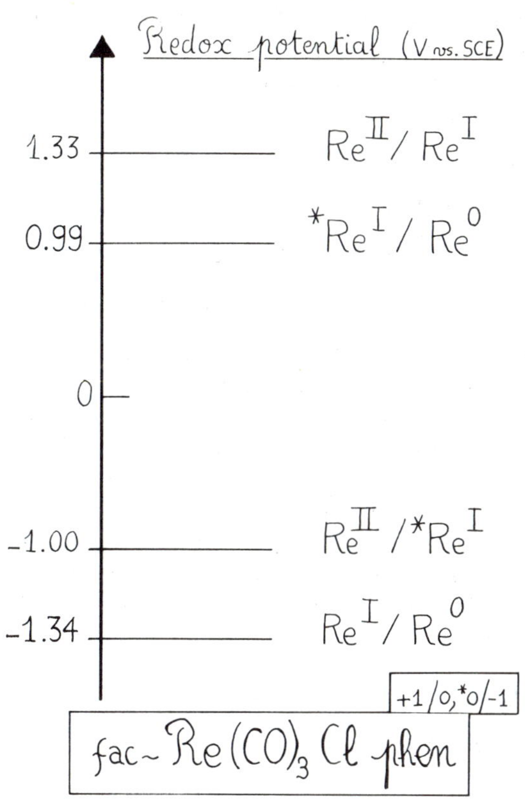

- $Os(bipy)_3^{2+}$ displays properties similar to those of $Ru(bipy)_3^{2+}$ [10]; however, the excited state has a smaller lifetime and the redox properties are quite different from those of $Ru(bipy)_3^{2+}$. In particular, the MLCT state is a more powerful reductant than that of its ruthenium equivalent. It is noteworthy that the redox parameters can be adjusted by introducing electron withdrawing or electron donating substituents at the appropriate positions of the diimine ligands. Copper(I) complexes are d^{10} and some of them show intense charge transfer transitions in the visible [11]. In particular, using properly substituted phenanthrolines, it has been possible to obtain long lived MLCT excited states of tetrahedral complexes [12], able to transfer an electron or energy to an appropriate acceptor [13].

4. Electron Transfer Reactions of the Excited State

Without entering into the details of the bimolecular quenching processes, it is necessary to know that the lowest excited state of a given complex, *A, can act as an energy donor, an electron acceptor (oxidant) or an electron donor (reductant):

$$^*A + Q \xrightarrow{k_{en.t.}} A + Q^* \quad \text{energy transfer}$$

Q = quencher

$$^*A + Q \xrightarrow{k_r} A^- + Q^+ \quad \text{reductive quenching}$$

$$^*A + Q \xrightarrow{k_{ox}} A^+ + Q^- \quad \text{oxidative quenching}$$

Depending on the redox properties of the quencher and on the energy level of its lowest excited state, one or several processes mentionned above can be operative. However, in almost all cases, after an electron transfer has taken place, the recombination reaction (back electron transfer) is extremely fast. In other words, if no external way for separating the photochemically generated charges can be found, the electrochemical energy produced for a short time is waster by charge recombination. As an illustration, let us consider the reaction between $Ru(bipy)_3^{2+}$ and MV^{2+} (MV^{2+} = methyl viologen: *N,N'*-dimethyl-4,4′-bipyridinium dication) under light irradiation [14, 15]:

$$Ru(bipy)_3^{2+} \xrightarrow{h\nu} {}^*Ru(bipy)_3^{2+}$$

MLCT excited state

oxidative quenching:

$$^*Ru(bipy)_3^{2+} + MV^{2+} \xrightarrow{k_{ox}} Ru(bipy)_3^{3+} + MV^{+\cdot}$$

Knowing the redox potentials of $Ru(bipy)_3^{3+}/{}^*Ru(bipy)_3^{2+}$ ($E^0 = -0.84$ V), $Ru(bipy)_3^{3+}/Ru(bipy)_3^{2+}$ ($E^0 = +1.26$ V) and $MV^{2+}/MV^{+\cdot}$ ($E^0 = -0.44$ V), it is easy to calculate the driving force of the oxidative quenching:

$$\Delta G^0_{ox} = -0.84 - (-0.44) = -0.40 \text{ eV}$$

Therefore, the oxidative quenching is a thermodynamically favourable reaction; it is thus not surprising that this same reaction is very fast: $k_{ox} \sim 10^9\ M^{-1}\ s^{-1}$ in H_2O at room temperature. One can also imagine that the back reaction (recombination) is even more favourable, either thermodynamically or kinetically:

recombination or back reaction

$$Ru(bipy)_3^{3+} + MV^{+\cdot} \xrightarrow{k_b} Ru(bipy)_3^{2+} + MV^{2+}$$

$$\Delta G_b^0 = -0.44 - 1.26 = -1.70 \text{ eV}$$

$$k_b = 8 \times 10^9 \text{ M}^{-1} \text{ s}^{-1} \quad \text{in } H_2O \text{ at room temperature}$$

As a consequence, the amount of energy stored in the redox products $Ru(bipy)_3^{3+}$ and $MV^{+\cdot}$ is important (1.70 eV, i.e. ~39 kcal) but this electrochemical storage is only very short lived. In order to create a practical system, it will be necessary to transform the oxidized *and* the reduced compounds formed as transient species into stable and storable products; a promising means would of course be the generation of a fuel like H_2, by simultaneously oxidizing and reducing H_2O:

$$Ru(bipy)_3^{3+} + \tfrac{1}{2} H_2O \longrightarrow Ru(bipy)_3^{2+} + \tfrac{1}{4} O_2 + H^+$$

$$MV^{+\cdot} + H^+ \longrightarrow MV^{2+} + \tfrac{1}{2} H_2$$

Numerous examples of photoinduced electron transfer reactions can be found in the literature, making use of $Ru(bipy)_3^{2+}$ as photoactive component [6], the redox partner being an inorganic cation, a transition metal complex or an organic species.

5. Photochemical Conversion and Storage of Light Energy

5.1. PRINCIPLE

An excited state contains a certain amount of energy that must be converted to a storable form prior to deactivation of the excited state. A general scheme is represented below:

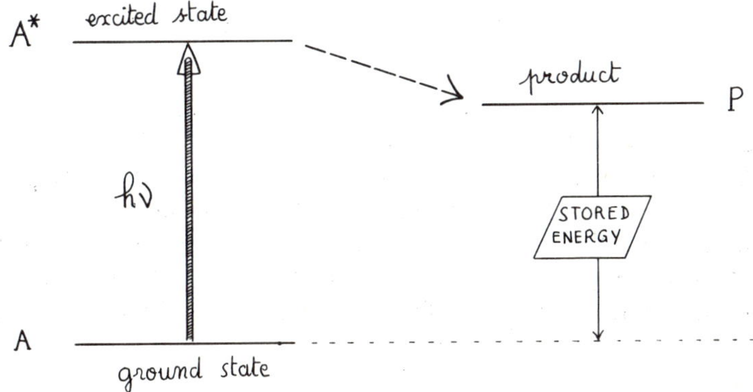

Such a diagram corresponds to the *stoichiometric* conversion of A into an *energy rich product.* It might be envisaged that P liberates an important fraction of the stored energy, in a thermal and possibly catalysed process with regeneration of the starting material. This would open the possibility of recycling A, making the device practically more feasible. The principle of a photoredox system is given below:

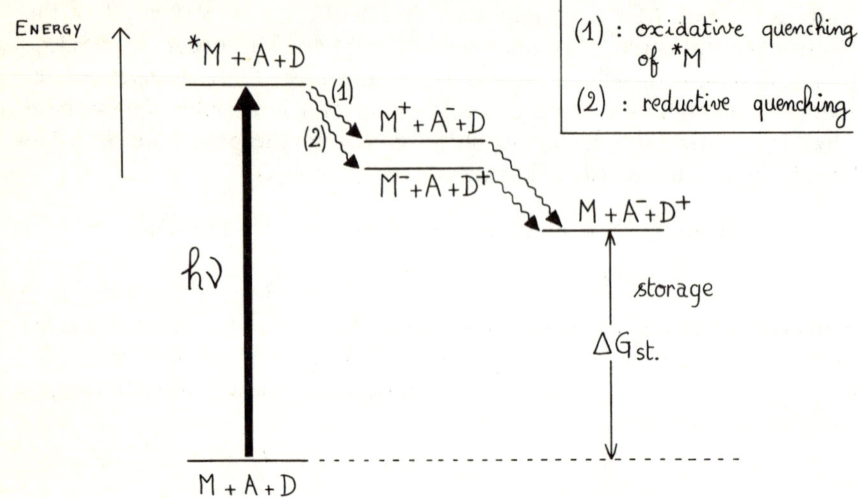

M is a transition metal complex; A and D are an electron acceptor and an electron donor, respectively. In order to store ΔG_{st} for a reasonable period of time, one must either avoid the reaction between A^- and D^+ (energy wasting process) or rapidly convert those two species into A and D with simultaneous reduction and oxidation of an appropriate substrate (H_2O, for instance).

5.2. A NON-REDOX EXAMPLE: ISOMERISATION OF NORBORNADIENE

An attractive and well studied reaction [16] is the photoisomerisation of nornadiene (NBD) into quadricyclane (Q).

NBD is almost transparent to *solar radiation.* However, in the presence of an appropriate inorganic sensitizer (CuCl, for instance), it undergoes valence isomerisation to Q when exposed to sunlight. Q is a highly strained molecule but it is reasonably stable and can be stored at room temperature. Under certain conditions, Q can liberate its stored energy as

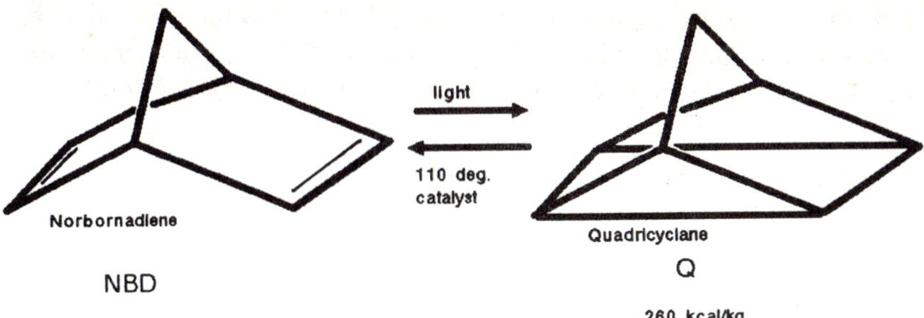

heat, with regeneration of NBD. Although many fundamental and practical problems remain to be solved, the NBD/Q system represents a possibility for converting light energy into storable thermal energy.

5.3. PHOTOCHEMICAL REDUCTION OF WATER

The photochemical cleavage of water into H_2 and O_2 is certainly the most actively investigated energy storage reaction:

$$H_2O(l) — H_2(g) + \tfrac{1}{2}\, O_{2(g)} \qquad \Delta G^0 = 57\ \text{kcal/mole}^{-1}$$

The advantages of this reaction are numerous, and will not be discussed here The most obvious one is that H_2 is, potentially, a non-polluting fuel with an extremely high energy content.

The splitting of H_2O can be considered as a set of two reactions:

$$\begin{array}{llll} \text{reduction:} & H_2O + e^- & \text{———} \ \tfrac{1}{2}\, H_2 + OH^- & E'^0 = -0.41\ \text{V at } pH = 7 \\ \text{oxidation:} & \tfrac{1}{2}\, H_2O & \text{———} \ \tfrac{1}{4}\, O_2 + e^- + H^+ & E'^0 = +0.82\ \text{V at } pH = 7 \\ \text{splitting:} & \tfrac{1}{2}\, H_2O & \text{———} \ \tfrac{1}{2}\, H_2 + \tfrac{1}{4}\, O_2 & \Delta G^0 = 1.23\ \text{eV} \end{array}$$

Clearly, the energy required for splitting H_2O into its elements depends very much on the number of electrons involved in the redox process [17]. For instance, a *mono*electronic reaction, leading to highly unstable radicals ($H^{\cdot}$ and $OH^{\cdot}$) necessitates an enormous energy input:

$$H_2O \longrightarrow H^{\cdot} + OH^{\cdot} \qquad \Delta G^0 \sim 5\ \text{eV}.$$

Such an endergonic monoelectronic reaction cannot be brought about by *solar light*, the thermodynamic requirements being inconsistent with the energy supplied by *visible* photons. On the other hand, 1.23 eV per electron (the ΔG^0 value for the tetraelectronic process) corresponds to a photon of high wavelength (λ ~ 1040 nm), showing that in principle, a large fraction of the solar spectrum might be used for photodissociating H_2O.

The first systems capable of evolving H_2 from water by visible light irradiation were developed in 1977 [18, 19]. Since this period, many other systems have been proposed, based on the following reaction scheme:

$Ru^{II} = Ru(bipy)_3^{2+}$

$Rh^{III} = Rh(bipy)_3^{3+}$

PS = $Ru(bipy)_3^{2+}$ D = electron donor

R = relay ($Rh(bipy)_3^{2+}$; MV^{2+}; etc...)

C_{red} = heterogeneous catalyst

The redox properties of $Ru(bipy)_3^{n+}$ (ground state and excited state) have been taken advantage of: $^*Ru(bipy)_3^{2+}$ is able to transfer an electron to a relay (MV^{2+}, or a rhodium(III) complex or another electron acceptor) whose reduced form reacts with water to yield hydrogen; the latter reaction might be accelerated by the presence of a heterogeneous redox catalyst. The ruthenium(II) complex is regenerated in the reaction between $Ru(bipy)_3^{3+}$ and an electron donor D. This compound D is irreversibly converted to an oxidation product. An ideal system would, of course, use H_2O as electron donor, with formation of O_2. This remains to be done, but model systems for H_2O oxidation have also been proposed [20]:

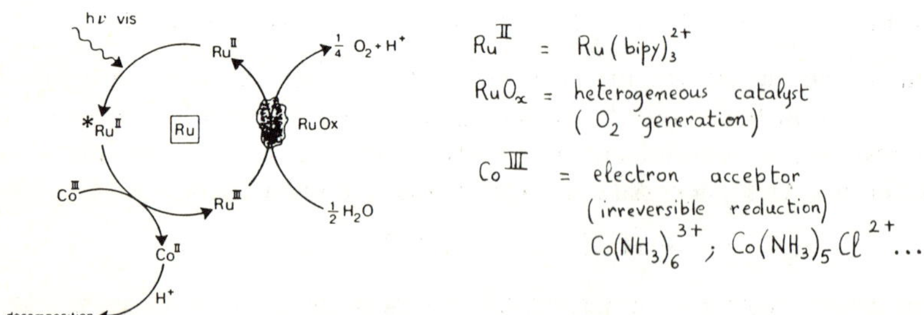

Quantum yields of 10 to 20% have been obtained for the photochemical H_2-generating systems, the turnover numbers on the various components being very high. Although an impressive amount of work has been devoted to those systems and in spite of the considerable progress performed in the field, the practical applications of these photochemical reactions are probably not for tomorrow. The main reason is that *coupling* oxidative and reductive systems has revealed extremely difficult.

Indeed, the diagram represented below corresponds to a line of research but its experimental reality has still to be proven:

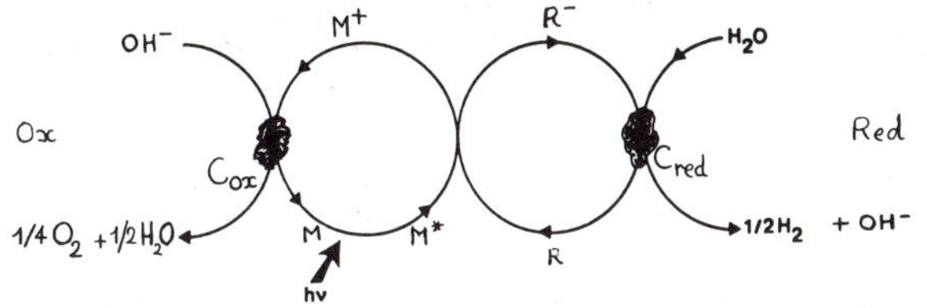

$M = Ru(bipy)_3^{2+}$ R = Relay (acceptor : MV^{2+}...)

C_{red} and C_{ox} = heterogeneous catalysts (specificity)

6. Concluding Remarks

Coordination photochemistry is clearly a rapidly expanding area, in particular with respect to photoredox catalysis. The excited state can act as an extremely powerful electron transfering agent. It is only recently that $Ru(bipy)_3^{2+}$ has been considered as a promising redox photosensitizer, the photodissociation of water being an appealing target. Artificial photosynthesis is at its beginning and many photoactive molecular species or photosynthetic systems will be discovered in the future. For instance, metalloporphyrins, chromium or copper complexes have turned out to be as efficient as $Ru(bipy)_3^{2+}$ in several photoredox processes. In addition, many other reactions than water splitting are expected to be found and studied.

Laboratoire de Chimie Organo-Minérale
Institut de Chimie
1 rue Blaise Pascal
F-67008 Strasbourg Cédex
France.

References

1. V. Balzani and V. Carassiti: *Photochemistry of Coordination Compounds*, Academic Press, London (1970).
2. *J. Chem. Ed.*, **60** (October), pp. 785—887 (1983).

3. V. Balzani, F. Bolletta, M. T. Gandolfi and M. Maestri, *Topics Curr. Chem.*, **75**, 1 (1978).
4. N. Sutin and C. Creutz, *Adv. Chem. Ser.*, No. 168, 1 (1978); N. Sutin, *J. Photochem.*, **10**, 19 (1979); N. Sutin and C. Creutz, *Pure Appl. Chem.*, **52**, 2717 (1980).
5. R. A. Marcus, *Discuss. Faraday Soc.*, **29**, 21 (1960).
6. K. Kalyanasundaram, *Coord. Chem. Rev.*, **46**, 159 (1982).
7. A. J. Lees and A. W. Adamson, *J. Am. Chem. Soc.*, **102**, 6874 (1980); D. M. Manuta and A. J. Lees, *Inorg. Chem.*, **22**, 572 and 3825 (1983).
8. M. Wrighton and D. L. Morse, *J. Am. Chem. Soc.*, **96**, 998 (1974).
9. J. C. Luong, L. Nadjo and M. S. Wrighton, *J. Am. Chem. Soc.*, **100**, 5790 (1978).
10. J. N. Demas, E. W. Harris, C. M. Flynn and D. Diemente, *J. Am. Chem. Soc.*, **97**, 3838 (1975).
11. D. R. McMillin, M. T. Buckner and B. T. Ahn, *Inorg. Chem.*, **16**, 943 (1977); M. W. Blaskie and D. R. McMillin, *Inorg. Chem.*, **19**, 3519 (1980).
12. C. O. Dietrich-Buchecker, P. A. Marnot, J. P. Sauvage, J. R. Kirchhoff and D. R. McMillin, *J. Chem. Soc. Chem. Commun.*, 513 (1983).
13. A. Edel, P. A. Marnot and J. P. Sauvage, *Nouv. J. Chim.*, **8**, 495 (1984).
14. C. R. Bock, T. J. Meyer and D. G. Whitten, *J. Am. Chem. Soc.*, **96**, 4710 (1974); R. C. Young, T. J. Meyer and D. G. Whitten, *J. Am. Chem. Soc.*, **98**, 286 (1976); C. R. Bock, J. A. Connor, A. R. Gutierrez, T. J. Meyer, D. G. Whitten, B. P. Sullivan and J. K. Nagle, *Chem. Phys. Letters,* **61**, 522 (1979).
15. E. Amouyal, B. Zidler, P. Keller and A. Moradpour, *Chem. Phys. Letters,* **74**, 314 (1980).
16. C. Kutal, *Adv. Chem. Ser.*, No. 168, 158 (1978) and references cited therein.
17. V. Balzani, L. Moggi, M. F. Manfrin, F. Bolletta and M. Gleria, *Science* **189**, 852 (1975).
18. B. V. Korriakin, T. S. Dzhabiev and A. E. Shilov, *Dokl. Akad. Nauk. SSSR* **223**, 620 (1977).
19. J. M. Lehn and J. P. Sauvage, *Nouv. J. Chim.*, **1**, 449 (1977); M. Kirch, J. M. Lehn and J. P. Sauvage, *Helv. Chim. Acta,* **62**, 1345 (1979).
20. J. M. Lehn, J. P. Sauvage and R. Ziessel, *Nouv. J. Chim.*, **3**, 423 (1979).

J. M. BASSET

AN INTRODUCTION TO THE FIELD OF CATALYSIS BY MOLECULAR CLUSTERS AND BY SUPPORTED MOLECULAR CLUSTERS AND COMPLEXES

The purpose of this chapter is not to make an extensive survey of the field of molecular clusters in coordination chemistry, a field which is expanding very rapidly in various directions, but to present a short introduction of the field of catalysis by clusters for the scientific community of surface catalysis. This community is not always aware of the progresses in molecular chemistry and more specifically in molecular clusters which are at the borderline between the solid state and the molecular state. The examples which will be given will be taken from reviews in the field of molecular clusters [1—5] in the field of catalysis by molecular clusters [6—7] as well as in the field of surface organometallic chemistry [8—14]. For a deeper approach of the fields considered, it is suggested to refer to those specialized review articles.

1. An Introduction to Molecular Clusters

1.1. DEFINITION

Molecular metal clusters represent a very large family of compounds in which some elements form a framework structure that has a polyhedral form or can be considered a fragment of a polyhedron. In these polyhedra or polyhedral fragments, the framework elements are generally bonded to each other in a multicenter form, that is, the molecular orbitals that contribute largely to the framework bonding, typically involve orbitals from more than two framework atoms.

The various classes of clusters include [5]:

a. Polyhedral boranes (plus, possibly, elements of Groups III, IV, V and transition metals.)
b. Lithium clusters, which form a much smaller class. Be and Mg tend to form chain-like structures.
c. The Naked clusters class, which contains no other element or ligands (they essentially include post transition metals or metalloidal elements like Ge, Sn, Pb, P, As, Sb, Se, Te, Hg).

A. Mortreux and F. Petit (Eds.), Industrial Applications of Homogeneous Catalysis, 293—333.

d. Transition metal clusters (plus Cu, Ag, Au) (*vide infra*).
e. Molecular metal oxides and metal sulfide clusters in which both the metal and the O or S atoms comprise the framework structure (early transition only).
f. A diverse class of clusters may be generated by expansion of gaseous molecules or atoms through a high speed nozzle. It is possible to generate in the gas phase M_x neutral molecules with M a metallic element, but M_x will typically have a range of integer x values.
g. Hypothetical M_x clusters represent an area much investigated by quantum mechanical calculations.

In this chapter we shall consider only transition metal clusters.

Transition metal clusters have the general composition $[M_xL_y]^Z$ where: x is an integer: the established values are 3, 4, 5, 6, 7, 8, 9, 10, 12, 13, 15, 17, 18, 19, 26, 30, 38, 52; y is an integer never less than x and typically is 2 to 3 times larger than x; L is a ligand that may be a molecule like CO, NO, R_3P, etc., or may be a radical like H, halide, alkyl, etc.; the formal charge Z of the cluster may be zero or may be positive or negative (anionic clusters are more common than cationic clusters in the carbonyl subclass); and, depending upon cluster size there may be one or more atoms enclosed or encaged by the cluster polyhedron. If the enclosed atoms are non-metallic, e.g. H, C, N, P, As, these clusters are refered to as interstitial complexes. With large clusters a transition metal atom can be enclosed or many transition metal atoms can be enclosed. Those clusters represent a fragment of close packed metallic arrays.

The following figures will illustrate examples of molecular clusters of increasing nuclearity and (or) complexity which correspond to the above definitions:

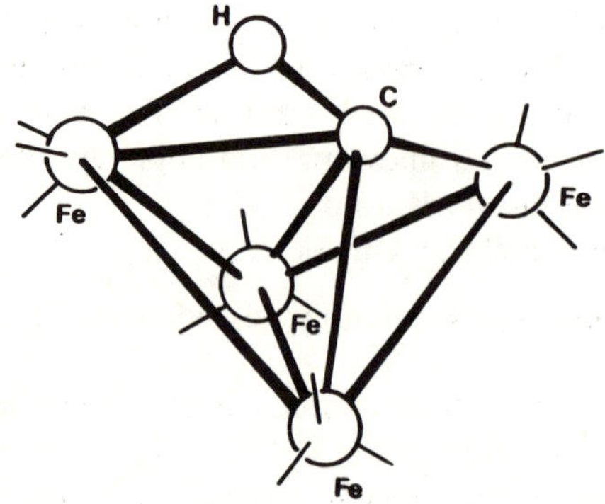

Fig. 1. X ray structure of $HFe_4(\eta^2\text{-}CH)(CO)_{12}$ According to [7].

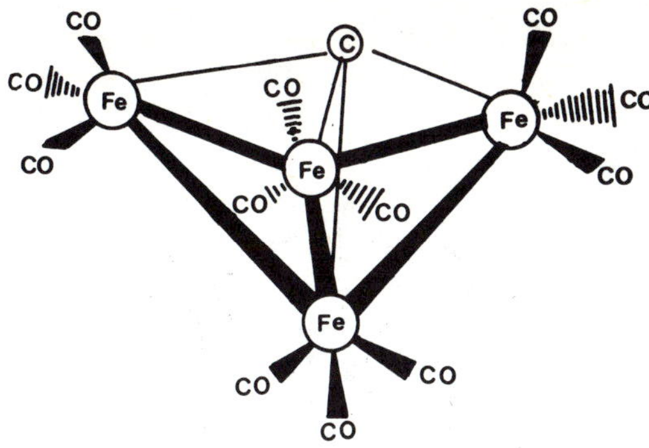

Fig. 2. X ray structure of $Fe_4C(CO)_{12}$ According to [7].

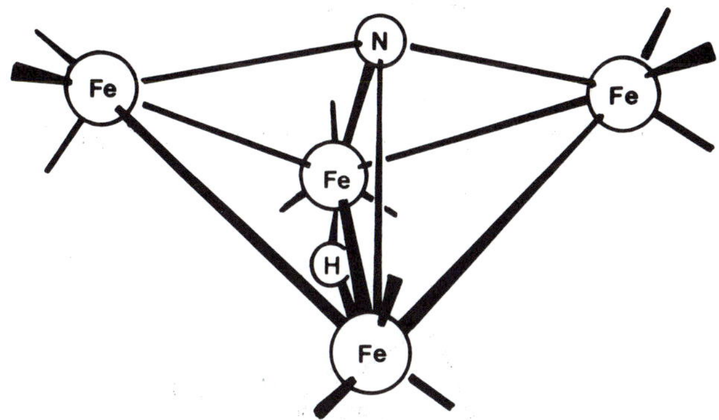

Fig. 3. X ray structure of $HFe_4N(CO)_{12}$

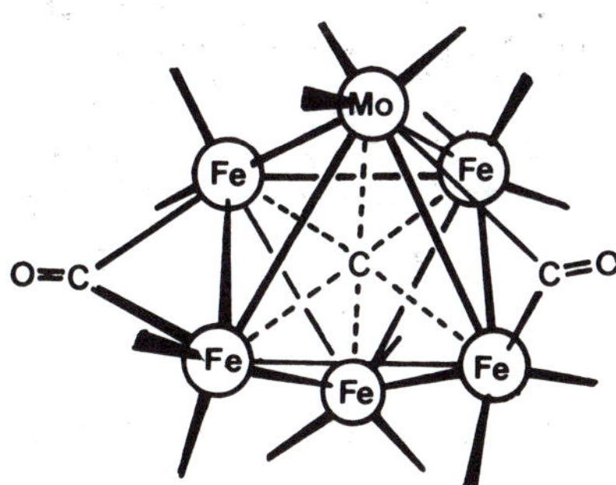

Fig. 4. Molecular structure of $[MoFe_5(CO)_{17}]^{2-}$ (According to [7]).

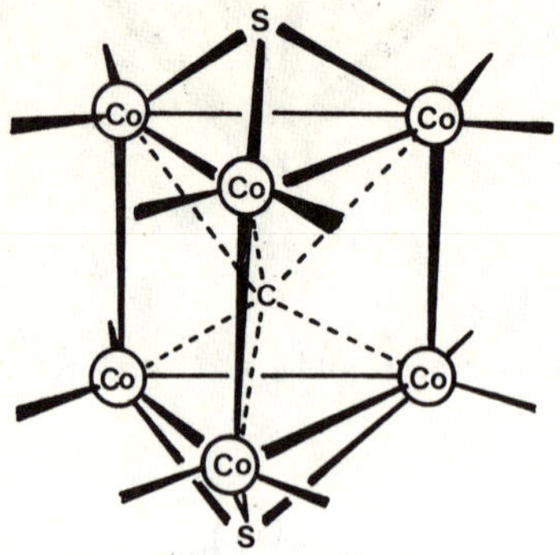

Fig. 5. Molecular structure of $Co_6C(CO)_{12}S_2$ (According to [7]).

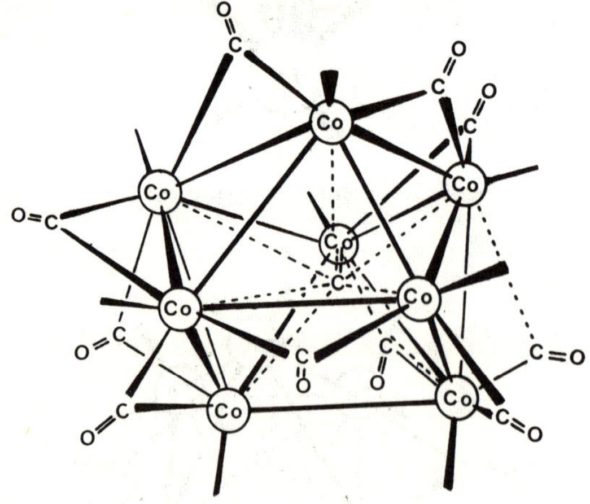

Fig. 6. Molecular structure of $[Co_8C\,(CO)_{18}]^{2-}$ According to [7].

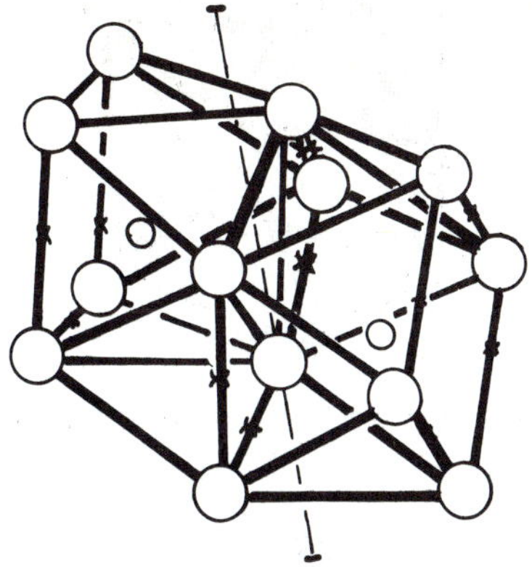

Fig. 7. Molecular structure of $[Co_{13}(CO)_{12}(\mu\text{-}CO)_{12}C_2H]^{n-}$ According to [1].

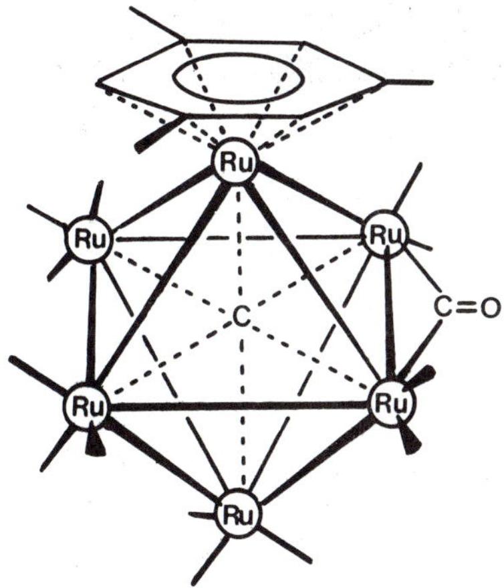

Fig. 8. Schematic representation of $Ru_6\,C(CO)_{14}$ (mesitylene) (According to [7]).

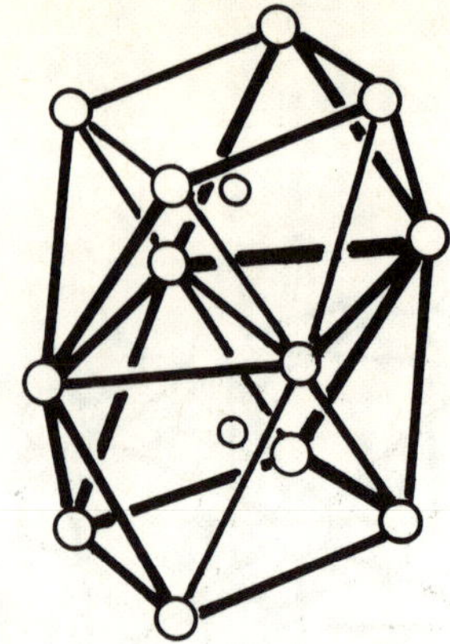

Fig. 9. Schematic representation of $[Rh_{12}(CO)_{16}(\mu\text{-}CO)_8C_2]^{2-}$ (According to [1]).

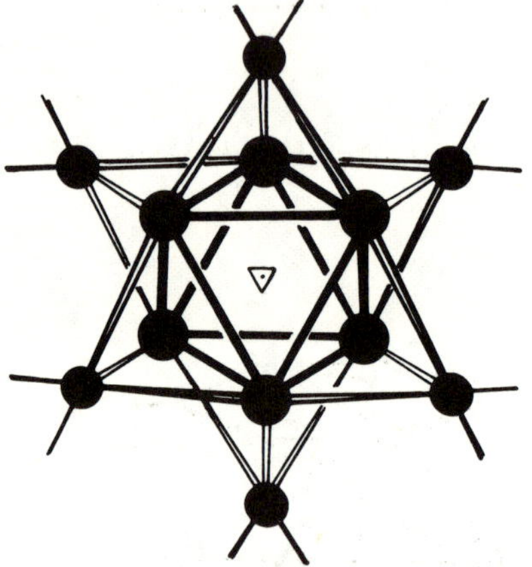

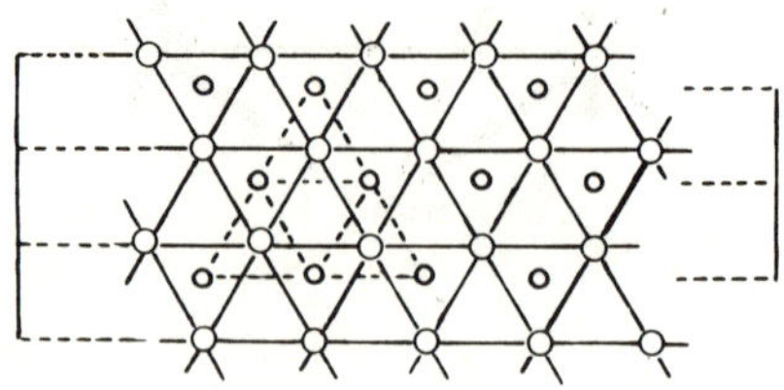

Fig. 10. Above: Molecular structure of $[Fe_6Pd_6(CO)_{24}H]^{3-}$ (According to [1]). Below: The corresponding representation of two layers in M'_3M'' alloy such as Cu_3Au (ccp) and Ni_3Sn (hcp) (see [1]).

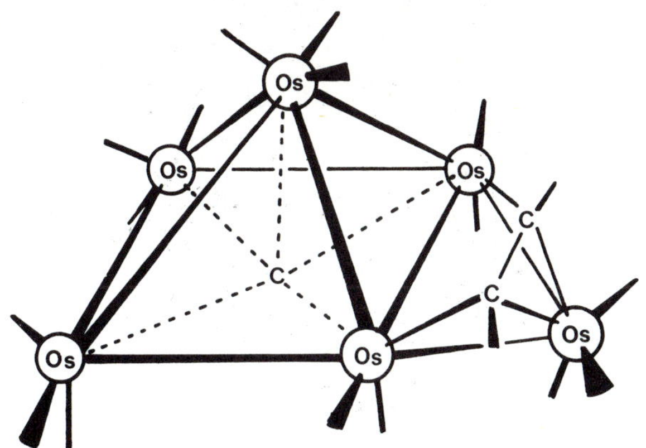

Fig. 11. Molecular structure of Os_6 C $(CO)_{16}$ $(CH_3C_2CH_3)$ according to [7].

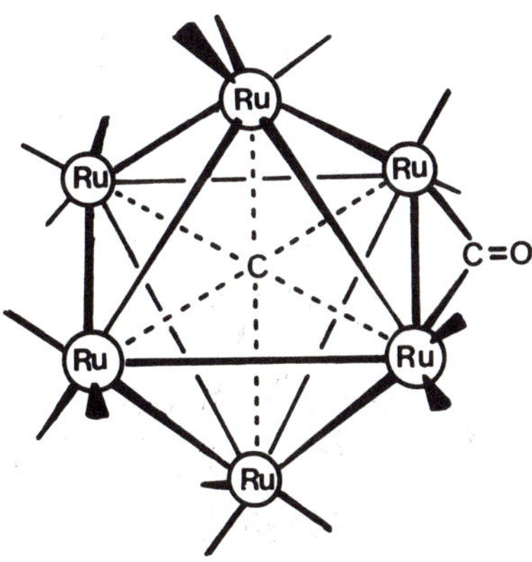

Fig. 12. Molecular structure of Ru_6 C $(CO)_{17}$ (According to [7]).

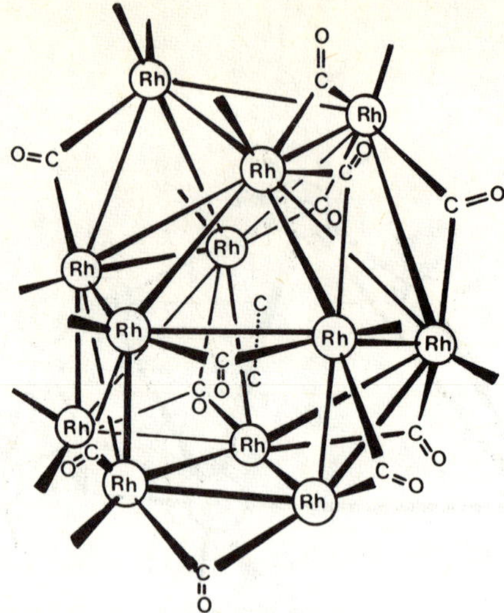

Fig. 13. Molecular structure of $Rh_{12}C_2(CO)_{25}$ (see [1]).

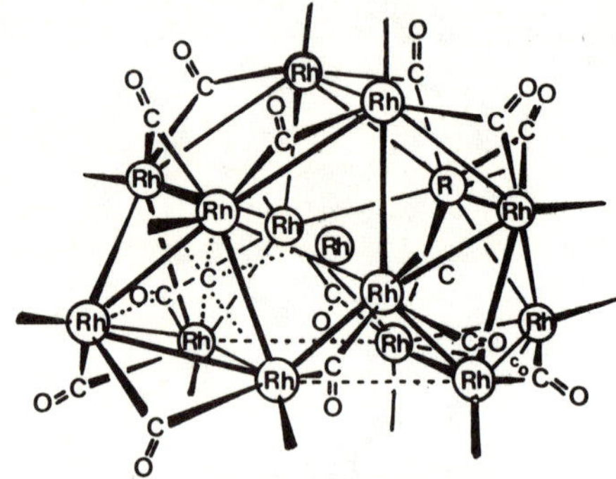

Fig. 14. Molecular structure of $[Rh_{15}C_2(CO)_{28}]^-$ (see [1]).

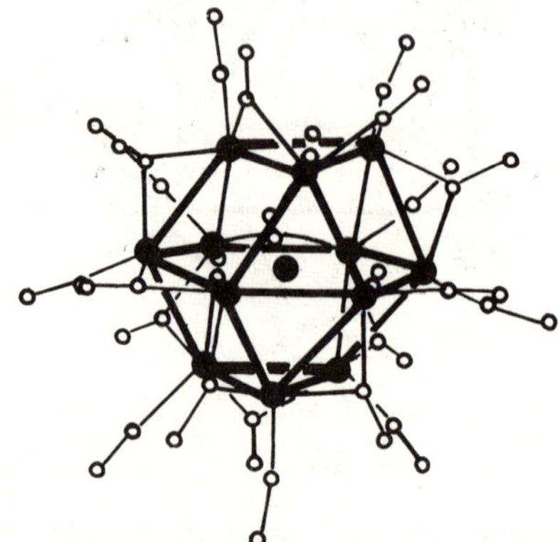

Fig. 15. X ray structure of $[H_{5-n}Rh_{13}(CO)_{24}]^{n-}$ (n = 2, 3, 4) (anticubo-octahedral metal skeleton [1]).

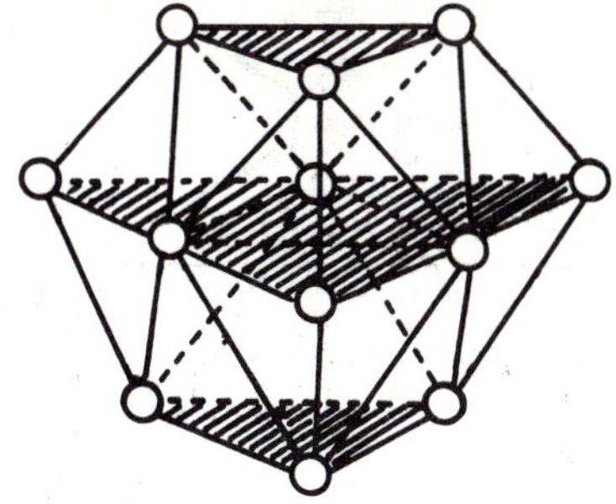

Fig. 16. X ray structure of $[Ni_{12}(CO)_2H_{4-n}]^{n-}$ ($n = 2, 3, 4$) [1].

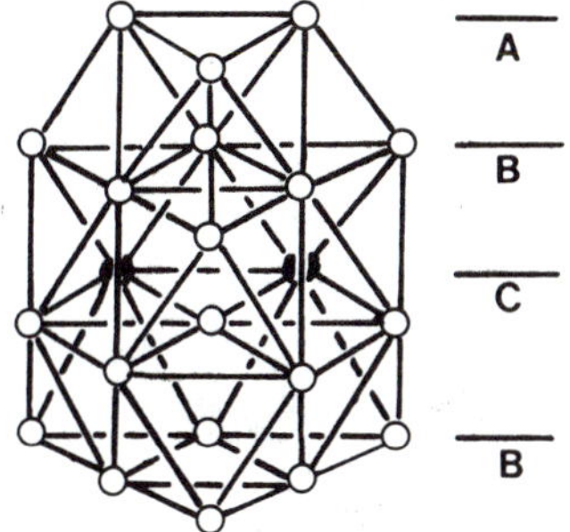

Fig. 17. X ray structure of $[Rh_{22}(CO)_{12}(\mu_2\text{-}CO)_{18}(\mu_3\text{-}CO)_7]^{4-}$ (according to [1]).

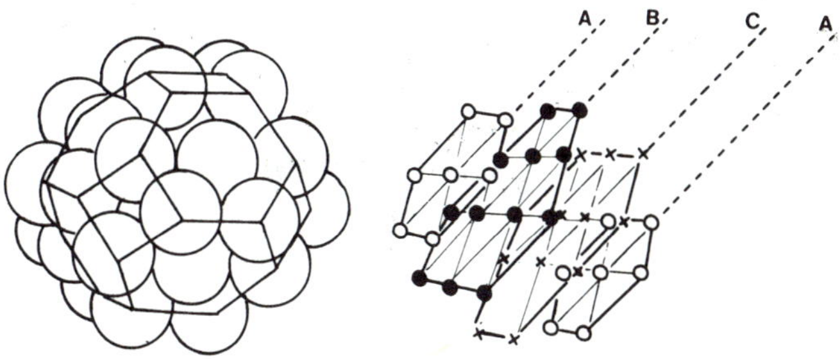

Fig. 18. The truncated octahedron of platinum atoms found in the $[Pt_{38}(CO)_{44}]^{2-}$ and the sequence of cpp layers [1].

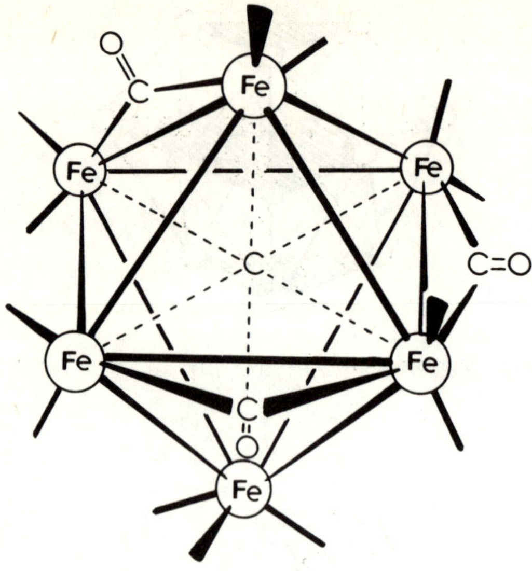

Fig. 19. Schematic representation of $[Fe_6\,C\,(CO)_{16}]^{2-}$ (According to [7]).

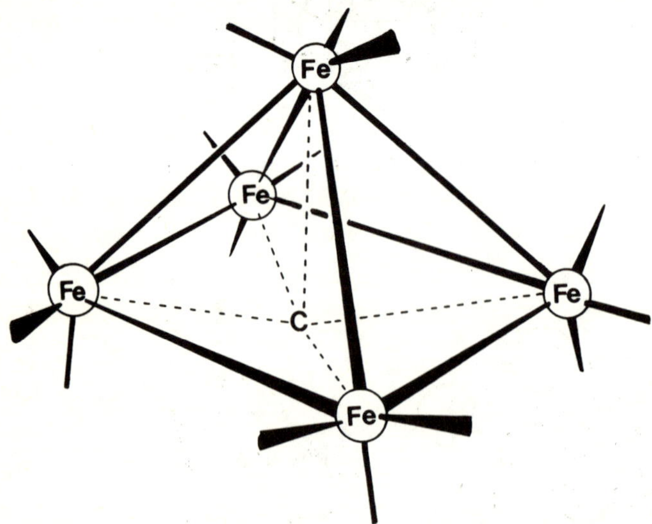

Fig. 20. Schematic representation of $Fe_5C(CO)_{15}$ (According to [7]).

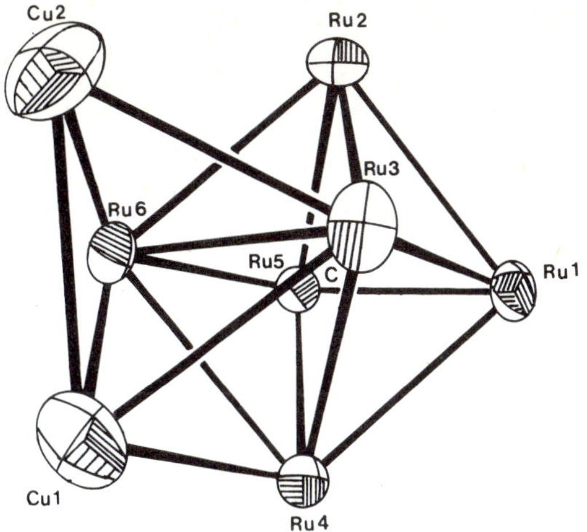

Fig. 21. X ray structure of $(CH_3CN)_2Cu_2Ru_6C(CO)_{16}$ according to Bradley [20].

1.2. BONDING IN MOLECULAR CLUSTERS

1.2.1. *The metal-metal bond in clusters*

Two main approaches have been used to rationalize metal-metal bonding in transition metal carbonyl clusters: The 18 electrons rule and the polyhedral skeletal electron pair theory (PSEPT). (For a complete survey of this field it is necessary to consider the reviews of K. Wade [3], B. F. G. Johnson [2], M. Mingos [33] and R. Hoffmann [34]). In this lecture we will consider only the 18 electrons rule which accounts for the geometry of the metallic frame of simple clusters of small nuclearity. In this rule one assumes that the skeletal atoms are held together by a network of two electron pair/two center ($2e/2$ center) bonds and that each individual cluster atom utilizes its *nine* atomic orbitals to accomodate both metal valence electrons and ligand electron pairs and also to form two electrons metal-metal bonds. Therefore each transition metal can accomodate a total of 18 electrons. Most of the low oxidation diamagnetic complexes obey this rule. However exceptions occur especially with d^8 metal ions with square planar geometry for which the high lying p_z orbital is found to be non-bonding and empty. The application of the 18 electrons rule to cluster carbonyl systems is in general restricted to small nuclearity clusters. Three hypotheses are made:

— The metal—metal bond will be a $2e/2$ center bond
— The metal—metal bond corresponds to polyhedral edges
— Ligands are used as $2e$ pair donors only, leading to the view that the same metal polyhedron will be derived irrespective of whether electrons are present as anionic charge or ligand pairs.

We will consider a few examples which illustrate the application of the 18 electron rule to account for the geometry of simple low nuclearity clusters.

a. In $Mn_2(CO)_{10}$, each manganese has 7 valence electrons and one considers that each CO donates 2 electrons to the cluster frame. The total number of electrons available is $(2 \times 7) + (10 \times 2) = 34$. The number of valence orbitals available is $9 \times 2 = 18$, which require, to be fully occupied, $18 \times 2 = 36$ electrons. Therefore the number of Mn—Mn $2e/2$ center bonds is 1.

b. $Ru_3(CO)_{12}$: Each ruthenium has eight valence electrons.

1.	Total number of electrons	= 48
2.	Valence orbitals available	= 27
3.	Required number of electrons to fill 27 A.O.	= 54
	Difference item 3 − item 1	= 6

Therefore the number of Ru—Ru $2e/2$ center bonds = 6/2 = 3, which corresponds to a triangular geometry.

c. $Rh_4(CO)_{12}$: Each cobalt has nine valence electrons

1.	Total number of electrons: $(9 \times 4) + (2 \times 12)$	= 60
2.	Valence atomic orbitals available: 9×4	= 36
3.	Required number of electrons to fill 36 A.O.	= 72
	Difference item 3 − item 1	= 12

Therefore the number of Co—Co $2e/2$ center bonds = 12/2 = 6 which corresponds to a tetrahedral arrangement.

d. $Os_5(CO)_{16}$: Each Osmium has eight valence electrons

1.	Total number of electrons: $(8 \times 5) + (16 \times 2)$	= 72
2.	Valence orbitals available: 9×5	= 45
3.	Required number of electrons to fill 45 A.O.	= 90
	Difference item 3 − item 1	= 18

It follows that the number of Os—Os $2e/2$ center bonds is equal to 9 corresponding to a trigonal bipyramidal (9 edges) unit.

The 18 electrons rule may also be applied to electron rich or electron poor clusters, which are of great interest in catalytic processes.

e. $H_2Os_3(CO)_{12}$:

1. Total number of electrons: $(3 \times 8) + (12 \times 2) + (2 \times 1)$ $= 50$
2. Valence orbitals available: 9×3 $= 27$
3. Required number of electrons to fill 27 A.O. $= 54$
   Difference item 3 − item 1 $= 4$

Therefore the number of Os—Os $2e/2$ center bonds = 2 corresponding to a linear or open triangle geometry.

f. $H_2Os_3(CO)_{10}$:

1. Total number of electrons: $(3 \times 8) + (10 \times 2) + (2 \times 1)$ $= 46$
2. Valence orbitals available: $= 27$
3. Required number of electrons to fill 27 A.O. $= 54$
   Difference item 3 − item 1 $= 8$

As a consequence the number of Os—Os $2e/2$ center bonds is 4.

The cluster may be represented as a triangle with two Os—Os single bond and one Os═Os double bond (Figure 22).

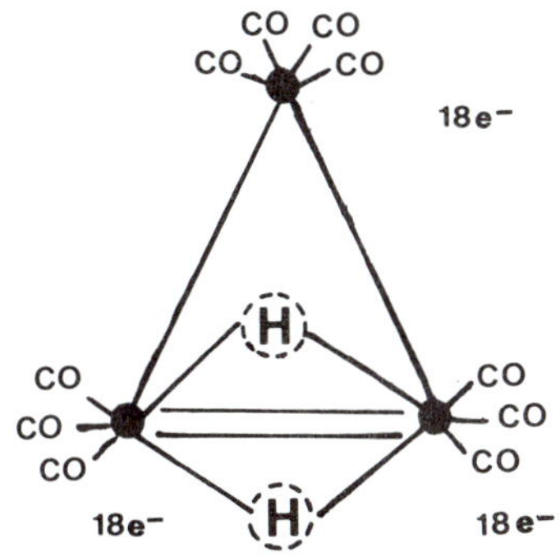

Fig. 22. $H_2Os_3(CO)_{10}$ structure deduced from the application of the 18 electrons rule.

1.2.2. *The metal-ligand bond*

We shall consider here only the CO ligand, which is the most commonly observed ligand in low valent cluster compounds. The main ways in which a carbonyl ligand can coordinate to one or more of the metal atoms of a cluster are as follows [3]:

1. Terminal attachment to one metal atom by a linear of nearly linear M—CO unit. (Usually the angle MCO ranges between 165 and 180°):

   M—C=O

2. Doubly bridging (M_2) between two metal atoms coordinating exclusively through the carbon atom (η'), the CO axis being perpendicular or nearly perpendicular to the M –M axis:

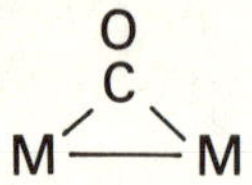

   O
   C
   M——M

3. Doubly bridging in such a way as to imply a dihapto coordination to one of the metal atoms with some . . . O bonding.

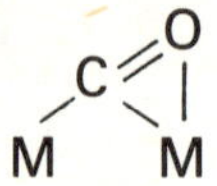

   C=O
   M M

4. Triply bridging (M_3) between three metal atoms the CO axis being perpendicular or nearly perpendicular to the M_3 plane.

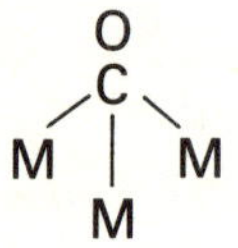

   O
   C
   M | M
   M

5. Triply bridging (M_3) between three metal atoms while coordinating dihapto to a fourth.

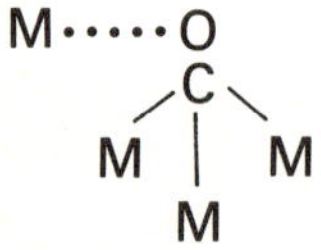

   M·····O
   C
   M | M
   M

When a CO molecule is coordinated to a single metal atom, the metal—carbon distance is shorter than a metal—carbon distance of a metal—alkyl bond. Besides the C—O distance is longer than in free CO. These observations can be rationalized in terms of resonance between the canonical forms M=C=O and M^-—C^+=O indicating an M—C bond order between 1 and 2 and a C—O bond order between 2 and 3. The more commonly bonding description is represented in Figure 23. The lone pair electrons on carbon can be donated to a suitable vacant metal orbital (e.g. the d_{z^2} A.O.) while the ligand—metal electron drift is compensated by a back donation from metal to ligand (e.g. from filled metal d_{xz} and d_{yz} A.O. into the ligand π^* orbitals). (Figure 23).

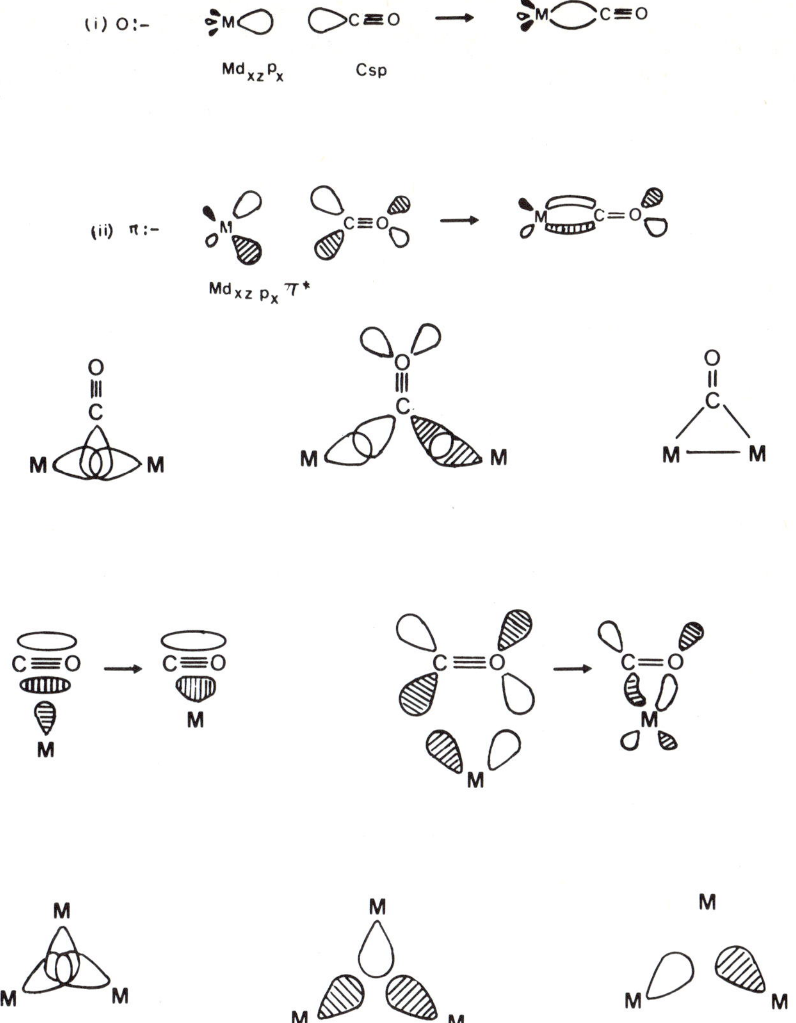

Fig. 23. Molecular orbitals involved in CO bonding. A. Linear CO; B. μ_2 bridging CO; C. η^2-μ_2, bridging CO; D. $\eta^1\mu_3$ bridging (according to Wade [3]).

A monohapto doubly bridging carbonyl still functions as a 2 electrons ligand: the lone pair orbital on carbon can overlap with a suitable in phase combination of filled metal orbitals, while an out of phase combination of filled metal orbitals can interact with the CO π orbital (Figure 23). The net result is to reduce the CO bond order to about 2.

A doubly bridging dihapto carbonyl ligand is a more unusual type of bonding. Nevertheless it can be observed in the dinuclear manganese complex $Mn_2(CO)_5(Ph_2PCH_2PPh_2)_2$. In this complex a carbonyl ligand is terminally bound to one manganese atom while coordinating η^2 to the other. As a result this the carbonyl ligand behaves as a poor electron ligand. (Figure 23).

In a triply bridging carbonyl ligand the lone pair orbital on carbon can combine with a suitable in phase combination of metal orbitals. Similarly to the other cases a π back-donation from suitable E symmetry combinations of metal orbitals into the π^* orbitals of CO. The metal ligand bonding can thus be regarded as involving three pairs of electrons, enough to allow the alternative description of three metal—carbon single bonds holding the carbon atom to the metal triangle, with the CO bond order formally reduced to one (Figure 23).

1.3. DYNAMIC BEHAVIOUR OF MOLECULAR CLUSTERS

In a molecular cluster the migration of metal atoms or ligands over or inside the cluster frame is an area of very intensive investigation. Not only do these kind of studies try to elucidate the mechanistic pathways of these fluxional processes in cluster chemistry but also it gives new concepts in surface science about the mobility of metal atoms as well as chemisorbed molecules on or inside a metallic lattice. Three different classes of fluxional processes have been observed in molecular clusters [2, 7]

— Ligand migration over the cluster surface.
— Arrangement of the metal cluster polyhedron.
— Ligand migration within the metallic cluster unit.

1.3.1. *Ligand migration over the cluster surface*

These kinds of studies have been restricted to carbon monoxide ligands, hydride ligands and some organic ligands as isocyanides. With carbon monoxide the ligand migration may occur between:

— two metal centres
— three metal centres
— four metal centres

In Figure 24 is represented the pairwise exchange mechanism in which ligand bridges are opened and closed in a pairwise fashion. In bridged dinuclear complexes this process involves the formation of a non bridged

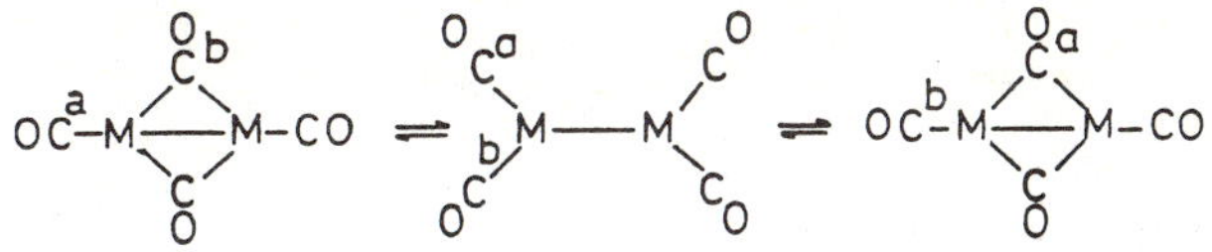

Fig. 24. The pairwise exchange mechanism according to B. F. G. Johnson [3].

intermediate. This intermediate may be of sufficient life time to allow rotation about the metal—metal bond prior to bridge reformation to bring about *cis-trans* isomerisation.

The simplest mechanism for CO migration in a tetranuclear carbonyl is represented on Figure 25. This process involves a $C_{3v} \rightleftharpoons T_d$ interconversion of the $M_4(CO)_{12}$ clusters. There is a simple bridge-break, bridge-make mechanism of the type employed in binuclear complexes.

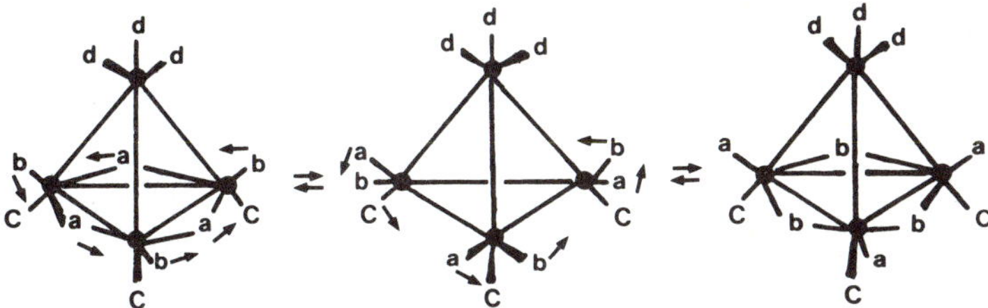

Fig. 25. The interconversion of the C_{3v} and T_d forms of the tetrametal dodecacarbonyls according to E. Muetterties [7].

The cluster $Ir_4(CO)_{11}PPh_2Me$ exhibits three distinct fluxional processes. The process of lowest activation energy is a cyclic permutation of the three bridging and three terminal carbonyl ligands about the basal plane. At higher temperatures all carbonyls but one undergo exchange. Finally at higher temperatures all eleven carbonyls equilibrate (Figure 26).

1.3.2. *Structural rearrangement within the metal core*

Two types of such structural rearrangement occur which may be relevant to surface science:

— rearrangement with retention of the ligand arrangement.
— rearrangement with simultaneous ligand mobility.

In the cluster $[Pt_9(CO)_{18}]^{2-}$ (structure shown in Figure 27), it has been shown by ^{195}Pt NMR spectroscopy that the outer Pt_3 triangles rotate about the pseudo three fold axis with respect to the inner one. The larger dianions $[Pt_9(CO)_{18}]^{2-}$ and $[Pt_{12}(CO)_{24}]^{2-}$ have been shown to exchange

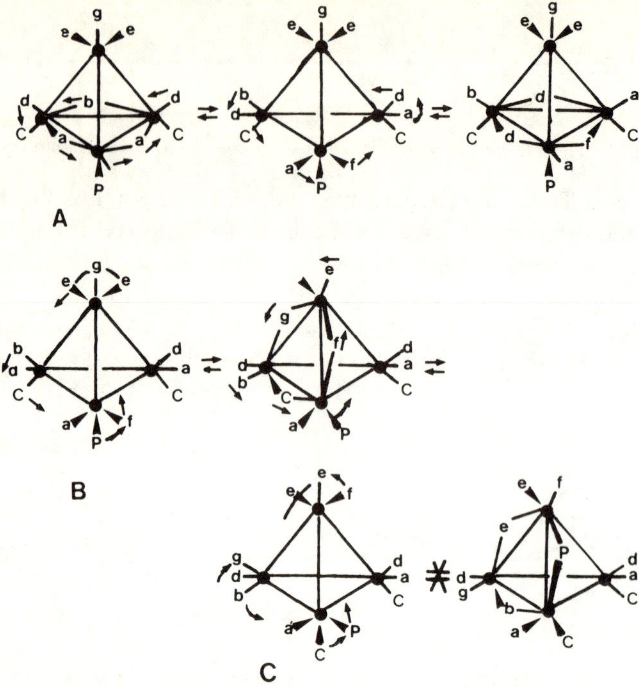

Fig. 26. Fluxional behaviour of $Ir_4(CO)_{11}PPh_2Me$.

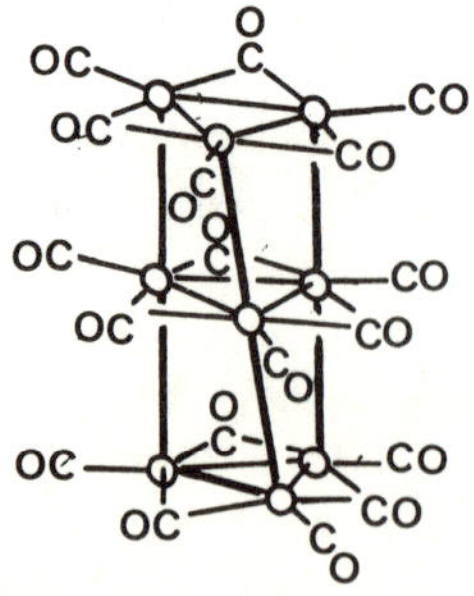

Fig. 27. X ray structure of $[Pt_9(CO)_{18}]^{2-}$.

$Pt_3(CO)_6$ units in solution. With the mixed metal cluster $H_2FeRuOs_2(CO)_{13}$ the CO and H migration occur in concert with the deformation of the quasi tetrahedral metal framework (Figure 28).

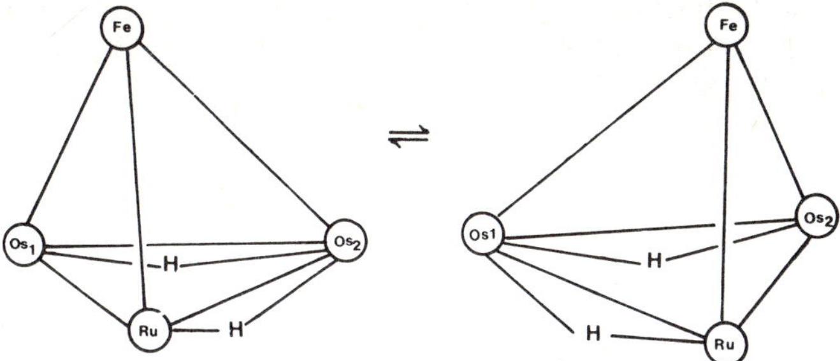

Fig. 28. Skeletal rearrangement in $H_2FeRuOs_2(CO)_{13}$.

1.3.3. *Ligand migration within the metal cluster unit*

Hydride ligands are able to occupy interstitial sites inside the metallic clusters. The carbonyl hydride anions $[H_2Rh_{13}(CO)_{24}]^{3-}$ and $[H_3Rh_{13}(CO)_{24}]^{2-}$ have an anticubo octahedral metal skeleton with the thirteenth rhodium atom occupying the interstitial site; this structure represents a fragment of hexagonal close packing. In these clusters the hydride ligands migrate about the whole cluster and couple to all thirteen rhodium nuclei. In this migration process the hydride occupy both octahedral and tetrahedral interstitial sites.

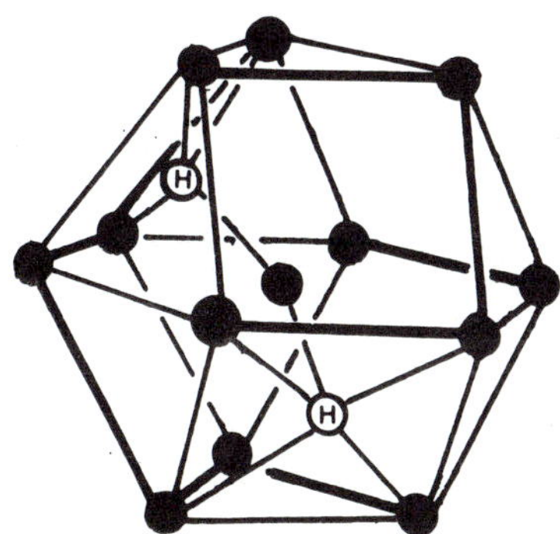

Fig. 29. X ray structure of $[H_2Rh_{13}(CO)_{24}]^{3-}$: fluxional behaviour of the hydride ligands.

1.4. REACTIVITY OF MOLECULAR CLUSTERS

A recent review by Deeming [4] has considered in detail and in a very comprehensive way, the large field of the reactivity of molecular clusters. One can classify reactions of molecular clusters in two areas depending on the fact that the reaction is followed or not by a major modification of the metal skeleton. One can also consider the reactivity in terms of electrophilic or nucleophilic attack, oxidative addition, etc.

1.4.1. *Electrophilic attack*

The electrophilic attack may occur on the metal center, which is the usual case since the metals have usually the highest electron density and can therefore be protonated. In most cases the resulting hydrido ligand will occupy a bridging position:

$$[Re_3(CO)_{12}]^{3-} \underset{-H^+}{\overset{+H^+}{\rightleftharpoons}} [HRe_3(CO)_{12}]^{2-} \underset{-H^-}{\overset{+H^+}{\rightleftharpoons}} [H_2Re_3(CO)_{12}]^- \rightleftharpoons H_3Re_3(CO)_{12}$$

In some cases the electrophilic attack can occur at the oxygen atom of a bridging carbonyl. Thus $AlBr_3$ may react with a linear carbonyl ligand of $Ru_3(CO)_{12}$ which becomes bridged:

$$Ru_3(CO)_{12} + AlBr_3 \longrightarrow Ru_3(CO)_{11}CO \longmapsto AlBr_3$$

Similarly a proton can attack at the oxygen atom of a bridging carbonyl:

$$[HFe_3(CO)_{11}]^- + H^+ \longrightarrow HFe_3(CO)_{10}(C\,O\cdots\cdots H)$$

1.4.2. *Nucleophilic addition*

On the metallic frame. This type of attack increases the number of electrons associated with the cluster. As a result a modification of the geometry or of the nuclearity of the starting cluster. There are however electron deficient clusters as $H_2Os_3(CO)_{10}$ where the addition of an electron pair does not modify the overall geometry of the cluster. This cluster will therefore be able to be a catalyst (*vide infra*) for olefin isomerisation and (or) hydrogenation. On Figure 30 are given two examples of nucleophilic addition with and without modification of the overall geometry of the cluster.

1.4.3. *Nucleophilic attack at the ligands*

It usually occurs at the carbon atom of a linear carbonyl ligand. For example the nucleophilic attack of an OH^- group on a coordinated CO gives rise, via β-H elimination, to the formation of anionic hydrido clusters:

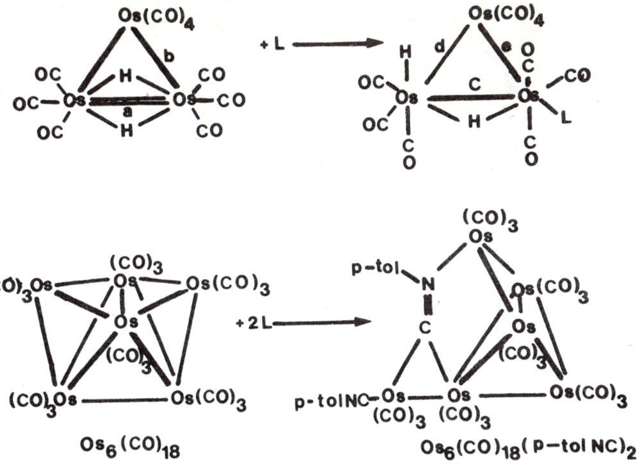

Fig. 30. Nucleophilic addition to $H_2Os_3(CO)_{10}$ (upper) and $Os_6(CO)_{18}$ (lower).

$$M_x(CO)_y + OH^- \longrightarrow M_x(CO)_{y-1}(CO_2H) \longrightarrow M_x(CO)_{y-1}H^- + CO_2$$

This kind of attack is frequently encountered in water gas shift reactions catalyzed by molecular clusters.

Another possibility of nucleophilic attack may occur with carbido cluster where the carbonyl can be coordinated to the "surface" carbon atom. Typical examples are given on Figure 31 where the nucleophile can be an alcohol, an amine or an hydride.

1.4.4. *Oxidative addition*

With mononuclear complexes one can define the oxidative addition of an X—Y molecule to a transition metal M_T by the simple equation

$$L_nM^n + X\text{—}Y \rightleftharpoons L_nM^{n+2}\,(X)\,(Y).$$

When applied to molecular clusters oxidative addition may concern more than one metal atom. In fact this kind of activation may concern cleavage of H—H bond, C—H bonds of olefinic or acetylenic compounds and various types of ligands. One of the unique aspects of oxidative addition in clusters is the ability of clusters to cleave bonds of a ligand in a very close vicinity to the original point of attachment of this ligand (Figure 32). Thus, the first oxidative addition of ethylene to an Os_3 cluster is likely to be as in (A) (Figure 32) and that of pyridine as in B. Other mechanisms of vinylic

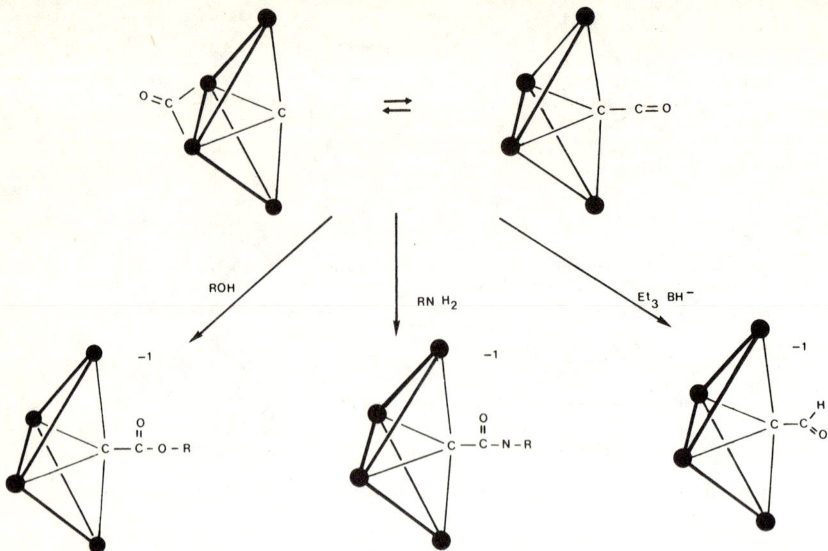

Fig. 31. Examples of nucleophilic attack on CO coordinated to an exposed "surface" carbon (according to J. Bradley [20]).

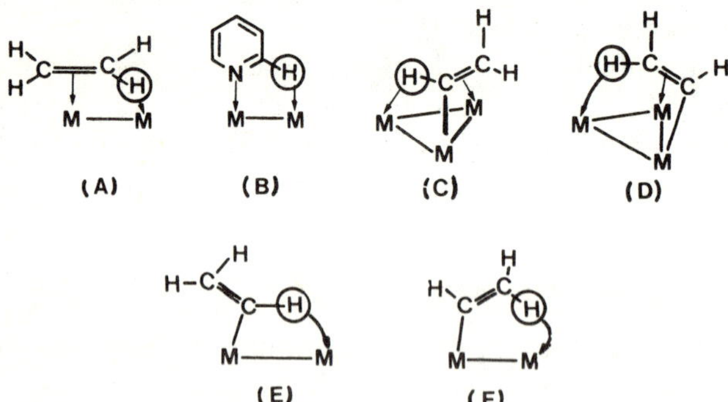

Fig. 32. Oxidative additions with the cluster $Os_3(CO)_{12}$ after Deeming [3].

C—H activation are represented on Figure 33. Obviously surfaces may activate olefinic bonds in the same way.

1.5. MOLECULAR CLUSTERS AS STRUCTURAL MODELS OF INTERMEDIATES OR CHEMISORBED SPECIES IN SURFACE SCIENCE

One of the most important aspects of cluster chemistry is the possible

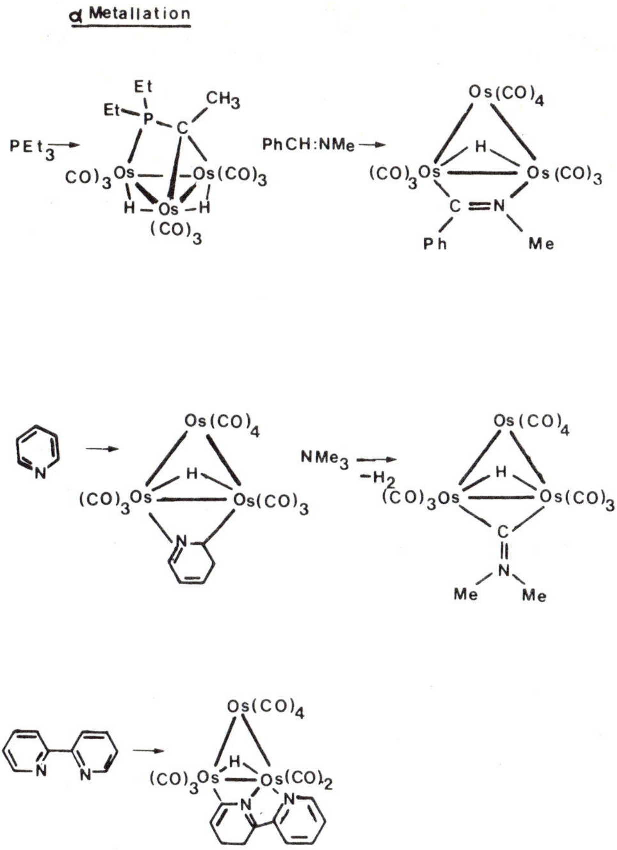

α Metallation

PEt$_3$ → $(CO)_3Os$–Os$(CO)_3$ (H–Os–H, $(CO)_3$; Et–P(Et)–C–CH_3) PhCH:NMe → $Os(CO)_4$, H, $(CO)_3Os$–Os$(CO)_3$, C=N, Ph, Me

pyridine → $Os(CO)_4$, H, $(CO)_3Os$–Os$(CO)_3$, N NMe$_3$ → $-H_2$ $Os(CO)_4$, H, $(CO)_3Os$–Os$(CO)_3$, C=N, Me, Me

bipyridine → $Os(CO)_4$, H, $(CO)_3Os$–Os$(CO)_2$, N, N

Fig. 33. Activation of C—H or N—H bond by $Os_3(CO)_{12}$ in close vicinity to the original point of attachment (after Deeming [3]).

relationship existing between the structure of chemisorbed species and that of ligands coordinated to the metallic cluster frames. There are many examples where the type of bonding between a ligand and a cluster frame is also postulated to occur on a metallic surface. A general review in this field has been written by R. Ugo [8], which describes in a very comprehensive way this very important aspect of surface science. We would like to present here only a few examples of clusters structures which are relevant to surface science and to catalytic reactions which occur on metal surfaces.

The catalytic hydrocondensation of CO is already a very old reaction in the field of heterogeneous catalysis. It is generally assumed that such reaction occurs on metallic surfaces of group VIII transition metals (sometimes on carbides of the same metals). The mechanism which is the most commonly admitted, at the moment, implies a dissociation of CO, followed by a stepwise reduction of surface carbon into $\geqq$C—H, $>CH_2$, and $—CH_3$ surface species. Figure 34 illustrates the various intermediates in which CO can be activated and reduced on metallic surfaces. The *A* cluster, $[Fe_4(CO)_{13}]^{2-}$, illustrates the possibility for a CO ligand to occupy

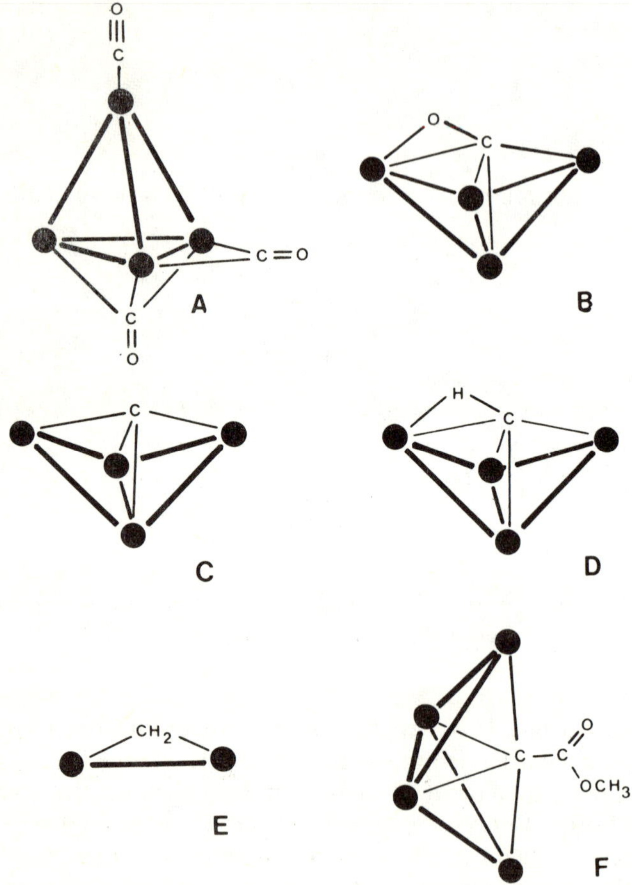

Fig. 34. Iron clusters representing some intermediates postulated in Fischer—Tropsch synthesis.

a linear coordination, a doubly bridging type of coordination, and a triply bridging type of coordination. The *B* cluster, $[HFe_4(CO)_{13}]^-$, illustrates the possibility for a CO ligand to be coordinated both through the carbon and the oxygen atom onto a metallic frame. This is a kind of precursor state to the CO dissociation process. The cluster *C*, $[Fe_4(CO)_{12}C]^{2-}$, can be considered as a possible model for a surface carbido species arising from the CO dissociation. The *D* cluster, $[HFe_4(CH)(CO)_{12}]^{2-}$, could represent a partially hydrogenated form of surface carbon. The bridging CH_2 ligand in *E*, $(Fe_2(CO)_8(CH_2))$, is probably one of the final states of activation and reduction of CO and many mechanistic paths require either $\gtrless CH_2$ coupling or $\gtrless CH_2$ reduction to methane:

$$2\ M\overset{CH_2}{\diagup\ \diagdown}M \longrightarrow C_2H_4$$

$$M\overset{CH_2}{\diagup\ \diagdown}M \xrightarrow{H} CH_4$$

Finally the presence of a C—CO bond in the cluster $[HFe_4(CO)_{12}CCO_2CH_3]^-$ may illustrate a new way of making an oxygenated hydrocarbon starting from a surface carbide (Figure 34).

On Figure 35 is represented another type of mechanism for C—C bond formation starting from a molecular cluster having already an internal carbon atom $[Fe_6C(CO)_{16}]^{2-}$. By oxidation of such cluster with tropylium cation the internal carbon becomes exposed in the cluster $Fe_4(CO)_{13}C$. This cluster can coordinate CO probably by an intramolecular fluxional process. In the presence of methanol a nucleophilic attack may occur at this carbide-coordinated CO. The resulting $\equiv$C—C(=O)—OMe may then by hydrogenated to CH_3COOCH_3. The other steps proposed by J. Bradley indicate a possible pathway for returning to $[Fe_6C(CO)_{16}]^{2-}$.

The reduction of nitroaromatic compounds as well as that of nitriles can occur at metallic surfaces. Various steps have been proposed although no one has been really identified so far. On Figure 35 is represented the stepwise reduction of C$\equiv$N triple bond at the "surface" of the anionic hydrido cluster $[HFe_3(CO)_{11}]^-$. The various steps observed by H. D. Kaesz are expected to be good models of the intermediates in these surface reactions (Figure 36).

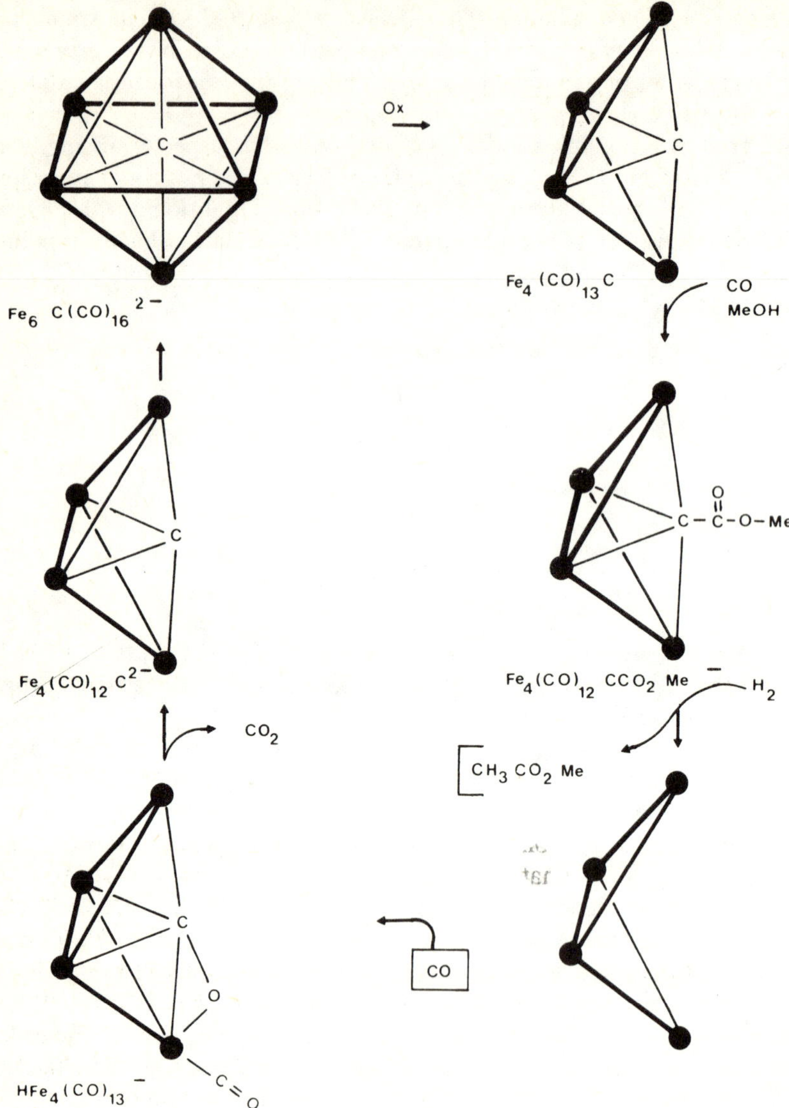

Fig. 35. Various possible steps for making CH_3COOCH_3 from $CO + H_2$ with the cluster $[Fe_6C(CO)_{16}]^{2-}$. Some steps necessary to return to the starting cluster are hypothetical. (According to J. Bradley [20].)

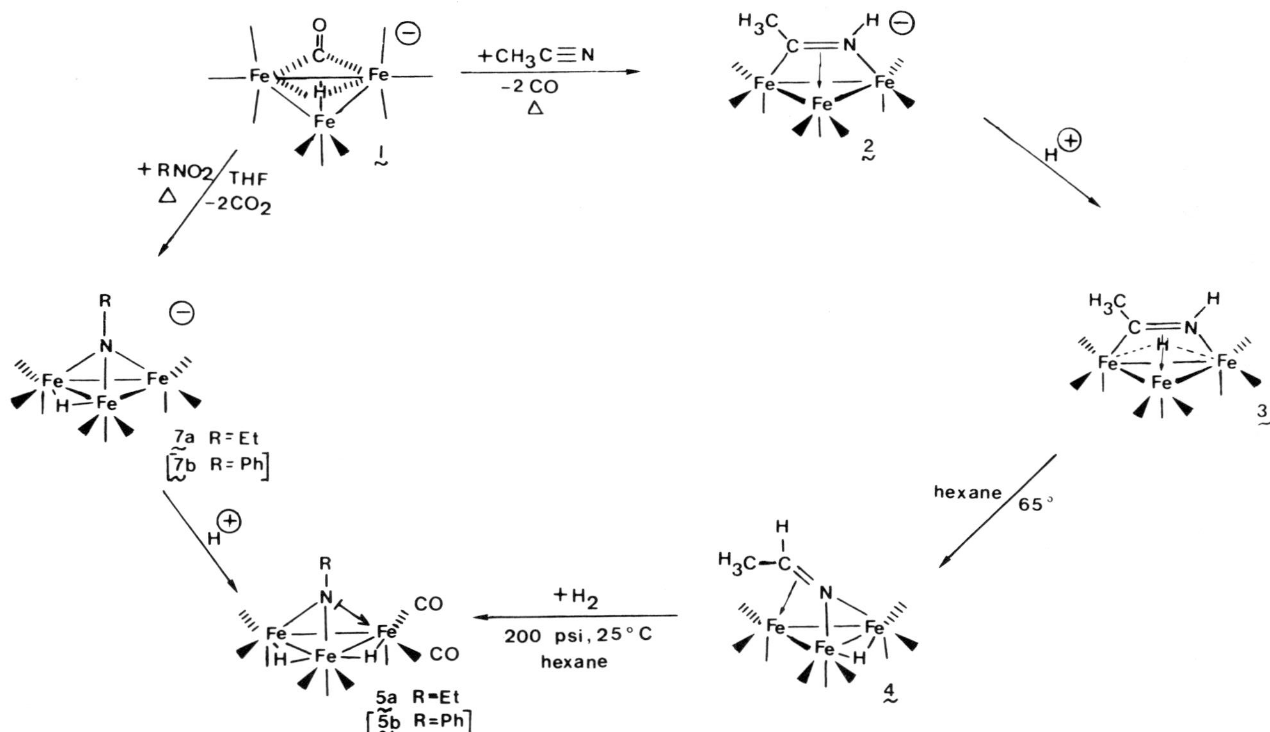

Fig. 36. Stepwise reduction of a C≡N triple bond or R—NO_2 group with the $[HFe_3(CO)_{11}]^-$ cluster according to H. D. Kaesz.

2. Catalysis by Molecular Clusters

2.1. THE RELATIONSHIP BETWEEN MOLECULAR CLUSTERS AND SMALL METAL PARTICLES

The most interesting aspect in the study of molecular clusters concerns the frontier situation that they occupy between the molecular state and the metallic state. This frontier situation is also expected in catalysis where the clusters are at the borderline between molecular catalysis and solid state catalysis. Heterogeneous catalysis on metals is not always well understood. Surfaces are not well defined at a microscopic level and they contain corners, faces, edges etc., so that the selectivity of a given reaction may depend on the respective amount of such geometric parameters. In the last twenty years homogeneous catalysis has been developed considerably. Usually the catalytic reaction occurs in the coordination sphere of a single transition metal atom surrounded by a variety of well defined ligands which may orientate the reaction in the desired direction depending on the electronic or steric effects of those ligands (Figure 37). Due to this frontier situation, the molecular cluster presents a considerable interest in catalysis and this for many reasons:

- Its metallic frame presents a geometry very close to that encountered in small metallic particles encaged in the cavities of some zeolithes. *A priori* one might expect that it will be possible to carry out in a molecular cluster frame the catalytic reaction which occur on a surface but with a better control of the activity and of the selectivity due to the presence of known ligands. The objective is a rather ambitious one since it is well known that the rigidity of the metallic frame of many clusters is weak and the geometry of the metallic frame will depend on the number of electrons brought by the ligands, and on the nature of such ligands.
- The presence of many metals in a mixed metal cluster offers the possibility of a cooperative effect in reactivity: one may speculate about the possibility of two different metals being responsible for two different types of activations necessary for the overall catalytic reaction.

2.2. HOMOGENEOUS CLUSTER CATALYZED REACTIONS

The number of publications related to catalytic reactions using molecular clusters has increased considerably over the last five years. In 1977, the first review [6] dealing with this aspect of catalysis contained only 50

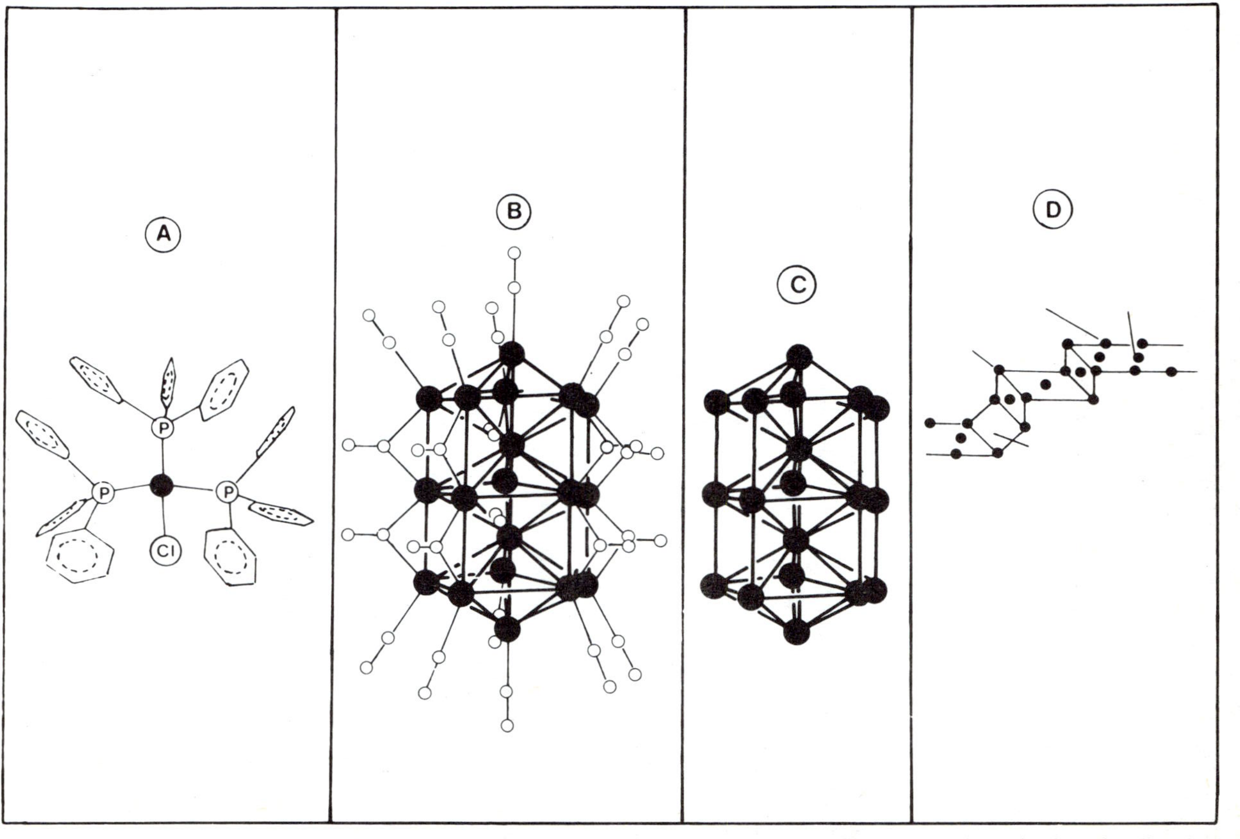

Fig. 37. The molecular clusters at the boarder line between molecular state and metallic state and at the boarder line between homogeneous catalysis and heterogeneous catalysis.

references. At the moment one can estimate the number to be close to 300. It is not the purpose of this paper to deal with such numerous examples.

Two aspects deserve to be briefly discussed:

— Is there any reaction which can be catalyzed by molecular clusters and which cannot be catalyzed by mononuclear complexes or metal particles?
— Is there any example of a catalytic reaction which occurs on a molecular cluster framework?

Regarding the first question, it is important to mention the works of Union Carbide related to the synthesis of ethylene glycol from syn-gas with Rh or Ru complexes under drastic conditions of pressure (above 500 atm.). With rhodium complexes it has been established that an anionic cluster $[Rh_5(CO)_{15}]^-$ was present in the reaction medium under catalytic conditions.

The same kind of observation was made recently by D. Dombeck [31] of Union Carbide for the same reaction using ruthenium clusters. In the case of ruthenium there is, under catalytic conditions, evidence for the presence of $[HRu_3(CO)_{11}]^-$ $HRu(CO)_4^-$ and $Ru(CO)_3I_3^-$. An almost complete catalytic cycle has been established by Dombeck. It appears that the hydrido anionic cluster $[HRu_3(CO)_{11}]^-$ or $HRu(CO)_4^-$ make a nucleophilic attack at CO coordinated to the mononuclear carbonyl Ru(II) complex to give a formyl species. The reaction, here, would obey a very complex mechanism involving both mononuclear and polynuclear species. This phenomenon seems to be a general rule in many reactions involving CO.

Regarding the second question the number of examples showing unambiguously cluster catalyzed reaction is still rare and is typically of mechanistic character. We would like to give only two examples dealing with the isomerisation of olefins with $H_2Os_3(CO)_{10}$ (Figure 38) and with the water gas shift reaction catalyzed by $Ru_3(CO)_{12}$ (Figure 39). The first reaction, studied in detail by Deeming [26] and Shappley [35], occurs in the triangle of $H_2Os_3(CO)_{10}$ due to the presence of an electronically unsaturated cluster allowing coordination of the olefin to an osmium atom without loss of a CO ligand. The second reaction of water gas shift, occurs in alkaline solution with $Ru_3(CO)_{12}$ as starting cluster. Three mechanisms have been postulated by P. Ford [25] which involve in each case anionic hydrides arising from nucleophilic attack of H_2O or OH^- at CO coordinated to the molecular cluster frame. Another mechanism has been proposed by Shore [36] based on the reactivity of KH with $Ru_3(CO)_{12}$.

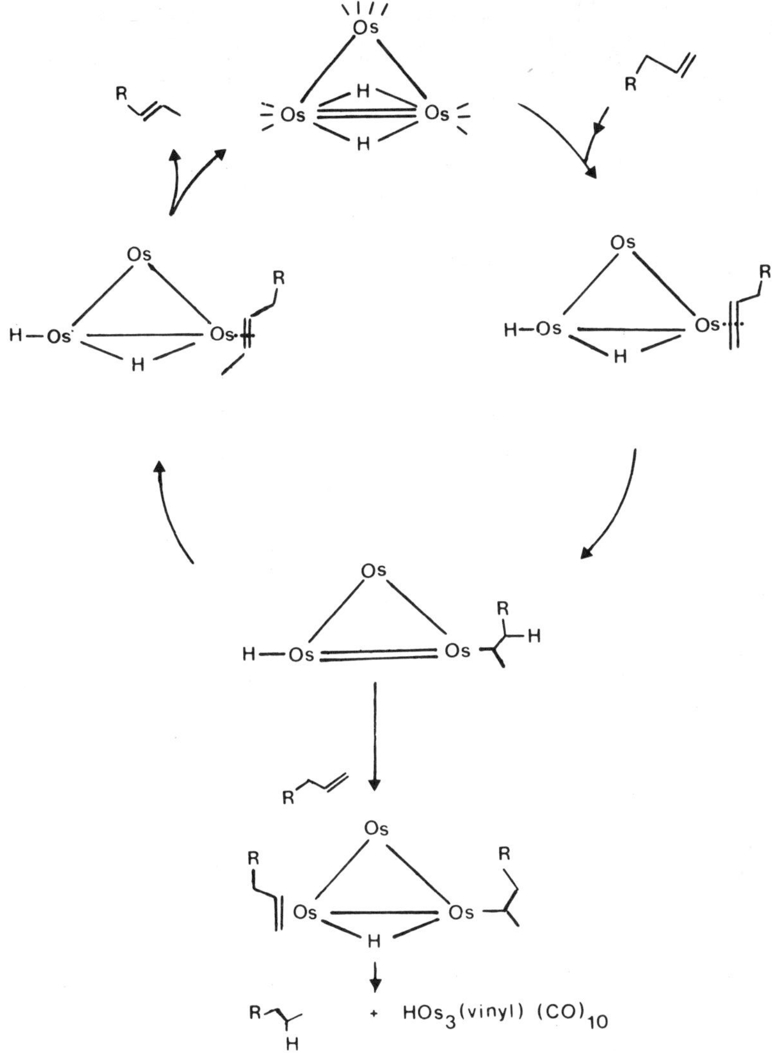

Fig. 38. Isomerization of olefin with $H_2Os_3(CO)_{10}$ according to Deeming [26].

2.3. CATALYSIS BY SUPPORTED MOLECULAR CLUSTERS

In the field by supported clusters many cases may occur. In some cases the grafted molecular cluster may remain intact during the complete catalytic cycle. In other cases the molecular cluster may be involved in some steps

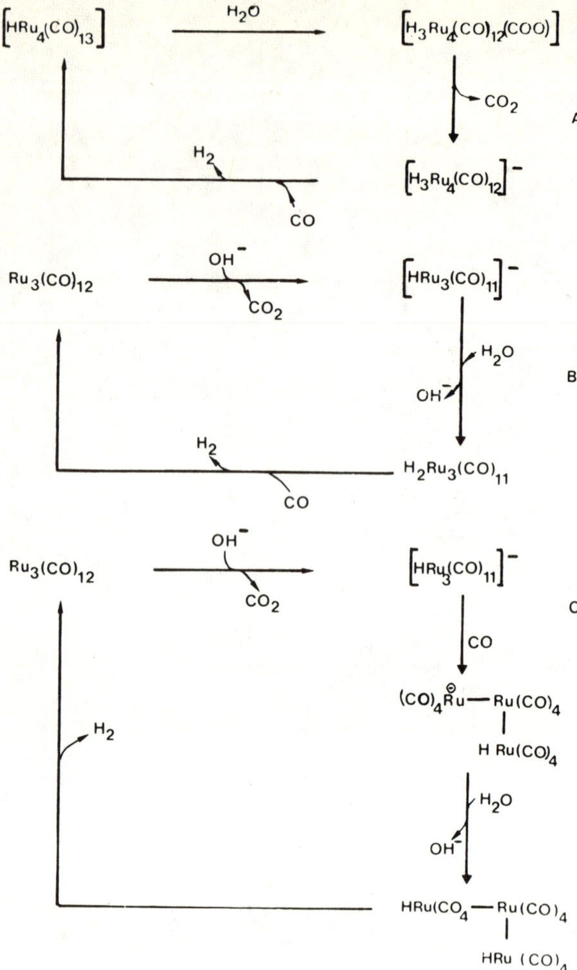

Fig. 39. Three possible mechanisms for the water gas shift reaction catalyzed by $Ru_3(CO)_{12}$ according to P. Ford [25].

of the catalytic cycle. Finally, and this is probably the most important aspect of cluster catalysis, the cluster may be decomposed into a small metal particle which is the active species in the catalytic reaction.

2.3.1. *The molecular cluster frame remains intact*

The reaction of $Os_3(CO)_{12}$ with the silanol groups of silica gives the grafted cluster $(H)Os_3(CO)_{10}(OSi\equiv)$. The grafting occurs by oxidative addition of the silanol group, the Os—Os bond. This grafted cluster contains a bridging hydride and a bridging $3e^-$ oxygen ligand which can be considered as good candidates for giving catalytic properties. Effectively the cluster is a catalyst for the reactions of olefin hydrogenation [27]. The mechanism of such reactions has been deduced from spectroscopic as well as kinetic studies (Figure 40). The first step of such mechanism seems to be the opening of one oxygen Os bond which occurs at *ca.* 80 °C. This opening favors the coordination of ethylene which is a reversible process. Then the ethylene would reversibly insert into the metal-hydride bond giving a σ-alkyl group. As a result the Os_1 atom would become coordinatively unsaturated and oxidative addition of hydrogen would occur. Finally the last step would be the reductive elimination of ethane with regeneration of the starting cluster. There are other examples in the litterature where catalytic cycles have been shown to occur in a molecular cluster frame supported on an inorganic oxide or polymer. (See p. 326.)

2.3.2. THE SUPPORTED MOLECULAR FRAME IS INVOLVED IN SOME STEPS OF THE CATALYTIC CYCLE

When $Rh_6(CO)_{16}$ is supported on alumina, the resulting solid is a catalyst for the water gas shift reaction. The reaction occurs between 25° and 100 °C and the mechanism has been studied in detail by labelling experiments as well as by infrared studies [28]. The first step of the mechanism is the destruction of the cluster frame by oxidative addition of $\geq$Al—OH groups to the Rh—Rh bond of the cluster, forming $Rh^{I}(CO)_2(OAl\leq)$ and $Rh^{III}(H)(H)(OAl\leq)$ surface species. The next step is the expulsion of H_2 by molecular CO with formation of $Rh^{I}(CO)_2(OAl\leq)$ surface mononuclear complex. The final step is the regenation of the cluster $Rh_6(CO)_{16}$ under $CO + H_2O$ with formation of CO_2 and H^+. It appears therefore that the mechanism of the water gas shift reaction occurs by stepwise destruction and regeneration of the cluster in the two key steps of the mechanism (Figure 41). The involvement of molecular clusters in some catalytic steps seem to be quite general in many reactions which involve molecular CO [23]. The reason for this is rather simple. The cluster state is a rather stable zero-valent state when CO is present. This does not necessarily means that the cluster is involved in a complete catalytic cycle but rather in some steps of the catalytic cycle.

2.3.3. *The molecular cluster is decomposed into very small particles of metal*

Preparation of heterogeneous catalysis can use molecular clusters as

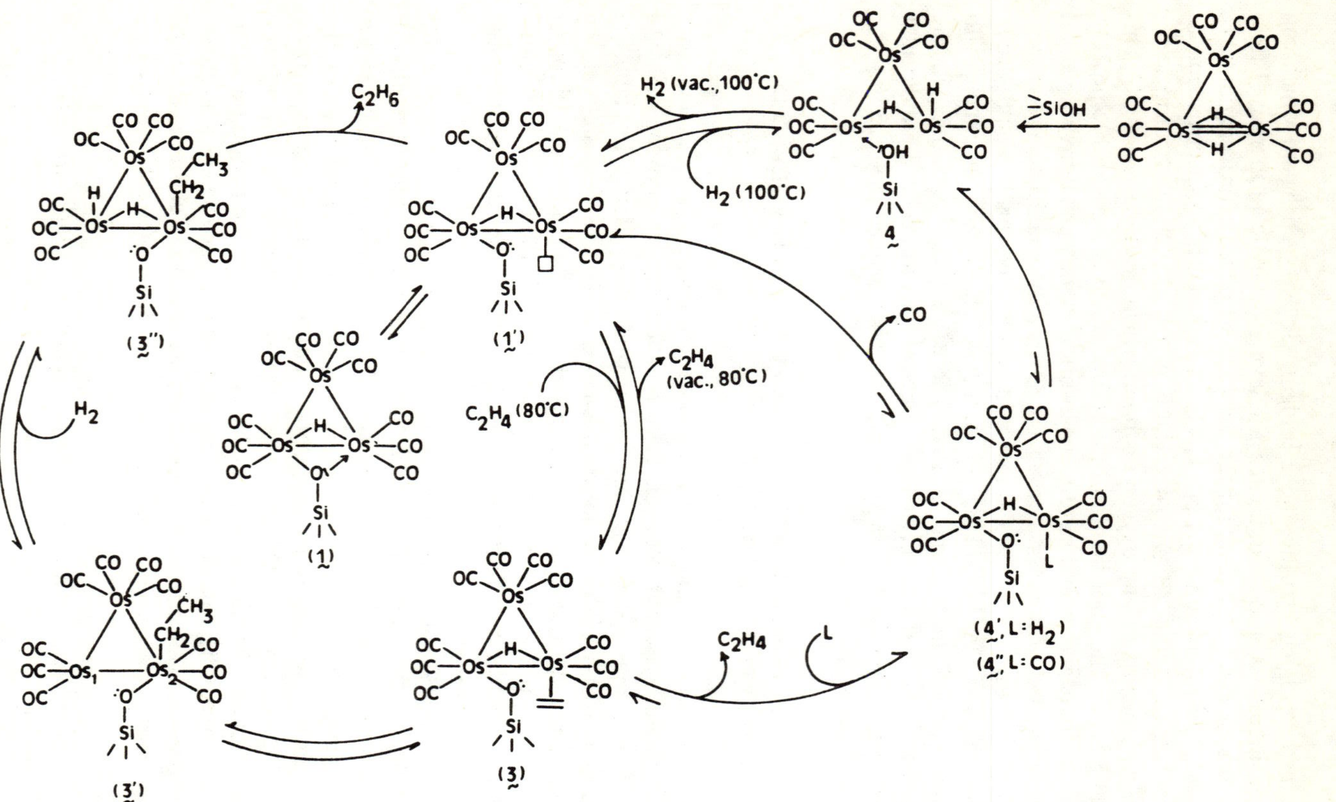

Fig. 40. Mechanism of ethylene hydrogenation with (μ-H) (μ-OSi≡) $Os_3(CO)_{10}$(ref. 27)

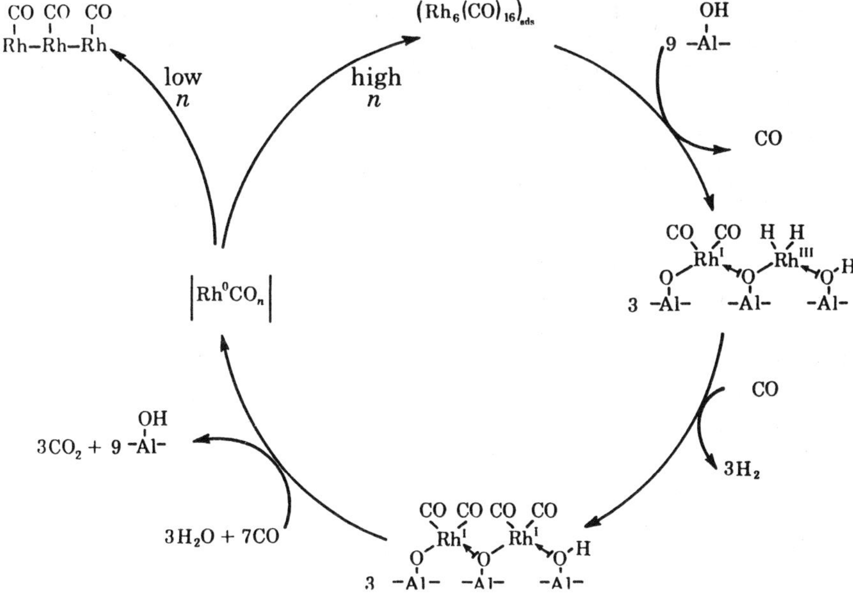

Fig. 41. Possible mechanism of water gas shift with $Rh_6(CO)_{16}$ supported on alumina [28].

starting material. As far as the development of heterogeneous catalysts derived from cluster compounds is concerned [23] such catalysts will be valuable only if they exhibit activities or selectivities that differ from those afforded by material prepared conventionally, that is halide impregnation or ion exchange followed by hydrogen reduction. In the case of ruthenium, cluster derived catalysts are shown to display greatly enhanced activity for the complete hydrogenation of straight chain aliphatic hydrocarbons to methane and provide a temperature advantage of 150 °C relative to conventionally prepared ruthenium catalysts where only moderate hydrocarbon conversions are noted. The increased activity superficially correlates with the smaller metal crystallite sizes (15—20 Å) reproducibly obtained with metal cluster compounds as catalysts precursors [23].

In the case of Fe, the thermal decomposition of $[HFe_3(CO)_{11}]^-$ absorbed on alumina, leads to the formation of very small particles of Fe (10 Å) which cannot be obtained by any other route. When these catalysts are used in Fischer—Tropsch synthesis, those supported particles are selective for low molecular weight olefins and the selectivity appears to be much larger than that obtained with conventionally prepared catalysts [29].

2.4. SUPPORTED CLUSTERS AND HETEROGENEOUS CATALYSIS: SURFACE ORGANOMETALLIC CHEMISTRY

One can define surface organometallic chemistry (S.O.M.C.) as the study of the reactivity of organometallic complexes with the surface of divided oxides and by extension of zeolites [37]. It is thus possible to make new catalysts which have no equivalent in homogeneous catalysis and in heterogeneous catalysis. These new surface complexes may be mononuclear, dinuclear, polynuclear (or metal particles). The strategy followed in this area is much closer to that followed in homogeneous catalysis than to that followed in heterogeneous catalysis. Besides, the study of the reactivity, stoichiometric reaction of organometallic complexes with surfaces introduces the concepts of coordination chemistry to heterogeneous catalysis. The field of S.O.M.C. has started with the study of the reactivity of mononuclear complexes of the type $Zr(CH_2—C_6H_5)_4$ with functional groups of silica and alumina [30]. It has resulted from this approach a new generation of highly active polymerisation catalysts. This field has broadened recently thanks to the use of molecular clusters which give rise to a very interesting reactivity of their metal—metal bond as well as metal ligand bonds with functional groups of surfaces.

We would like to illustrate S.O.M.C. by the following examples related to the behaviour of $Os_3(CO)_{12}$ adsorbed on silica and on alumina (Figure 42 and 43).

The first reaction which occurs during the thermal decomposition of $Os_3(CO)_{12}$ on hydroxylated silica is the oxidative addition of a silanol group in the metal—metal bond of $Os_3(CO)_{12}$ with departure of two moles of CO. This grafted cluster has been characterised by infrared spectroscopy, Raman spectroscopy, ^{13}C NMR and EXAFS. A model compound of the type $(H)Os_3(CO)_{10}(OSiPh_3)$ has also been synthesised; it exhibits spectroscopic data similar to that of the grafted cluster. Due to the presence of a bridging oxygen and bridging hydride, this grafted cluster exhibits various catalytic properties in reactions such a olefin hydrogenation or hydroformylation.

Thermal decomposition of the grafted cluster $(H)Os_3(CO)_{10}(OSi\lessgtr)$ results in a total destruction of the molecular cluster frame. This destruction probably results from a multiple oxidative addition of silanol groups on 3 metal atoms with formation of $(H)Os(CO)_3(OSi\lessgtr)$ which can be reversibly decarbonylated to $HOs(CO)_2(OSi\lessgtr)$. These mononuclear osmium hydrido species are also catalysts for hydroformylation.

It is only by treatment under vacuum above 250 °C that small metal particles of osmium are formed. Those particles have a narrow size

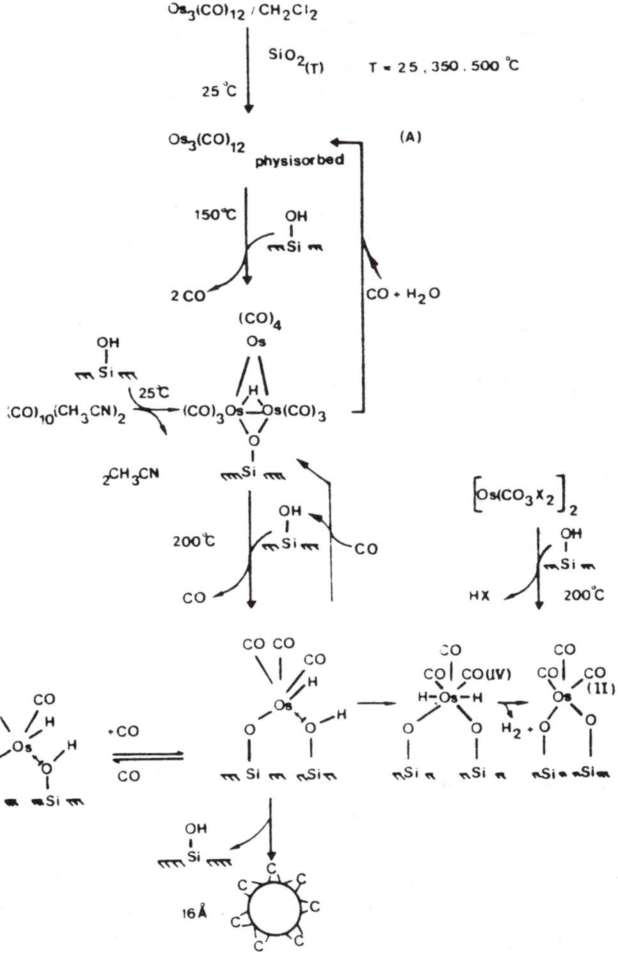

Fig. 42. Surface organometallic chemistry of $Os_3(CO)_{12}$ on silica [32].

distribution around 16 Å. The mechanism of such formation is the reductive elimination of a ⋛Si—OH from $HOs(CO)_3(OSi⋚)$:

$$(H)Os(CO)_3(OSi⋚) \longrightarrow Os(CO)_3 + ⋛Si—OH$$

Above 250 °C the equilibrium could be shifted towards the left whereas below 250 °C this equilibrium could be shifted to the right and lead to an aggregation phenomenon. Those metallic particles of Os are active in the reaction of methanisation of carbon monoxide.

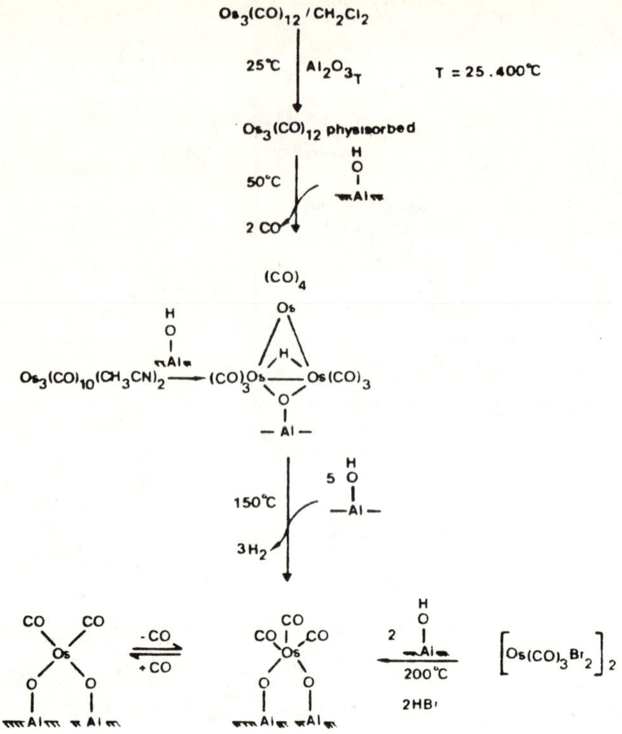

Fig. 43. Surface organometallic chemistry of $Os_3(CO)_{12}$ on alumina [32].

The reaction of $Os_3(CO)_{12}$ with the alumina surface is different from that observed with the silica surface. The first step is the oxidative addition of an $\geq$ Al—OH group into the Os—Os bond which gives the species $(H)Os_3(CO)_{10}(OAl\leq)$. However it is much less stable than on silica; it decomposes above 150 °C to give mononuclear Os(II) species with liberation of 3 moles of H_2. In contrast to what is observed on silica, these surface species are linked to silica by two covalent bonds $Os(OAl\leq)_2$. Those osmium(II) carbonyl complexes are not easily reduced by H_2 to metallic osmium. In contrast to $(H)Os(CO)_3(OSi\equiv)$, the $Os(CO)_3(OAl\leq)$ species is not a catalyst for olefin hydroformylation. In order to achieve a reduction to Os metal, it is necessary to reduce Os(II) with H_2 at 400 °C for 20 hours. The metallic particles which are thus obtained are much smaller than on silica since their size is situated around 8Å.

Surface organometallic chemistry has already been at the origin of new concepts in surface science on oxides that can be summarized as follows:

— reaction of a metal-alkyl bond with an OH group:

$$M_t\text{—}R + \underset{\overset{|}{S}}{OH} \longrightarrow RH + S\text{—}O\text{—}M_t$$

— reaction of a Π allyl group with an OH group

$$\widehat{M_t} + \underset{\overset{|}{S}}{OH} \longrightarrow + S\text{—}O\text{—}M_t + C_3H_6$$

— oxidative addition of an OH group into a single metal species

$$M_t + \underset{\overset{|}{S}}{OH} \longrightarrow S\text{—}O\text{—}M_t\text{—}H$$

— oxidative addition of an OH group into a metal-metal bond

$$M_3 + \underset{\overset{|}{S}}{OH} \longrightarrow M_3(\mu\text{-}H)(\mu\text{-}O\text{—}S)$$

— multiple oxidative addition with reductive elimination of H_2:

$$M_3 + \underset{\overset{|}{S}}{OH} \longrightarrow 3\ (S\text{—}O)_2M + 3\ H_2$$

— nucleophilic addition at coordinated CO followed by β-H elimination:

$$M_3\text{—}CO \quad \underset{\overset{|}{S}}{OH} \longrightarrow CO_2 + M_3(\mu\text{-}H)$$

— nucleophilic attack by molecular water with construction of a metallic frame:

$$3\,H_2O + 7\,CO + 6\,\text{O(S)}\cdots\text{Rh(CO)}_2 \rightarrow 3\,CO_2 + 6\,\text{OH–S} + Rh_6(CO)_{16}$$

— complexation of a carbonyl ligand into a Lewis center of the surface:

$$(CO)_5W\!-\!C=O \longmapsto Al;\ (H)Fe_3(CO)_{10}CO^- \longmapsto \overset{+}{Al}$$

— nucleophilic attack of an O^{2-} ligand into a coordinated CO

$$Fe_3(CO)_{12} + \underset{|}{O^{2-}}\!\!-\!Mg \longrightarrow (CO)_4Fe\!-\!C(O)_2\!\cdots Mg$$

— oxidative addition into a coordinatively unsaturated metal atom:

$$\text{Rh(O–Al)}_2 + H_2 \longrightarrow \text{H}_2\text{Rh(O–Al)}_2$$

References

1. P. Chini. *J. Organomet. Chem.*, **200**, 37—61 (1980).
2. B. F. G. Johnson. *Transition Metal Clusters* (B. F. G. Johnson Ed.) John Wiley and sons (1980).
3. K. Wade. *Transition Metal Clusters* (B. F. G. Johnson Ed.) John Wiley and sons 193—263 (1980).
4. A. J. Deeming. *Transition Metal Clusters* (B. F. G. Johnson Ed.) John Wiley and sons 391—469 (1980).
5. E. L. Muetterties. *La Recherche* **117** (1980), 1364—1372 (John Wiley and sons, 1981) 203—238.

6. A. K. Smith and J. M. Basset. *J. Mol. Cat.,* **2**, 229 (1977).
7. E. Band and E. Muetterties. *Chem. Rev.,* **78**, 639—658 (1978).
8. R. Ugo. *Cat. Rev.,* **11**, 225 (1975).
9. G. Martino. *Models and Precursors for Metallic Catalysts* (J. Bourdon Ed.) 399 (1980).
10. J. M. Basset and R. Ugo. *Asp. Homog. Cat.* (R. Ugo Ed.) **3**, 136 (1978).
11. E. L. Muetterties, T. N. Rhodin, E. Band, C. F. Brucker and W. R. Pretzer. *Chem. Rev.* **79**, 91 (1979).
12. D. C. Bailey and S. H. Langer. *Chem. Rev.,* **81**, 109—148 (1981).
13. T. L. Brown, **10**, 159—180 (1981).
14. J. Evans. *Chem. Soc. Rev.,* **10**, 159—180 (1981).
15. H. Vahrenkamp. *Phil. Trans. Roy. Soc. Lond.* **A.308**, 17—26 (1982).
16. R. E. Colborn, A. F. Dyke, S. A. R. Knox, K. A. Macpherson, K. A. Mead, A. G. Orpen, J. Roue and P. Woodward. *Phil. Trans. Roy. Soc. Lond.* **A.308**, 67—73 (1982).
17. A. Fusi, R. Ugo, R. Psaro, P. Braunstein and J. Dehand. *Phil. Trans. Roy. Soc. Lond.* **A308**, 125—130 (1982).
18. G. Longoni, A. Ceriotti, R. D. Pergola, M. Manassero, M. Perego, G. Piro and M. Sansoni. *Phil. Trans. R. Soc. Lond.* **A.308**, 47—57 (1982).
19. B. F. G. Johnson and J. Lewis. *Phil. Trans. R. Soc. Land.* **A.308**, 5—113.(1982).
20. J. Bradley, *Phil. Trans. R. Soc. Lond.* **A.308**, 103—113 (1982).
21. B. Heaton. *Phil. Trans. R. Soc. Lond.* **A.308**, 95—102 (1982).
22. D. M. P. Mingos. *Phil. Trans. R. Soc. Lond.* **A308**, 75—83 (1982).
23. R. Whyman. *Phil. Trans. R. Soc. Lond.* **A.308**, 131—140 (1982).
24. B. F. G. Johnson and J. Lewis. *Phil. Trans. R. Soc. Lond.* **A.308**, 5—15 (1982).
25. P. Ford. *Acc. Chem. Res.* **14**, 31 (1981).
26. Deeming. *J. Organomet. Chem.,* **114**, 313 (1976).
27. B. Besson, A. Choplin, L. D'Ornelas and J. M. Basset. *J.C.S. Chem. Commun.,* 842 (1982).
28. J. M. Basset, B. Besson, A. Choplin and A. Theolier. *Phil. Trans. R. Soc. Lond.* **A.308**, 115—124 (1982).
29. D. Commereuc, Y. Chauvin, F. Hugues and J. M. Basset. *J.C.S. Chem. Commun.,* 154 (1980).
30. Ballard. *Adv. Cat/Rel. Subj.* **23**, 263 (1973).
31. Dombeck B.D. *Adv. Catal.* **32**, 325 (1983); *Organometallics* **4**, 1707 (1985).
32. R. Psaro, R. Ugo, G. M. Zanderighi, B. Besson, A. K. Smith, and J. M. Basset. *J. Organomet. Chem.* **213**, 215 (1981).
33. D. M. P. Mingos. *Nature (London) Phys. Sci.* **236**, 99 (1972).
34. R. Hoffmann, *Angew. Chem. Int. Ed. Engl.* **21**, 711 (1982).
35. J. R. Shapley, J. B. Keister, M. R. Churchill , B. G. de Boer. *J. Am. Chem. Soc.* **97**, 4145 (1975).
36. J. C. Bricker, C. C. Nagel and S. Shore. *J. Am. Chem. Soc.* **104**, 1444 (1982).
37. J. M. Basset and A. Chorlin, *J. Mol. Cat.* **21**, 95 (1983).

W. KEIM

FUTURE TRENDS IN HOMOGENEOUS CATALYSIS

Most industrial reactions are catalytic, and many processes and process improvements result from new catalysts. It is estimated that about 60—70% of all chemicals, at one or another time, see a catalyst. Catalytic reactions can be divided as shown in Figure 1.

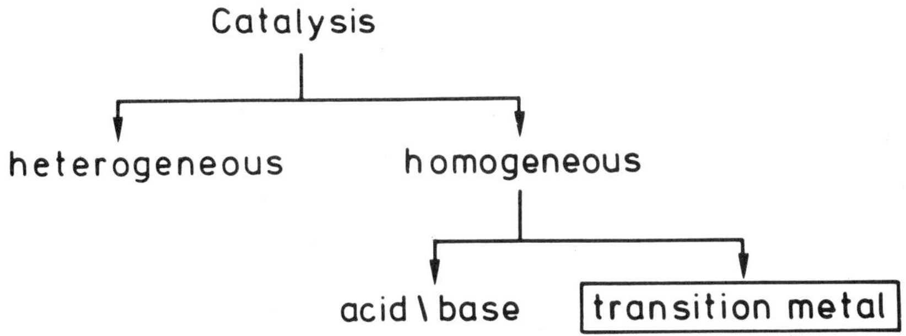

Fig. 1. Classification of catalysts.

By far most important are heterogeneous catalysts. The market share of the homogeneous type is estimated at 10—15%. Among homogeneous catalysts the last 30 years have seen prodigious growth. Many new processes emerged, many new products became available [1]. Assessment of the merchant catalyst value for homogeneous catalysts is hardly possible. Other than in heterogeneous catalysis homogeneous catalysts are frequently captively used and in situ preparation by the manufacturer is quite common. In addition, it is very difficult to distinguish between homogeneous and heterogeneous systems because of the problem of particle size determining homogeneity. Often the catalyst solutions are clear to the naked eye, but light scattering experiments suggest the presence of aggregates. An additional complication may arise when, as in polymerization, the catalyst is deposited on the polymeric particles.

A. Mortreux and F. Petit (Eds.), Industrial Applications of Homogeneous Catalysis, 335—347.

1. Industrial Applications of Homogeneous Catalysis

Applications of homogeneous catalysis are found in all major chemical reactions as is elucidated in Figure 2.

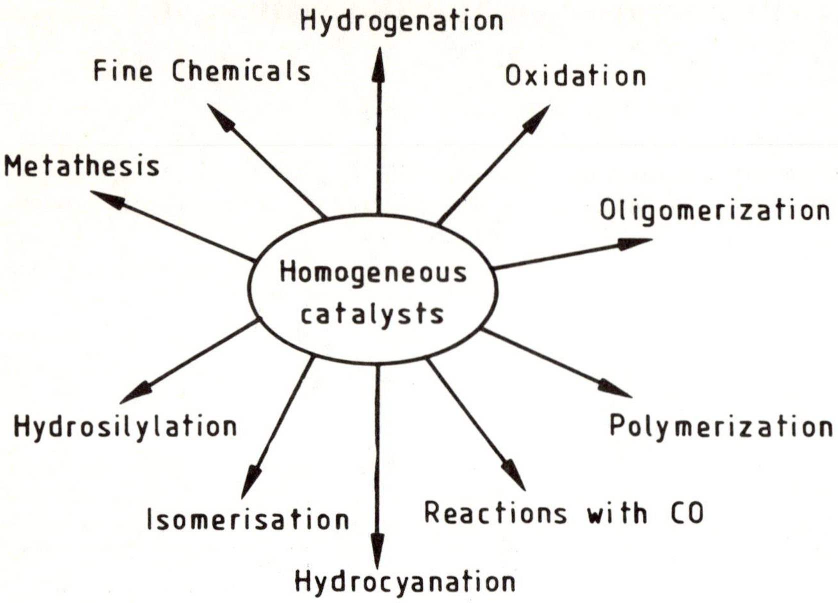

Fig. 2. Reactions catalyzed by homogeneous transition metal catalysts.

Applications of homogeneous hydrogenation are found in polymer synthesis, hydrogenation of aldehydes to alcohols (oxo-process) and asymmetric hydrogenation (synthesis of *l*-dopa by Monsanto Comp.).

The largest scale of application of homogeneous catalysis is found in oxidation of hydrocarbons by molecular oxygene or peroxides. Mechanistically homogeneous metal catalyzed oxidation reactions can be divided into homolytic and heterolytic type of reactions. Heterolytic systems are more in line with coordination chemistry. The organic substrate and oxygen/or the oxidising agent are activated by the metal centre. The catalytic cycle can involve a series of two electron steps (Wacker process, Halcon process).

Oligomerization processes practized embrace monoenes and dienes. Here the synthesis of α-olefins from ethylene (Shell Higher Olefin Process), the oligomerization of propylene/butene (Dimersol Process), and the cyclic dimerization and trimerization of butadiene to cyclooctadiene and cyclododecatriene (Wilke-Chemistry) are prestigious examples.

Mechanistically related to oligomerizations are polymerization reactions. Organometallic catalysts, both soluble and insoluble, are used commercially to polymerize and copolymerize ethylene, propylene, butadiene and isoprene to high-density polyethylene, to polypropylene, ethylene/propylene/diene rubber, *cis*-1,4-polybutadiene and *cis*-1,4-polyisoprene. The use of transition metal complexes yields ordered polymers whose physical properties differ from those of free radical polymers. Special mention must be made of the recent results by J. A. Ewen and W. Kaminsky [2] who obtained atactic polypropylene with homogeneous catalysts. Reactions with CO belong to very important examples of homogeneous catalysis. They are associated with the names of W. Reppe and O. Roelen and belong to the earliest homogeneous processes practized in industry. They can be classified in carbonylation and hydroformylation reactions.

The commercial introduction of hydrocyanation has been pioneered by Du Pont. The nickel complex catalyzed addition of two molecules of HCN to butadiene yielding in high regioselectivity adiponitrile is practized in two plants.

Isomerization via homogeneous catalysts occurs, for instance, as an intermediate step in catalytic processes. Thus in Shell's hydroformylation route, which converts internal olefins to primary alcohols, isomerization takes place prior to CO-insertion. Homogeneous isomerization of 2-methyl-3-butenenitrile to the linear nitrile is an essential step in du Pont's hydrocyanation. Noteworthy is the recent commercial asymmetric isomerization of neryl and geranyl amines [3].

Metathesis can be catalyzed homogeneously and heterogeneously. The biggest applications of metathesis such as the SHOP process [4] and Phillips Triolefin process use heterogeneous catalysts. Norbornene (Norsorex by CdF Chimie), cyclooctene (Vestenamer by Hüls AG), and dicyclopentadiene (Hercules) practice homogeneous catalysis.

Hydrosilylation of allyl chloride, acetylene and siloxanes is carried out to produce intermediates in silicone chemistry.

2. Advantages and Disadvantages of Homogeneous Catalysis

To elucidate the advantages and disadvantages of homogeneous and heterogeneous catalysis, Figure 3 exhibits a comparison.

Most noteworthy is the selectivity with which homogeneous system can operate. For instance, the Monsanto process to manufacture acetic acid exhibiting 99% selectivity is the world's best route to this chemical. One of

	homogeneous	heterogeneous
new products	in both possible	
selectivity	often better	
activity	metal atoms are used	only surface atoms are used
reaction conditions	mild (20-200°C)	severe (>250°C)
Diffusion	practically unknown	given
reproducibility	given	often difficult (know-how)
understanding	in situ spectroscopic methodes (NMR,IR,ESR, etc.) can be applied	difficult, often impossible under reactions conditions
catalyst recycle	full of problems	easy

Fig. 3. Comparison of homogeneous and heterogeneous catalysis.

the greatest drawbacks in applying homogeneous catalysis are difficulties to recycle the catalyst. Spurred by these problems a new research discipline "Support of homogeneous catalysts" is emerging [5]. Often the use of two phases is also feasible: one phase (e.g. polar solvent) contains the catalyst, the products form the second phase [6].

3. Future Applications of Homogeneous Catalysis

Regarding future applications many favorable positions can be seen. The field of coordination chemistry, the backbone of homogeneous catalysis, has developed to an advanced level. It can be generalized that probabilities for technological jumps need a broad foundation of fundamental knowledge prior to breakthroughs. A major impulse of the past can be traced back to the Ziegler—Natta catalysis, which in its beginning was more an art, but which has developed into a science, in many regards well understood. This understanding has led to many new catalytic systems, many technical applications.

It also can be anticipated that in the future more metals of the periodic table will find entrance in homogeneous catalysis. As is evident from Figure 4, only 12 metals, so far, have found broader industrial applications.

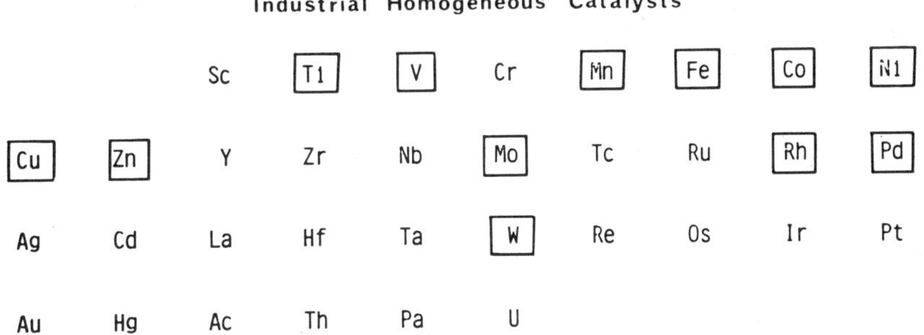

Fig. 4. Transition metals used in homogeneous industrial processes.

The number of publications dealing with lanthanides or actinides is steadily growing indicating new possibilities. In the framework of this chapter four driving forces for future homogeneous catalysis will be discussed:

(a) Impacts from changes in raw material supply;
(b) Impacts steming from engineering requirements;
(c) Technological drives;
(d) Impacts driven by society needs.

3.1. CHANGING RAW MATERIALS SUPPLY

The increase in oil prices, which we experienced between 1974—1983 has given rise to a growing interest in alternative feedstocks such a coal, methane, biomass and CO_2. Of course, mineral oil will be the backbone for the chemical industry for a long time to come, yet opportunities for alternative feedstocks can be seen from the following points:

— *Innovation*: The chances in searching for a competitive advantage are greater where new technology exists. Using crude oil derivatives narrows the opportunities for new processes, because similar technological paths exist for all competitors.
— *Independence*: An erratic behaviour of feedstock prices and a shortage in availability make planning and forecasting very difficult. A reliable, secure feedstock supply and long term contracts at calculable prices are the basis of a prosperous industry. Independence also embraces concern for national security aimed at using own available feedstocks.
— *Availability of own carbon resources*: Countries with substantial own carbon resources such as coal, natural gas or biomass may find it beneficial to develop their own resources, thus improving their balance of payments.

Opportunities for alternative raw materials applying homogeneous transition metal catalysis can be seen in: synthesis gas chemistry, alkane chemistry, and CO_2 chemistry.

3.1.1. *Synthesis gas chemistry*

Synthesis gas offers many routes to industrial chemicals, which can be classified into a direct and indirect path as outlined in Figure 5 [7].

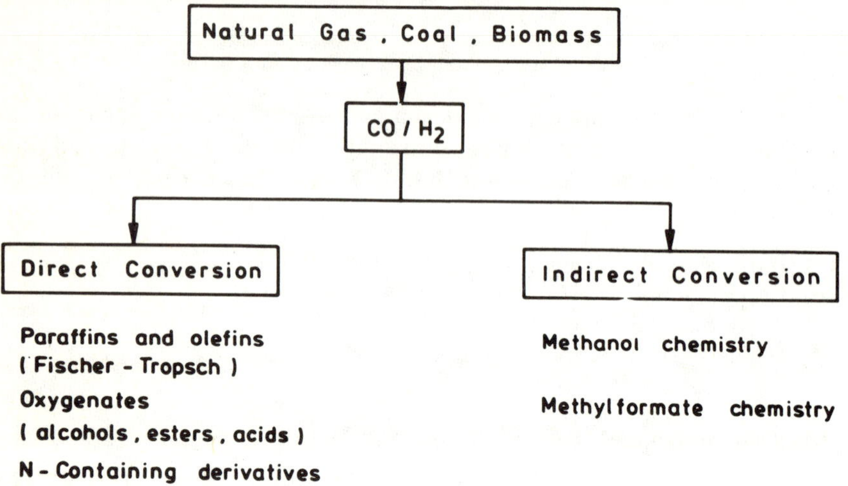

Fig. 5. Direct and indirect conversion of CO/H_2.

Synthesis gas can be derived from many different sources such as crude oil, coal, biomass, oil shale, tar sand. With the recent global decline in crude oil prices, the attractiveness of syngas as an alternate feedstock has ebbed bringing many efforts and considerations in this field back to the research stage. But there is general consent that coal gasification will continue to be developed. In addition, experts also see a greater conversion of natural gas to CO/H_2 coming. Long range synthesis gas will compete with mineral oil and the basic research and development for its future application must be done today to have it available when it is needed.

The use of homogeneous transition metal catalysis in syngas chemistry can be seen for various reasons:

— Homogeneous catalysts often show significant advantages in selectivity, activity and ease of modification. Hence, a key factor in synthesis gas chemistry is better selectivity, homogeneous systems could play an important role. Homogeneous reactions occur under rather mild reaction conditions, thus offering economic benefits in the saving of energy.

- The hydrogenation of CO is exothermic and heterogeneous systems are plagued by problems of heat removal. Homogeneous systems operating in the liquid phase offer advantages.
- Homogeneous transition metal catalysts, in a broad sense, are off-springs of organometallic complex chemistry. Here complexes and reactions with CO and H_2 are among the most studied areas. A wealth of information and understanding is available providing the nutrient for applications.

In the direct conversion homogeneous catalysts could be applied to provide oxygenates. Potential applications exist for alcohols, acids and esters. Alcohol synthesis is a favorable way to upgrade synthesis gas because of its positive thermodynamics. A number of homogeneous catalysts are known to convert CO/H_2 to alcohols. Most remarkable is the high-pressure homogeneous reduction of CO in the work by Union Carbide aimed at an industrial synthesis of ethylene glycol [8].

The indirect path offers many opportunities for homogeneous catalysts, especially related to methanol chemistry via carbonylation. Among organic chemicals, methanol belongs to industry's most significant products amounting to about 16 million tons in 1986. In addition to the traditional markets, a very significant demand for methanol in new areas, such as energy or single cell proteins may emerge in the future. If energy prices develop in such a way that it becomes economically or industrially attractive to use methanol in gasoline or as a fuel for power production, the market potential at low prices will be excellent.

Two processes of direct methanol carbonylation are well established already: The direct carbonylation of methanol yielding acetic acid, the Monsanto process and carbonylation of methyl acetate giving acetic anhydride, a technology commercialized by Tennessee Eastman Kodak. By adjusting the $CO:H_2$ ratio, catalytic systems for the reductive carbonylation of methyl acetate can be tuned to the production of acetic anhydride, ethylidene diacetate or acetaldehyde.

A direct carbonylation reaction demonstrating in an impressive way the potential of homogeneous catalysis is the double carbonylation [9]. Phenyl pyruvic acid derivatives can be prepared directly from benzyl halides.

While the direct carbonylation is well accepted by industry, the reductive and oxidative carbonylations are still in the research and development stage. Using Texaco technology the combined synthesis of ethene and ethanol is feasible via homologation of acids [10].

The oxidative coupling of CO in the presence of an alcohol to yield oxalate esters is under study by Ube Industries and Union Carbide [11]. In a subsequent reaction, the oxalate can be hydrogenated to ethylene glycol. Oxalate esters can also be reacted with NH_3 giving oxamides, a fertilizer.

The oxidative carbonylation of methanol to dimethylcarbonate has found small industrial application [12]. Great promise is offered by the homologation of methanol to yield ethanol and/or acetaldehyde [13].

Finally, methyl formate must be mentioned as a potential C_1 intermediate [14]. It can be synthesized directly from CO/H_2 or indirectly via methanol. Around methyl formate various potential applications could revolve as is shown in Figure 6.

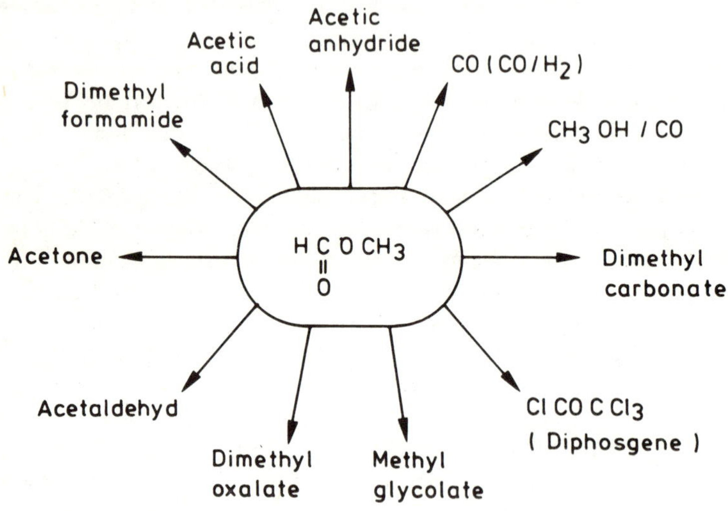

Fig. 6. Methyl formate as intermediate in syngas chemistry.

3.1.2. *Alkane chemistry*

When considering that alkanes are among the most abundant of naturally occuring hydrocarbons, the potential of CH-activation of alkanes by transition metals is quite obvious. Although chemical reactions of saturated hydrocarbons are industrially utilized (thermal conversions, chlorination, oxidation) generally, selectivities are low and energy requirements are high. Therefore, the chemical industry normally uses the building blocks ethene, propene, butadiene, benzene and xylenes, which amount to >90% of all base chemicals, to built up large volume products. Significant cost advantages can be anticipated with the direct use of saturated hydrocarbons as starting material thus circumventing the olefin route. In addition, the huge resources of methane and the difficulties to transport it to the places of consumption necessitate considerations to activate C—H-bonds and couple C—C-bonds. Here homogeneous catalysis could provide solutions. If it would be possible to economically insert metal complexes L_nM, L_nM_x, into CH-bonds of paraffins RH, reactions as outlined in equations (1) to (4) are conceivable:

$$\text{RH} \xrightarrow{L_nM} \text{R—R} + H_2 \quad \text{(growth)} \tag{1}$$

$$\text{RH} + O_2 \xrightarrow{L_nM} \text{aldehydes, acids, alcohols} \tag{2}$$

$$\text{RH} + \text{CO}/H_2 \xrightarrow{L_nM} \text{alcohols} \tag{3}$$

$$\text{RH} + \text{olefin} \xrightarrow{L_nM} \text{growth products} \tag{4}$$

In the last years remarkable success in CH-activation by metal complexes has been achieved [15]. Figure 7 exhibits two examples based on methane demonstrating the ease and the principle of CH-activation.

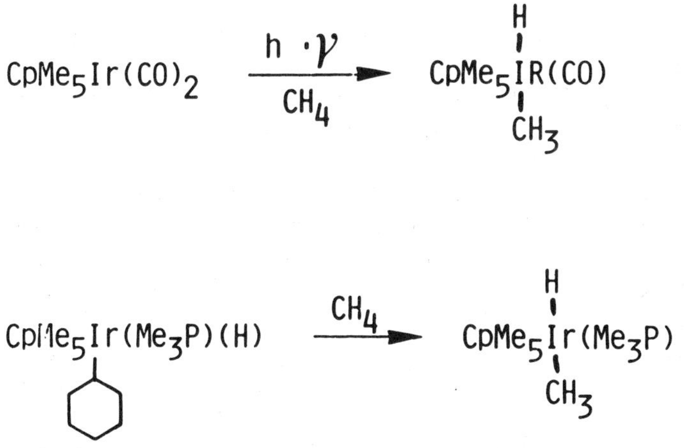

$$\text{CpMe}_5\text{Ir(CO)}_2 \xrightarrow[CH_4]{h \cdot \nu} \text{CpMe}_5\text{IR(CO)}(\text{H})(\text{CH}_3)$$

$$\text{CpMe}_5\text{Ir(Me}_3\text{P)(H)}(\text{C}_6\text{H}_{11}) \xrightarrow{CH_4} \text{CpMe}_5\text{Ir(Me}_3\text{P)}(\text{H})(\text{CH}_3)$$

Fig. 7. Methane activation.

The many examples of recent CH-activation give hope that activation of paraffins by homogeneousatalysts may be achievable in the near future.

3.1.3. *Carbon dioxide chemistry*

The practically unlimited amounts of carbon dioxide available in the atmosphere or bound as carbonates make this chemical an attractive feedstock for chemical synthesis. Recent findings to coordinate CO_2 in metal complexes and model reactions with CO_2 using transition metals do away with the myth that CO_2 is inert. A comparison may be drawn with H_2O, which easily reacts with energy rich olefins. Indeed, equations (5) and (6) demonstrate the ease with which CO_2 reacts with olefins in catalytic cycles [16]. Here more research is needed to explore the full potential.

$$2\ CH_3(CH_2)_3C{\equiv}CH + CO_2 \longrightarrow \text{(4,6-dibutyl-2-pyrone; Bu, Bu, O, O)} \quad (5)$$

$$2\ \text{(butadiene)} + CO_2 \longrightarrow \text{(lactone; O, C=O)} \quad (6)$$

3.2. IMPACTS BY ENGINEERING REQUIREMENTS

The price of a chemical is determined to a great extent by the cost of the raw material and the cost of operating a plant. With regard to the raw material two goals will emerge in the future: use of alternative feedstocks and optimization of existing feedstocks. In both, catalysis will have a major impact. High selectivities at high conversions are desirable for an industrial process. One of the most noteworthy virtues of homogeneous catalysis is selectivity. Operating at lower temperatures influences the activation energies thus eliminating competing reactions. Remarkable product selectivities have been reported as discussed already for acetic acid (Monsanto process).

Selectivity is also needed to produce pure products in high yields, a characteristic which is very important in the preparation of pharmaceuticals, intermediates for polymers and many other applications. Prices of chemicals vary drastically with its purity grade. Often high purity is a necessity as in polymer grade propene or olefins for metathesis. This is exemplified in the SHOP process, which consists of three reactions: oligomerization, isomerization and metathesis. The selectivity to linear olefins in the oligomerization section, which is carried out homogeneously using nickel complexes, is essential for metathesis [4]. Branched olefins react only slowly under metathesis conditions. Without the high selectivity of the homogeneous oligomerization catalyst integration of oligomerization and metathesis would have been impossible.

The formation of by-products can also lead to a market problem in selling or discarding the latter ones, thus impacting the economics of a process. No industry likes processes with two or more products because balance on the market normally is quite difficult. In addition, disposal of by-products could become more difficult in the future.

Besides selectivity, energy requirements will grow more essential in the future. One advantage of homogeneous catalysis lies in its mild reaction temperatures needed for reactions to occur. Also working in solution makes heat recovery — especially in exothermic reactions — often much easier. For instance, reflux of a solvent can be used to control the temperature in a reaction vessel.

Also benefits in diffusion control may be cited among the advantages. Heterogeneous catalysts are often plagued by diffusion limitations which are practically unknown or can be much easier solved in homogeneous systems.

Summarizing the above points, better process control for temperature, for mixing of reactants, and for circumventing diffusion — all lowering the cost of operating a plant — can be seen for homogeneous catalysts. There are various cases known where homogeneous systems excell when comparing engineering requirements.

3.3. TECHNOLOGICAL DRIVES

The last thirty years have seen an explosive development of organometallic chemistry. As mentioned already, organometallic chemistry provides the background for homogeneous catalysis. We can observe that more and more organic chemists utilize reactions steming from organometallic chemistry partly in stoichiometric partly in catalytic reactions. It can be anticipated that this development will continue, thus broadening the awareness of the new discipline and making it more applicable.

In highly industrialized countries there is a great drive for fine chemicals. Words like “value added”, “high tech” and “life science chemicals” are heard constantly. A drive for small volume chemicals such as pharmaceuticals, agricultural and electronic chemicals can be seen. Small tonnage, high-value chemicals are well suited for applications of homogeneous catalysts. Furthermore, in a multiple step batch process, as used for the synthesis for fine chemicals, cost savings are possible by reducing reaction steps. Here homogeneous catalysts hold great promise, because very complex molecules can be synthesized in one step reactions.

3.4. SOCIETY'S NEEDS

We observe that the requirements for safety and environmental control grow more stringent. Regarding safety, we will see a change to safer processes even a phasing out of certain technologies. Here homogeneous catalysis could find applications in providing better alternatives. For instance, hydrogenation in homogeneous systems, circumventing the use of fire hazardous Raney nickel, could lead to safer operations.

Regarding environmental constraints, we will see that processes burdening the environment will be phased out and that a change in products will occur. The inherent selectivity of homogeneous catalysis may prove advantageous in selecting processes with a minimum of disposable by-products. With regard to new or improved products, there is a great need for better control of stereoselectivity. Remarkable results have already been achieved by homogeneous catalysts. For instance, single optical isomers in greater 98% can be obtained. This is important in the pharmaceutical industry to maximize biological activity. An impressive example of stereoselectivity by homogeneous catalysis provides the polymerization of propene or butadiene. Depending on the catalyst used, different products can be obtained. This is shown for butadiene in Figure 8.

To demonstrate the impact of society needs, two examples should be mentioned: carbon dioxide and NO_x. There is growing concern regarding the greenhouse effect of CO_2. Shouldn't we increase our efforts at using

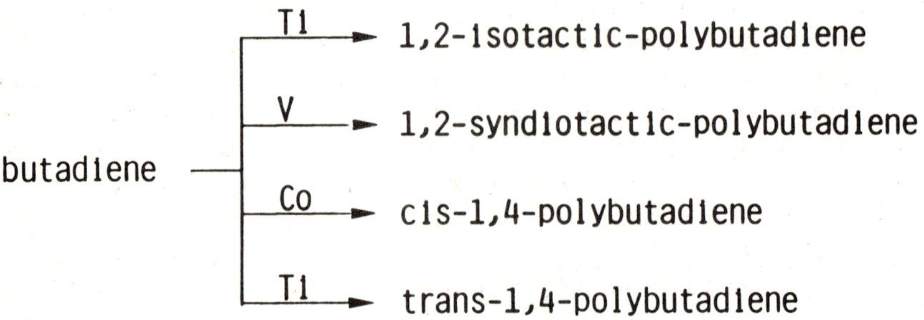

Fig. 8. Stereoselective polymerization of butadiene.

homogeneous catalysts to base more processes on CO_2? Problems with NO_x provide opportunities for homogeneous catalysis. A fair amount of studies with NO_x on a molecular level applying transition metal complexes have been carried out. This basic research may bear the nucleus for potential future applications.

Institut für Technische Chemie und Petrolchemie
RWTH Aachen
FRG

References

1. (a) W. Parshall in: *Homogeneous Catalysis*, J. Wiley & Sons, N.Y. (1980); (b) W. Keim: *Chem. Ind.* 397 (1984); (c) B. L. Goodall, R. Pruett, C. A. Tolman, F. F. Lutz, D. Forster, T. W. Dekleva, S. W. Polichnowski, B. D. Dombek, C. B. Murchison, R. L. Weiss, R. A. Stone, R. K. Grasselli, W. S. Knowles: *J. Chem. Ed.* 189—255 (1986).
2. J. A. Ewen: *J. Am. Chem. Soc.* **106**, 6355 (1984); W. Kaminsky, K. Külper, H. H. Brintzinger and F.R.W.P. Wild: *Angew. Chem.* **97**, 507 (1985).
3. K. Tani, T. Yamagata, S. Akutagawa, H. Kumobayashi, T. Taketami, T. Takaya, A. Miyashita, R. Noyori and S. Otsuka: *J. Am. Chem. Soc.* **106**, 5208 (1984).
4. (a) M. Peuckert, W. Keim: *Organometallics* **2**, 594 (1983); (b) W. Keim: *Chem. Ing. Techn.* **56**, 850 (1984).
5. F. R. Hartley in: *Supported Metal Complexes*, D. Reidel Publishing Company (1985).
6. A. Behr, W. Keim: *Erdöl, Erdgas, Kohle* **103**, March (1987).
7. (a) W. Keim in: *Chemistry for the Future* (Ed. H. Grünewald) Pergamon Press, Oxford (1984); (b) W. Keim: *Pure &Appl. Chem.* **58**, 825 (1986).
8. W. Keim in: *Catalysis in C_1 Chemistry*, D. Reidel Publishing Company (1983).
9. R. Perron (Rhone-Poulenc Ind.): *Be 837.401* (1975).
10. J. F. Knifton: *J. Catal.* **79**, 147 (1983).
11. F. J. Waller: *J. Mol. Catal.* **31**, 130 (1985).
12. R. Ugo, R. Tesei, M. M. Mauri, P. Rebora: *Ind. Eng. Chem. Prod. Res. & Dev.* **19**, 396 (1980).
13. (a) M. Röper: *Habilitationsschrift* RWTH Aachen (1985); (b) Union Rheinische Braunkohlen Kraftstoff AG: *Ger. Offen.* 3.343.519 (1983).
14. (a) W. Keim: *Am. Chem. Soc. Meeting N.Y.* (1986); (b) J. Haggin: *C & E News*, May 19, 7 (1986).
15. A. E. Shilov in: *Activation of Saturated Hydrocarbons by Transition Metal Complexes*, D. Reidel Publishing Company (1984).
16. (a) A. Behr: *Chem. Ing. Techn.* **57**, 893 (1985); (b) A. Behr: *Bull. Soc. Chim. Belg.* **94**, 671 (1985); (c) A. Behr: *Habilitationsschrift* RWTH Aachen (1987).

INDEX